Feature Selection and Feature Extraction on Omics Data

In today's data-driven world, biology and medicine are being transformed by the power of big data. Making sense of large, complicated biological datasets is a crucial problem that underlies every medical advancement and gene discovery. *Feature Selection and Feature Extraction on Omics Data* provides insight into this innovative area where biological science and computational science collide. This book, which is written in an approachable manner, explains the methods researchers employ to sort through vast amounts of multi-omics data to find insights that may result in better treatments, early disease diagnosis, and a greater comprehension of life at the molecular level. This volume provides a unique look at the technologies influencing the future of biological discovery and customized medicine, making it the perfect choice for anyone interested in learning more about how AI and data science are transforming biology and health.

This collection explores cutting-edge feature selection and extraction methods across a broad range of omics data formats, such as metagenomics, genomics, transcriptomics, epigenomics, and datasets. Readers will learn how these techniques can be used to improve disease classification, find promising biomarkers, uncover significant biological patterns, and aid in early diagnosis. The chapters discuss techniques designed to regulate sparsity, minimize dimensionality, and preserve biological interpretability while fusing fundamental ideas with practical applications. Case studies and real-world applications show how these methods enhance computational models' performance in tasks like disease prediction and gene identification. This book is a great resource whether you're new to omics data analysis or looking to improve your current workflows using sophisticated feature engineering techniques. It connects theory and application with contributions from subject matter experts to assist readers in converting unprocessed data into biologically significant insights, making it an essential resource in contemporary computational biology and precision medicine.

This book offers a comprehensive exploration of cutting-edge methodologies designed to address the complexities of high-dimensional biological datasets. This book serves as a practical and theoretical guide for researchers, data scientists, and students working at the intersection of bioinformatics and machine learning.

This book is a comprehensive and application-focused approach to one of the most pressing challenges in modern bioinformatics: making sense of high-dimensional omics data. While many resources touch on machine learning or biological datasets in isolation, this book bridges the two, offering a unified, practical guide that combines theoretical depth with real-world implementation across diverse omics domains—including genomics, metagenomics, transcriptomics, and epigenomics data.

Feature Selection and Feature Extraction on Omics Data

Edited by
Saurav Mallik, Zhongming Zhao, Soumita Seth, Aimin Li,
Kasmika Borah and Himanish Shekhar Das

CRC Press
Taylor & Francis Group
Boca Raton London New York

CRC Press is an imprint of the
Taylor & Francis Group, an **informa** business
A CHAPMAN & HALL BOOK

Designed cover image: Editor Ms. Kasmika Borah.

First edition published 2026
by CRC Press
2385 NW Executive Center Drive, Suite 320, Boca Raton FL 33431

and by CRC Press
4 Park Square, Milton Park, Abingdon, Oxon, OX14 4RN

CRC Press is an imprint of Taylor & Francis Group, LLC

ISBN: 978-1-032-96767-7 (hbk)
ISBN: 978-1-032-97000-4 (pbk)
ISBN: 978-1-003-59165-8 (ebk)

DOI: 10.1201/9781003591658

Typeset in Minion
by KnowledgeWorks Global Ltd.

Contents

About the Editors

Saurav Mallik is currently working as Research Scientist in the Department of Pharmacology and Toxicology, The University of Arizona, USA. Previously, he worked as Postdoctoral Fellow in Harvard T.H. Chan School of Public Health, Boston, MA, USA for more than three years (2019–2022), the Center of Precision Health, Department of School of Biomedical Informatics, University of Texas Health Science Center at Houston, Houston, Texas, USA (2018–2019), and in the Division of Bio-statistics, Department of Public Health Sciences, University of Miami Miller School of Medicine, Miami, Florida, USA (2017–2018). He obtained his PhD degree in the Department of Computer Science and Engineering (C.S.E.) from Jadavpur University, Kolkata, India in 2017 while his PhD work was carried out in Machine Intelligence Unit, Indian Statistical Institute, Kolkata, India as Junior Research Fellow and Visiting Scientist. He has obtained the award of Research Associateship from CSIR (Council of Scientific and Industrial Research), MHRD, Govt. of India in 2017. Dr. Mallik has more than 150 research papers in high-impact-factor peer-reviewed international journals, conferences, and book chapters. He published several books and patents. He is a member of the Institute of Electrical and Electronics Engineers (IEEE), USA, ACM and American Association for Cancer Research (AACR), USA, and Bioclues, India. He has also worked with section editors and reviewers affiliated with several well-reputed high-impact journals. His research interests include computational biology, knowledge retrieval and data mining, bioinformatics, biostatistics, and machine learning/deep learning.

Zhongming Zhao, Ph.D., M.S., is a chair professor at McWilliams School of Biomedical Informatics, The University of Texas Health Science Center at Houston (UTHealth). He holds University Chair for Precision Health and is the founding director of the Center for Precision Health. Before he joined UTHealth in 2016, he was Ingram Endowed Professor of Cancer Research, Professor (tenured) in the Departments of Biomedical Informatics, Psychiatry, and Cancer Biology at Vanderbilt University Medical Center, Chief Bioinformatics Officer of the Vanderbilt-Ingram Cancer Center (VICC), and Director of the VICC Bioinformatics Resource Center. Dr. Zhao has broad interests in bioinformatics, genomics, population genetics, precision medicine, and machine learning and has co-authored over 500 total publications in these areas (cited >25,000 times, H-index = 80). Dr. Zhao has served as the Editor-in-Chief, Associate Editor, or editorial board member of 22 journals. Dr. Zhao is the founding president of The International

Association for Intelligent Biology and Medicine (IAIBM, 2018). He was elected as a fellow in the American College of Medical Informatics (ACMI, 2021), the American Medical Informatics Association (FAMIA, 2022), and the American Institute for Medical and Biological Engineering (AIMBE, 2023).

Soumita Seth is currently serving as an Assistant Professor in the Department of Computer Science and Engineering of Future Institute of Engineering and Management, Kolkata, India, Affiliated to MAKAUT, Kolkata. Additionally, she has recently submitted her PhD thesis in the Department of Computer Science & Engineering (CSE) from a state-government university, Aliah University (AU), Kolkata, India. Previously, she completed M.Tech. and B.Tech. from the departments of CSE and IT, respectively. She is conducting her PhD research with The University of Texas Health Science Center at Houston (UTHealth), Houston, TX 77030, USA. She has academic experience of almost 7 years, full-time research experience of 2 years, and industrial experience of 2 years. Dr. Seth has more than 10 research papers in different top high-impact-factor peer-reviewed international journals, conferences, and book chapters. Her forthcoming books are with Scopus-indexed publishers. She has also worked with section editors and section reviewers affiliated with several well-reputed high-impact journals. Her research interests include computational biology, data mining, bioinformatics, pattern recognition, biological regulatory networks, biostatistics, and machine learning/deep learning.

Aimin Li is an assistant professor of Xi'an University of Technology, China. He got his master's degree from Xi'an University of Technology, and doctoral degree from Xidian University. He previously worked as a visiting scientist at the University of Texas Health Science Center, Houston, Texas, USA. His current research applications are in the areas of machine learning, bioinformatics, and regulatory networks. He has published 20+ research papers. He is also an editor of *International Journal of Computational Biology and Drug Design*, PC member of ICIBM (International Conference on Intelligent Biology and Medicine), and co-chair of BIBM IWRI 2020.

Kasmika Borah is a PhD research scholar in the department of Computer Science and IT, Cotton University, Assam, India. She completed her master's in Bioinformatics from Dibrugarh University, Assam, India. She has over seven years of experience doing research in the fields of drug design, machine learning of next-generation sequencing data, network pharmacology, molecular dynamic simulation, drug repurposing and medical dataset annotation, and image processing. She has also worked for a startup company on AI-based software development. She has published 14 articles in reputed international journals and books.

Himanish Shekhar Das received the B.Tech degree in Computer Science and Engineering from the Department of Computer Science and Engineering, National Institute of Technology Silchar, Assam, India, in 2012, the M.Tech. degree from the Department of Computer Science and Engineering, National Institute of Technology Durgapur,

Assam, India, in 2015, and the Ph.D. degree from the Department of Computer Science and Engineering, National Institute of Technology Silchar, Assam, India, in 2021. He is currently working as an Assistant Professor with the Department of Computer Science and Information Technology, Cotton University, Assam, India. Before joining Cotton University, he worked as an Assistant Professor with the Department of Computer Science and Engineering, Jorhat Engineering College, Assam, India and Birla Institute of Technology Mesra, Ranchi, India. He has published more than 30 papers in international/national journals/conference proceedings, and book chapters. His research interests include speech processing (language identification); medical image processing; machine learning; deep learning; and bioinformatics.

Contributors

Selorm Adablanu
University of Education
Winneba, Ghana

A. Alan Steve Amath
A. Alan Steve Amath is a Temporary Demonstrator in the Department of Computer Science, Faculty of Applied Sciences
South Eastern University of Sri Lanka
Sammanthurai, Ampara, Eastern Province, Sri Lanka

Rabia Musheer Aziz
Researcher Officer (Technical), Planning Department
State Planning Institute (New Division),
Lucknow, Uttar Pradesh, India

Tuward Jade Dweh
Assam Down Town University
Assam, India

Anushka Gupta
Center for Computational Biology and Bioinformatics, Amity Institute of Biotechnology
Amity University
Noida, Uttar Pradesh, India

Mohit Kumar Halder
Department of Computer Science and Engineering
University of Kalyani
Kalyani, West Bengal, India

Ruifeng Hu
Dept of Neurology,
Harvard Medical School,
Boston, MA, USA

Atikul Islam
Department of Computer Science and Engineering
Seacom Engineering College
West Bengal, India

&

Department of Computer Science and Engineering
University of Kalyani
Kalyani, Nadia, West Bengal, India

Gaurav Jain
Department of Computer Science & Applications
Dr. Hari Singh Gour Vishwavidyalaya
Sagar, Madhya Pradesh, India

Anita Kachari
Technical Assistant, Biological Sciences and Technology Division
CSIR-North East Institute of Science and Technology
Jorhat, Assam, India

R. Vijaya Lakshmi
Assistant professor
Faculty of Management Studies
ICFAI University
Raipur, Chhattisgarh, India

Pranjali Malviya
Department of Computer Sciences
VIT Bhopal University
Bhopal, India

Kalyani Mali
Department of Computer Science and Engineering
University of Kalyani
Kalyani, West Bengal, India

Karan Mech
Department of CSE
Tezpur University
Assam, India

Abhay Mishra
Department of Pharmacy
Indira Gandhi National Tribal University
Amarkantak, Madhya Pradesh, India

M. M. Mohamed Mufassirin
Dr. M.M. Mohamed Mufassirin is a Lecturer in Computer Science at the Department of Computer Science, Faculty of Applied Sciences
South Eastern University of Sri Lanka
Sammanthurai, Ampara, Eastern Province, Sri Lanka

Dola Mukherjee
Amity Institute of Biotechnology
Amity University
Kolkata, India

Priyanka Narad
Indian Council of Medical Research Headquarters
Ansari Nagar East, New Delhi, India

Kamlesh Kumar Pandey
Department of Vocational Education (Software Development)
Indira Gandhi National Tribal University
Amarkantak, Madhya Pradesh, India

Deepranjan Pathak
Biological Sciences and Technology Division
CSIR-North East Institute of Science and Technology
Jorhat, Assam, India

Tejaswini Pradhan
Assistant Professor
Department of Mathematics
Kalinga University
Raipur, Chattigarh, India

Guimin Qin
School of Computer Science and Technology
Xidian University
Xi'an, China

GVV Jagannadha Rao
Department of Mathematics
Faculty of Science
The ICFAI University
Raipur, Chhattisgarh, India

Pinju Saikia
Research Scholar, Department of CSE
Tezpur University
Assam, India

Suvobrata Sarkar
Department of Computer Science and Engineering, Institute of Engineering and Management (IEM)
University of Engineering and Management
West Bengal, India

&

Department of Computer Science and Engineering
University of Kalyani
Kalyani, Nadia, West Bengal, India

Abhishek Sengupta
Center for Computational Biology and Bioinformatics
Amity Institute of Biotechnology
Amity University
Noida, Uttar Pradesh, India

Muskan Syed
Center for Computational Biology and Bioinformatics
Amity Institute of Biotechnology
Amity University
Noida, Uttar Pradesh, India

Navneet Kumar Verma
VIT Bhopal University's School of Advanced Science and Language
Kothrikalan, Sehore, Bhopal, India

Abrar Yaqoob
VIT Bhopal University's School of Advanced Science and Language
Kothrikalan, Sehore, Bhopal, India

Preface

THE ADVENT OF HIGH-THROUGHPUT Sequencing technologies has transformed the landscape of molecular biology as well as biomedical research by enabling the comprehensive measurement of molecular entities across multiple omics layers, including genomics, transcriptomics, proteomics, metabolomics, metagenomics, and epigenomics. These technological advances have resulted in unprecedented volumes of high-dimensional and heterogeneous data. While such datasets provide a unique opportunity to advance our understanding of complex biological systems, they also pose substantial analytical challenges. Chief among these are the issues of dimensionality, sparsity, redundancy, and noise, which complicate the identification of robust and biologically meaningful patterns.

Feature extraction and selection are essential techniques for overcoming these challenges. They not only make it easier to simplify data and increase model performance, but they also make results easier to understand, which is crucial for translational applications and biomedical research. Over the past decade, methodological advances in statistics, machine learning, and artificial intelligence have greatly enriched the toolbox available for analyzing omics data. However, the field remains dynamic, with ongoing developments continually reshaping best practices and opening new avenues for exploration.

This book provides a systematic and comprehensive treatment of state-of-the-art methodologies in omics data analysis. It encompasses theoretical foundations, algorithmic frameworks, and practical applications, thereby aiming to serve as both a scholarly reference and a resource for applied researchers. Particular emphasis is placed on methods that address the curse of dimensionality, accommodate the unique characteristics of different omics platforms, and integrate multi-omics data to support systems-level biological insights.

The intended readership includes academic researchers, clinicians, and professionals in bioinformatics, computational biology, data science, biotechnology, and related disciplines. By consolidating current knowledge and highlighting emerging trends, this book aspires to contribute to the advancement of computational strategies for omics analysis and to support the broader goal of translating molecular data into actionable biological and clinical insights.

We compiled the chapters written by various experienced researchers working in the molecular biology, biotechnology, computational biology, bioinformatics, computer science, and related areas. Chapter 1 will introduce the statistical and state-of-the-art machine learning approaches in omics data interpretation. Chapter 2 will introduce feature selection and extraction techniques, and integration of omics data. Chapters 3 and 9 introduce the role of bioinformatics and feature selection in metagenomics data. Chapters 4–6 describe the feature selection and extraction techniques of gene selection for specific diseases. Chapters 7 and 10 will introduce feature selection in machine learning and deep learning. Chapter 8 described hybrid feature gene selection techniques. We definitely hope that the present book will be beneficial for all the researchers, clinicians, and industrialists.

Editors
Saurav Mallik,
Zhongming Zhao,
Soumita Seth,
Aimin Li,
Kasmika Borah,
Himanish Shekhar Das

CHAPTER 1

Machine Learning and Statistic-Based Feature Selection and Extraction Approach for Omics Data

Kamlesh Kumar Pandey, Abhay Mishra and Gaurav Jain

1.1 INTRODUCTION

Omics sciences have seen a significant transformation in recent decades and have been propelled by swift progress in high-throughput sequencing and various omics technologies [1]. The Human Genome Project commenced in 1990 to delineate and analyse the entire sequence of human genes. It achieved 85% completion by 2003 at a cost of approximately \$3 billion and finalized its assembly by 2022. This endeavour has markedly improved the capacity for analysing a genome within one day at a reduced \$1,000–\$5,000 cost [1, 2].

The omics data includes diverse biological and bioinformatics data such as genomes, microarrays, transcriptomes, metabolomes, and proteomes [3]. The number and quality have substantially improved because the rapid advancement and innovation of technology in the medical area have produced high-quality data in large-scale dimensions [4, 5]. The variety of biological knowledge has significantly expanded through diverse omics methodologies. Sequential technology is used to generate omics data. The spatial transcriptomics, single-cell RNA sequencing (scRNAseq), and RNA sequencing have emerged as significant advancements in transcriptomic technologies during the past years [6]. Table 1.1 shows the variety of omics types, molecules, data, and their applications.

RNA sequencing offers a comprehensive perspective on transcriptional activity while obscuring cellular heterogeneity. Single-cell transcriptomes are analysed by scRNA-seq which reveals cellular variety at the expense of geographical context [7]. Spatial transcriptomics integrates gene expression profiling with spatial data. It maintains the tissue architecture to delineate the localization of gene expression within the tissue environment. It provides insights into the spatial structure of cellular processes and interactions.

DOI: 10.1201/9781003591658-1

TABLE 1.1 Omics Data and Their Molecules and Applications

Omics Type	Molecule	Data	Applications
Genomics	DNA	Genome sequence	Disease gene mapping, GWAS
Transcriptomics	RNA	mRNA levels	Gene expression profiling
Proteomics	Proteins	Protein levels	Drug discovery, pathway analysis
Metabolomics	Metabolites	Metabolite profiles	Biomarker discovery
Epigenomics	Epigenetic marks	DNA methylation	Cancer, gene regulation studies
Microbiomics	Microbes	16S rRNA data	Gut health, immunity research

The transformation of omics data into a multidimensional framework is feasible. It is driven by innovations that enhance data collection from various sources [8]. Omics data include mutation profiles, the number of variations, microRNA expression, DNA methylation, gene expression levels, and so on. It is obtained from a variety of tissues and donor populations. It is obtained from various tissues and donor populations related to biological and bioinformatics applications. Omics technologies facilitate extensive biological understanding by acquiring data across various molecular strata that encompass genomics (whole-genome sequencing), transcriptomics (RNA-seq, single-cell RNA-seq), proteomics (mass spectrometry, CITE-seq), metabolomics (NMR, LC-MS), epigenomics (ChIP-seq, ATAC-seq), and methylomics (bisulfite sequencing) [2, 9, 10]. The different types of omics data and their associated molecule and data are described in Figure 1.1.

The omics data comprises several samples and attributes. The features significantly exceed the number of samples [11]. The dimensionality reduction safeguards against potential dimensionality catastrophe, and feature selection (FS) has demonstrated exceptional efficacy in data preparation in high-dimensional data. The molecular connections and enduring regulatory network provide insights into the essential mechanisms that

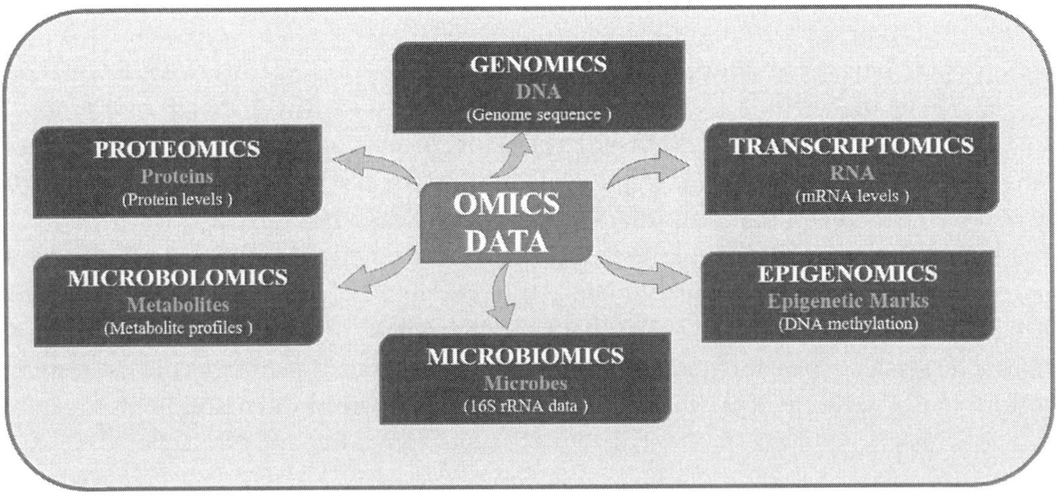

FIGURE 1.1 Different types of omics data.

regulate cellular functions and disease states. The simultaneous examination of several samples under diverse processes produces datasets by utilizing the multiple samples and conditions. It demonstrates another aspect of omics data's expanding multidimensional complexity [12]. The omics data set consists of several investigations, cell and tissue information, healthy and diseased tissues, pharmacological responses, cellular monitoring, etc.

The substantial increase of complex and multifaceted omics data presents novel difficulties and possibilities in omics data processing [10, 13, 14]. The data emergence underscores the essential requirement for strong analytical methodologies. It manages the complex nature of datasets and derives significant insights. The impressive computational resources have been allocated to the creation of advanced analytical tools that can effectively process and analyse high-dimensional omics data [9, 15].

Tensors are one of the multidimensional arrays that generalize matrices and help to provide a natural representation of multidimensional data. Tensors inherently express multidimensional interactions across a holistic perspective on biological processes. The two-dimensional tensors of omics data consist of samples and features of the biological processes. The three-dimensional tensor omics data uses multiple samples, features, and conditions [2].

The increase in dimensionality leads to an exponential rise in the volume of data required for accurate analysis and model development. The prevalent method of addressing high-dimensional datasets is to reduce the set of features that retains information to the greatest extent possible. Microarray data exemplifies this category of minor sample issues where each data point contains approximately 450,000 variables. It processes many data points, incurring significant computational expenses [16, 17].

The dimensionality of a dataset increases substantially, and it becomes increasingly challenging to demonstrate statistical significance due to the sparsity of relevant data within the dataset in issue. The overfitting is another challenging issue of the curse of dimensionality that misinterprets minor oscillations as significant variance in the data. Therefore, the analysis and model results are irrelevant to the decision-making approach. This challenge escalates due to extraneous features. Noise is an additional problem associated with the high-dimensional dataset that impacts the variance of measured variables. It arises from measurement inaccuracies and inherent variability of the high-dimensional omics dataset. The noise of high-dimension data affects the attribute and class [17, 18]. The essential challenges of the omics data are shown in Figure 1.2.

The FS and extraction methods are used to handle high-dimensional omics data and resolve the above issues. The FS ensures omics data is comprehensible for developing straightforward and practical models. It improves the effectiveness and precision of data mining, machine learning, sequence analysis, microarray analysis, spectrum analysis, and other modelling approaches [19]. The researcher has introduced numerous computational FS approaches to diminish the dimensionality of omics datasets [20]. These techniques require prior specification of the number of features as a prerequisite. This chapter aimed to examine the efficacy and effectiveness of the FS approach in the medical application according to biological or natural principles [5]

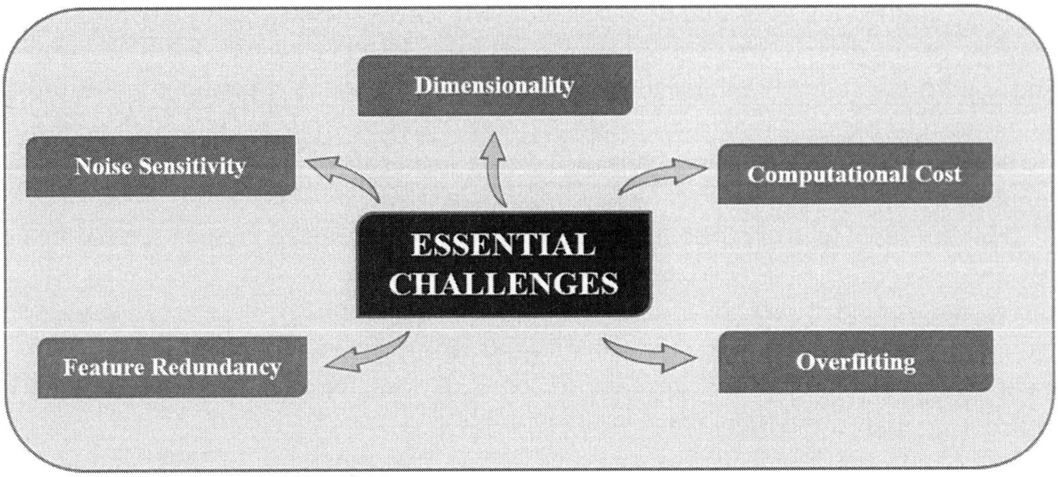

FIGURE 1.2 Essential challenges of the omics data.

The fundamental objective of this work is to analyse the FS and extraction methodologies in bioinformatics, biostatistics, and other medical fields that employ omics data for modelling and data analysis. The omics data research has challenges due to high dimensionality and small sample sizes. The remainder of the section is classified into seven distinct subsections. The second section delineates the challenges associated with high-dimensional omics data. The third section establishes the foundation of FS and its structure. The fourth section analyses the various FS classes and their corresponding techniques. The fifth segment illustrated the feature extraction (FE) framework and its corresponding methodologies. The sixth section establishes the stability metric for the FS and extraction methodology. The sixth section delineates the properties of stability measures for FS. The seventh section finishes the work and outlines future endeavours.

1.2 CHALLENGING FACTOR OF HIGH DIMENSIONAL OMICS DATA

Researchers face significant challenges in computing algorithms for assessing omics datasets due to the high dimensionality. The data dimensionality is characterized by small samples and significant variables. High dimensionality is a significant challenge in omics data, and it describes other issues related to omics data [21]. The challenging factor of the high-dimensional omics data is listed in Figure 1.3.

1.2.1 Small Sample Size

The predominant issue that several studies address in omics data is a small sample size, which frequently falls below 100. The primary challenge here is that the error estimate is significantly impacted by small samples [22]. The absence of adequate error estimation and the selection of an ineffective method contribute to the less-than-ideal performance of machine learning algorithms. The small number of samples is inadequate for machine

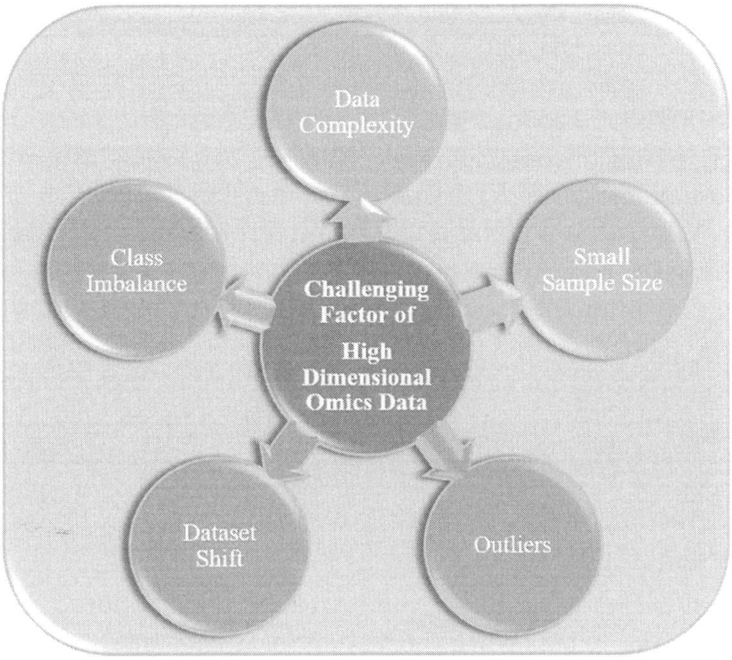

FIGURE 1.3 Challenging factor of the high-dimensional omics data.

learning, leading to imprecise classification results. The suitable validation technique determines the classification error. It presents an additional challenge to deal with small sample sizes [21].

1.2.2 Class Imbalance

Class imbalance presents a significant challenge in omics data. This scenario occurs when a dataset exhibits a bias in favour of a predominant class [23]. Classical machine learning algorithms and data analysis methods demonstrate a bias towards classes with small sample sizes. The learning principles for predicting these major situations are positively reinforced for performance accuracy. The recommendations for predicting minority class samples are frequently overlooked and occasionally viewed as noise. This illustrates that prediction algorithms typically favour more generic criteria. The minority class is frequently incorrectly classified in comparison to situations in other classes [21, 24].

The obstacle in the learning process cannot be ascribed to class imbalance. This difficulty stems from the limited sample size and the high complexity of omics data. The conventional preprocessing techniques frequently employed to tackle this challenge include under sampling and oversampling approaches. The previous method produces a subset from the original dataset by omitting specific instances from the majority class. This method involves forming a larger set by duplicating certain examples from the initial dataset and producing new examples derived from the original data. The Synthetic Minority

Oversampling Technique (SMOTE) is a resampling method that defines class imbalance issues in high-dimensional data [21]

1.2.3 Data Complexity

Data complexity refers to contemporary challenges that assess data attributes during the classification process, such as class overlap, decision boundaries, and separability [25]. The complexity pertains to the detailed structure of omics data incorporating gene analysis, RNA, DNA, and protein analysis. The complicated architecture generates a high degree of dimensionality, and this complexity is associated with the most significant prediction error. Thus, it is essential to reduce complexity and minimize prediction errors. The class overlapping evaluation assesses ability of an identified component to differentiate across classes. The researchers analysed the range and distribution of the weighted values within each category and aimed to detect overlaps across different categories [21, 25].

1.2.4 Dataset Shift

The dataset shift occurs when the original datasets are partitioned into training and testing segments. This occurs when the holdout data encounters a scenario where the distribution of classes among inputs and outputs differs between the training and testing stages [26]. The prevalent assumption that training and test data exhibit similar distributions is frequently contravened in real-world situations, possibly impacting omics data processing [21].

Machine learning algorithms often encounter substantial challenges in extracting the test and training subset from omics datasets. The test data demonstrates an estimated tenfold variation in total omics intensity relative to the training data [27]. The 26–74% test distribution significantly differs from the training data distribution of 49–51%. Insufficient FS procedures result in specific classifiers categorizing all samples into a singular class [27]. Dataset shift arises during the cross-validation method due to multiple data subsets [28].

1.2.5 Outliers

The outlier is a significant problem in omics data, adversely impacting the accuracy of the modelling method. High-dimensional data contain several outliers and predominantly outlier-associated dimensions irrelevant to the omics dataset [29]. The particular datasets comprise misclassified samples and labelled as potentially contaminated when they need to be categorized as outliers. They adversely affect the selection of valuable omics data from the sample. Outlier identification samples must be considered a preprocessing step in classification, adversely impacting FS and the resulting classification [30].

1.3 FEATURE SELECTION

FS is a fundamental preprocessing phase with omics data for executing tasks such as pattern recognition, classification, regression, and clustering [31]. This method presents a superior strategy for deriving optimal characteristics from datasets with high dimensionality [32]. It analyses high-dimensional next-generation sequencing (NGS) datasets and eliminates irrelevant, noisy expressions and redundant features to facilitate model creation [33].

The FS strategy reduces the computation expenses and improves model learning efficiency. It also diminished the algorithm's memory storage requirements and improved model accuracy comprehension [34, 35]. The primary objective of the FS procedure is to improve the performance and efficiency of models in regression, classification, and clustering within machine learning [35]. The applications of FS in NGS datasets consist of genes, proteins, DNA, RNA, and other omics data types. The omics dataset consists of class identification, biomarker discovery, mitigation of model overfitting, discerning pertinent information, and so on [36, 37]. The FS approaches are integrated with machine learning and deep learning algorithms. It evaluates the predicted efficacy of the selected features [38, 39]. Figure 1.4 illustrates the primary advantages of the FS method.

The FS technique constitutes a modelling and data analysis procedure component. The modelling technique initially pre-processed the omics dataset by addressing noise and outlier data, among other factors. The pre-processing stage transforms the data from one format to another for the straightforward interpretation of the pertinent feature set. It employed the pre-processed omics data to identify the most pertinent features from the high-dimensional omics dataset. It aligns with the specific application and objectives at hand. Figure 1.5 illustrates the conceptual foundation of the FS strategy.

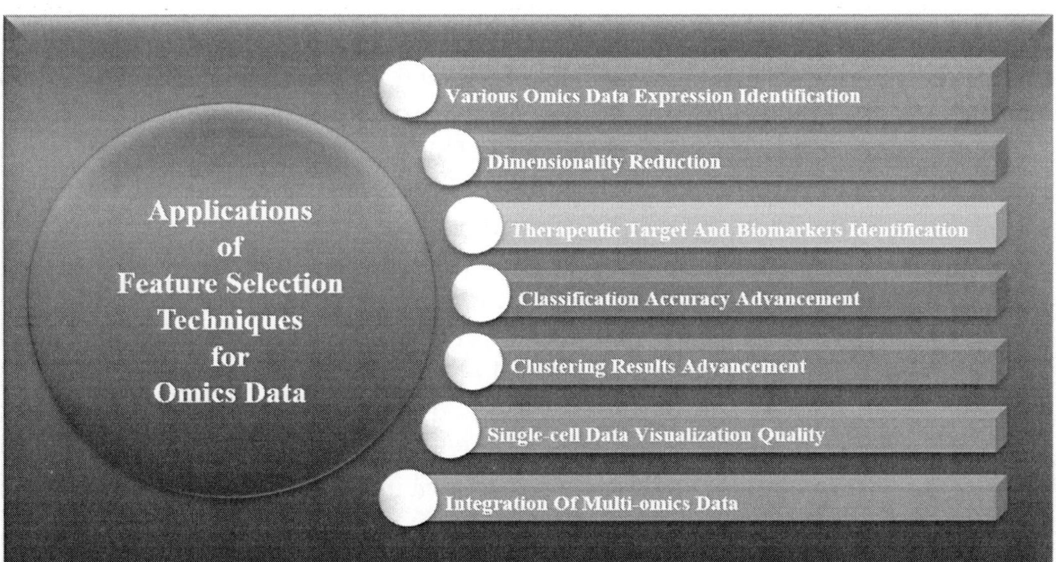

FIGURE 1.4 Fundamental advantages of the feature selection approach.

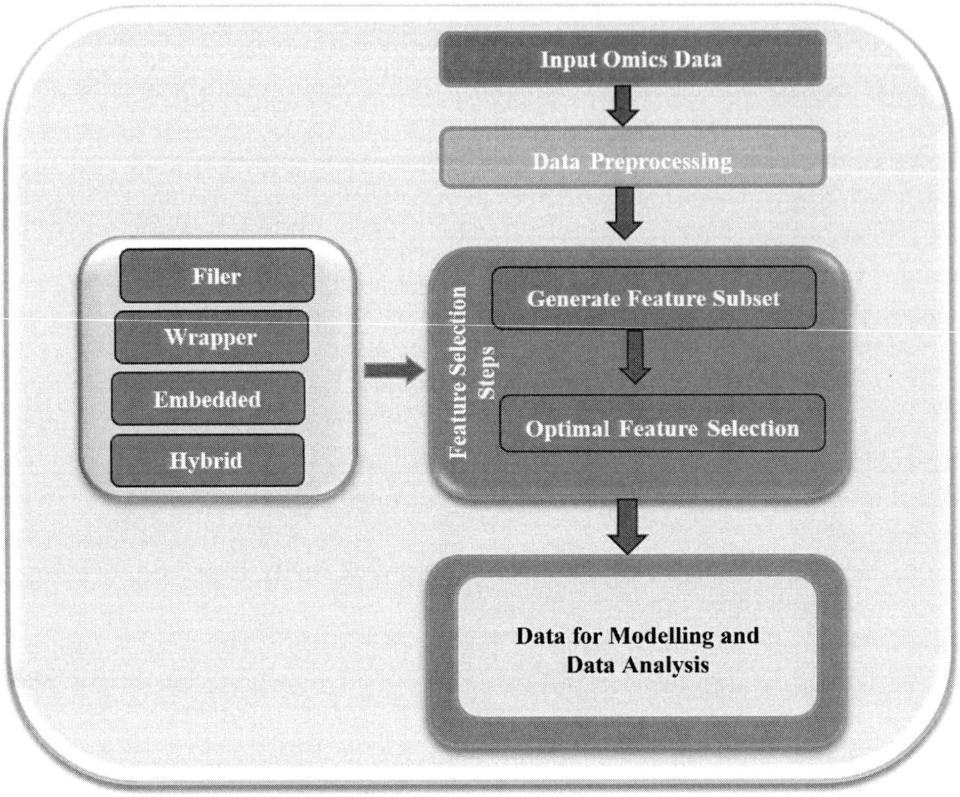

FIGURE 1.5 Conceptual framework of the feature selection process.

1.4 FEATURE SELECTION METHODS FOR HIGH-DIMENSIONAL DATA

The FS strategy is known as dimensionality reduction. FS is a fundamental strategy for identifying the most appropriate features in high-dimensional data. The supervised mutual information-based feature selection (MIFS) first selected a small feature subset from high-dimensional omics data. The second step selected the optimized feature set for the modelling using the wrapper and filter method. The wrapper methods utilize classification techniques to identify the significant features from the diminished feature set [40].

FS strategies are categorized into filters, wrapper, embedded, and hybrid classes. The FS techniques are divided into statistical and machine learning. The statistical FS is also known as bio-statistical, and it is classified as filter [35, 41]. The machine learning-based FS strategies are divided into unsupervised, supervised, and semi-supervised learning under the wrapper. The embedded class leveraged the merits of the wrapper and filter FS methodologies. The classification of the FS methods is shown in Figure 1.6, and their core details are described in Tables 1.2 and 1.3.

Table 1.2 presented herein elucidates the fundamental principles, concepts, and operational processes of the FS class as applied to omics data. The fundamental concept illustrates the essential notion of the FS category. The fundamental concept illustrates the foundational principle for the execution. The fundamental functionality illustrates the

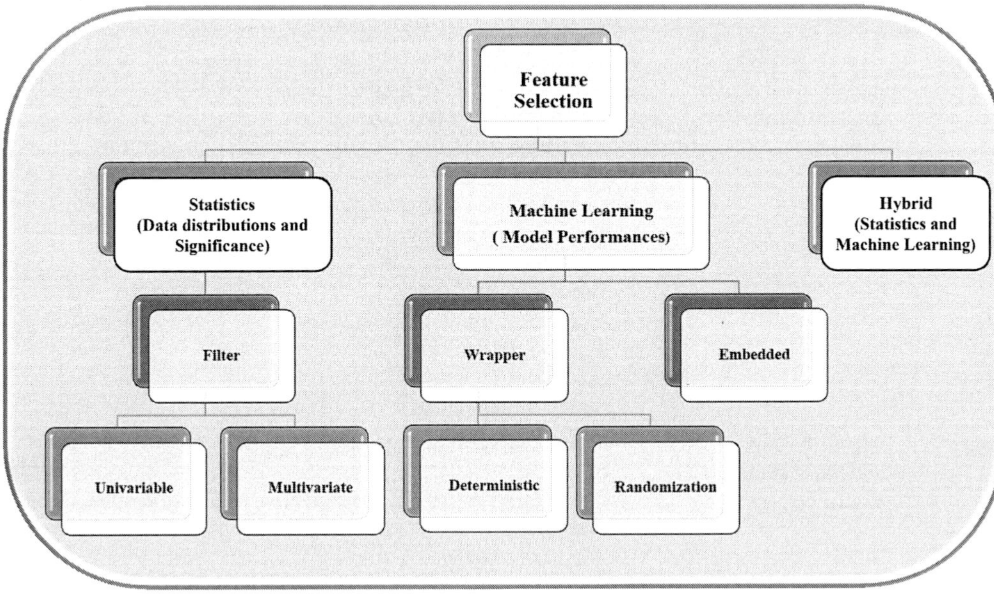

FIGURE 1.6 Classification of the feature selection.

essential operational process and primary capabilities of the various FS classes applied to omics data. This table also illustrates which model supports the machine learning classifier during the FE process.

Table 1.3 outlines the objectives, advantages, and limitations associated with the FS sort in omics data analysis. The objective delineates the function of the FS class within omics data. The advantages illustrate the strengths of the class, while the limitations highlight its weaknesses. Filter-based techniques are useful for preprocessing and exploratory analysis of omics data. Wrapper methods are particularly suitable for the ultimate

TABLE 1.2 Core Principle, Idea and Functionality of Feature Selection Class on Omics Data

Class	ML Model	Core Principle	Core Idea	Core Functionality
Filter	No	Statistical evaluation	Features are ranked based on statistical scores without involving any model training.	• Large datasets with many features • Early-stage feature reduction • Model-agnostic filtering is needed
Wrapper	Yes	Model performance	Evaluates subsets of features by training models and selecting those that yield the best performance.	• Small to medium datasets • Feature interactions matter • Accuracy is prioritized
Embedded	Yes	Model-inherent selection	Feature selection is inherently integrated into the model's training phase through the application of regularization techniques or by utilizing importance scores.	• Regularized models • Its need for efficiency with good accuracy • Medium-size data

TABLE 1.3 Objective, Advantages and Limitations of Feature Selection Class on Omics Data

Class	Objective	Advantages	Limitations
Filter	Quickly remove irrelevant and low-variance features before the modelling	• Fast and scalable for high-dimensional datasets • The classifier does not require feature extraction • It suitable for initial screening and dimensionality reduction • Easily interpretable statistical measures • It useful for limited labelled data • Easy to implement • No overfitting risk (model-independent)	• It ignores inter-feature dependencies • It retains irrelevant or redundant features • It cannot tailor selection to specific classifiers or tasks • It sensitive to data noise and feature scaling
Wrapper	Extract the feature subsets that maximize model performance	• It explores feature interactions with classifiers • It yields highly predictive feature subsets • It reduces irrelevant features effectively • It uses the heuristic and metaheuristic wrappers class to avoid local optima • It suitable for small to medium sample sizes • It captures interactions and high accuracy	• High computational cost, especially with large feature sets • High risk of overfitting without cross-validation • Performance is classifier-dependent • It does not well-suited for real-time or streaming data • It requires careful parameter tuning
Embedded	Perform feature selection during model learning for efficiency and accuracy	• It integrates selection during model training • It balances speed and accuracy better than wrappers • It captures non-linear feature dependencies • It works well with tree-based and regularized models. • It is efficient and integrates selection into training	• It heavily model-specific, and different models may select different features • It does not expose feature relationships clearly • It's slower than filters for large datasets • Selection criteria tied to model assumptions • It might miss important features not favoured by the model

enhancement of features when the precision of predictions is of utmost importance. The embedded approaches function to harmonize the omics data within structured modelling frameworks.

1.4.1 Filters Class (Statistical Feature Selection Techniques)

Filter FS methods are important for the dimensionality reduction in biological analysis of the different types of the omics data. The analysis highlights the strongest distinguishing attributes derived from their fundamental traits within the initial dataset. The filters class describes the quality of features group that solely focuses on the characteristics of that group and is independent of any learning methods. The fundamental characteristics of the information are employed to validate the pertinence of features regarding inclusion in the feature subset [42].

The statistical strategies utilized as a potent instrument for feature dimensionality reduction in diverse microarrays and NGS high-dimensional multi-omics datasets. The efficiency of statistical FS methods varies across distinct biological sequencing datasets [41]. It is essential to establish the criteria for selecting the most accurate and reliable FS methods from all available techniques for dimensionality reduction. Statistical FS strategies are categorized as filter and used inside the wrapper and embedding classes [35, 39, 43].

The filter methods perform FS based on the ranking and search space methods. The filter approach is categorized into Rank and Subset distinct categories to extract the relevant features from the omics dataset [44]. These class uses the univariate and multivariate statistical techniques to extract the features from the omics data. The implementation process of the filter-based FS process is shown in Figure 1.7.

The univariate method is classified as a parametric and non-parametric class. Some filter-based statistical FS strategies are detailed below and help reduce the dimensionality of omics datasets [39]. The rank-based FS approach employs univariate statistical method to validate the ranking of features, whereas disregarding the relationships among features [45]. This approach fails to find redundant features.

The multivariate class encompasses the entire search space. It assesses a complete feature subset based on its significance and redundancy [39]. The FS procedure utilizes multivariate

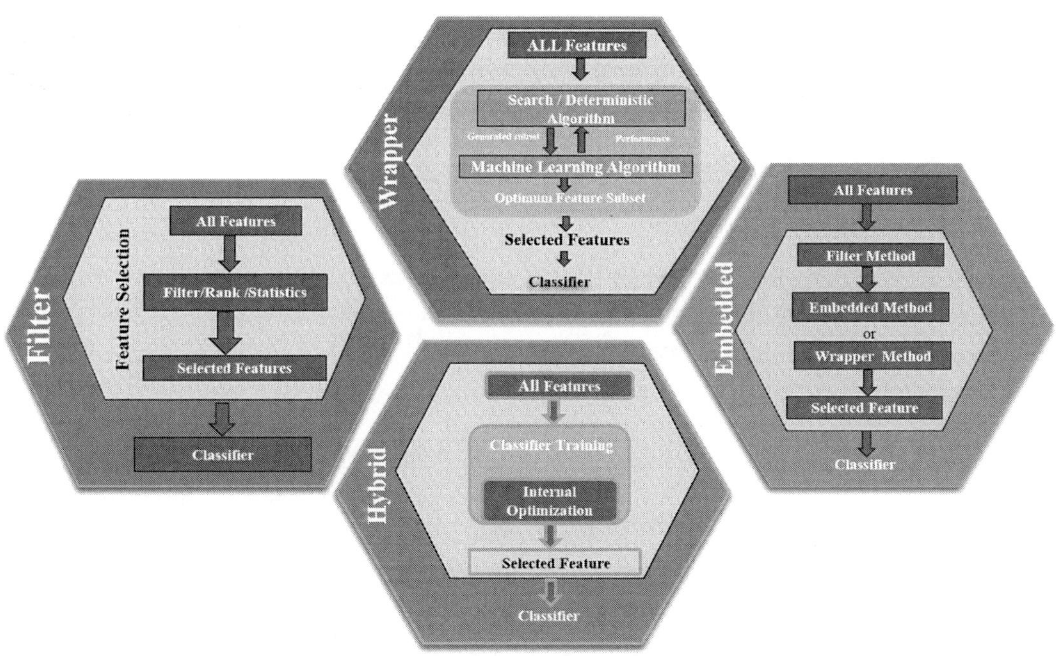

FIGURE 1.7 Implementation process of the feature selection methods.

statistical methods to assess the ranking of the complete collection of features. The benefit of employing a multivariate statistical approach lies in its ability to consider inter-feature relationships, which eliminates the need for a separate classification mechanism. It exhibits its superior efficiency compared to the wrapper method regarding processing complexity. The principal disadvantage of the multivariate strategy is its diminished speed and stability relative to the univariate ranking method. Several computationally efficient filter algorithms for high-dimensional data have emerged in recent years. This subsection elucidates various filter class-based methodologies for the selection of features in omics data [46].

1.4.1.1 MIFS (Mutual Information Feature Selection)

The MIFS approach eliminates replicate features in the omics dataset and is significant to the class variable. The MIFS formulation shown in Equation 1.1 [40].

$$MIFS = I(f_i, C) - \beta \sum_{f_s \in S} I(f_s, f_i) \tag{1.1}$$

The MIFS selects the significant m dimensions from a set of d dimensions. The S is predetermined set of features and candidate feature f_i is selected at each stage. It is based on its relevance to the class $I(f_i, C)$ without considering the joint mutual information (MI) of attribute set and class variable C. The parameter β significantly influences the selected feature set.

1.4.1.2 MIFS-U (Mutual Information Feature Selection-Uniform Prior)

The MIFS-U is an improved version of the MIFS. It is based on uniform information processing with designated features [47]. The MIFS-U is employed to identify the potential characteristic at each stage, and its formulation is shown in Equation 1.2 [40].

$$I(f_i; C) - \beta \sum_{f_s \in S} \frac{I(f_s; C)}{H(f_s)} I(f_i; f_s), \tag{1.2}$$

where $H(f_s) = \Sigma P(f_s) \, logP(f_s)$, *where* $f_s \in S$ and it is present as entropy. The assumption of present as uniform probability distribution and take as guarantees that the class C does not affect the ratio of the MI between f_i *and* f_s as well as entropy. The MIFS-U approach provides a more accurate evaluation of the MI measure compared to the MIFS method.

1.4.1.3 mRMR (Minimum Redundancy Maximum Relevance)

The minimum redundancy maximum relevancy (mRMR) criterion evaluates the redundancy and importance of the features. The redundancy is calculated using the MI between pairs of features. The MI assesses the significance of characteristics in relation to each attribute and the class labels [46, 48].

The mRMR FS algorithm is a parameter-free FS technique [48]. It identifies significant data subsets related to the variables and is termed optimum relevance. The mRMR exhibits several forms across multiple domains and identifies traits with the highest correlation with the transportation variable. The algorithm uses the heuristic methods of sequential backward, forward, and floating selections to determine the relevant features. The characteristics are identified as mutually distant and maintaining parallelism with the class variable. The mRMR technique substitutes the β in MIFS and MIFS-U by $\frac{1}{|S|}$. $|S|$ determines the cardinality of the features set. The FS formulation of the mRMR shown in Equation 1.3 [40].

$$I(f_i;C)-\frac{1}{|S|}\sum_{f_s\in S} I\left(f_i;f_s\right), \tag{1.3}$$

1.4.1.4 Normalized Mutual Information Feature Selection (NMIFS)

The approach of normalized mutual information feature selection (NMIFS) enhanced the accuracy of the FS process through iterative method of mRMR. The NMIFS method employed normalized MI to evaluate the redundancy among the selected characteristics. The entropy of the class variable attains its maximum value of one whenever its distributions of the class constant are uniform in binary classification. The entropy value is expected to be below one, resulting in a divergence in the selection criteria when comparing a single class to multiple classes. The disparity between these two variables may occasionally cause the FS method to prioritize a characteristic based on the maximum value of the left-hand variable, while disregarding the right-hand variable. The NMIFS method's selection criterion is Equation 1.4 [33, 49].

$$I(f_i;C)-\frac{1}{|S|}\sum_{f_s\in S} \hat{I}\left(f_i;f_s\right), \tag{1.4}$$

The $\hat{I}$ is the normalized MI, and it is formulated in Equation 1.5.

$$\hat{I}\left(f_i;f_s\right)=\frac{I\left(f_i;f_s\right)}{\min\left(H\left(f_s\right),H\left(f_i\right)\right)}. \tag{1.5}$$

The NMIFS method utilizes a normalized measure to tackle the disparity between the significance and overlap aspects of the selection criteria applied in the MIFS, mRMR, and MIFS-U procedures for feature–feature analysis.

1.4.1.5 Improved Normalized Mutual Information Feature Selection (INMIFS)

The INMIF is the improved approach of the NMIFS FS approach [50]. Normalized MI is utilized in class significance and feature-feature overlap. The selection requirement for the INMIFS approach is delineated in Equation 1.6 [40].

$$\hat{I}\left(C;f_i\right)-\frac{1}{|S|}\sum_{f_s\in S} \hat{I}\left(f_s;f_i\right) \tag{1.6}$$

The $\hat{I}$ determines the normalized MI. The INMIFS methodology aims to equalize the magnitudes of left and right-handed components of the selection criterion.

1.4.1.6 VWMRmR (Variable Weighted Maximal Relevance Minimal Redundancy)

The VWMRmR is the NMIFS approach. The features of the omics data are selected based on the weighted maximum relevance minimum redundancy method. The number of features already selected linearly augments the relative weight of redundancy. The criteria for selecting this strategy are presented as Equation 1.7 [40].

$$\hat{I}(C;f_i)-\left(1+\frac{w\cdot|S|}{k}\right)\cdot\frac{1}{|S|}\sum_{f_s\in S}\hat{I}(f_i,f_s) \tag{1.7}$$

The $\hat{I}(f_i,f_s)$ represents the normalized MI between f_i *and* f_s features. The variable w signifies a user-defined positive parameter that regulates the relative importance of redundancy in relation to relevance. The ***S*** is the selected feature set and k represents the number of features.

1.4.1.7 Term-Variance (TV)

The term variance (TV) is the unsupervised FS approach of the omics data. It extracts the features based on the rank [51]. The formulation of the TV is shown as Equation 1.8. The features X_i are ranked according to their sample variance.

$$\mathrm{TV}_i=\mathrm{var}(X_i)=\frac{1}{n}\sum_{j=1}^{n}\left(X_{ij}-\bar{X}_i\right)^2, \tag{1.8}$$

where $\bar{X}_i$ *is* the sample mean of feature Xi, $\bar{X}_i=\left(\frac{1}{n}\right)\sum_{j=1}^{n}X_{ij}$.

1.4.1.8 Laplacian Score (LS)

The Laplacian score (LS) posits that data belonging to the same class are proximate. The local geometric framework is essential for distinguishing feature identification. The significance of a feature is determined by its capacity for local preservation. The LS extracts the relevance feature as compared to TV [46].

The LS is a typical univariate unsupervised feature ranking technique. It uses the geometric structure of neighbourhood graph data to reduce dimensionality. The LS utilized the highest degree of locality-preserving capability to identify the relevant informative aspects. The purpose of the LS approach relies on the perspectives of two closely connected data points [52]. This approach employs the Laplacian matrix to define the constructed graph and the manifold structure of data that exhibits maximal weight characteristics. The LS methodology has the potential to diminish the dimensionality of various data type of omics datasets. The derivation of the LS method presents Equation 1.9 [39].

$$L_r = \frac{\tilde{j}_r^S L^u \tilde{j}_r}{\tilde{j}_r^S D^u \tilde{j}_r} \tag{1.9}$$

where L^u determines the graph Laplacian, D^u shows the diagonal matrix of the graph, and $\tilde{j}_r = j_r - \frac{j_r^S D^u 1}{1^S D^u 1} \cdot 1$

1.4.1.9 Locality Sensitive Laplacian Score (LSLS)

The Locality Sensitive Laplacian Score minimizes local within-class data while maximizing local between-class data. It helps classification by integrating discriminative information into a local geometric structure [53]. The LSLS methodology is based on the unsupervised technique. It integrates class label information into the graph through the utilization of the Laplacian matrix. The LSLS is a significant supervised method for gene subset selection in NGS data. The FS strategy that integrates the LSLS and wrapper techniques to identify significant feature subsets inside omics datasets [54]. The LSLS Equation 1.10 is defined as follows [39].

$$LS_r = \frac{\breve{g}_i^T \left(\alpha L_w - (1-\alpha) L_b\right) \breve{g}_i}{\breve{g}_i^T D \breve{g}_i} \tag{1.10}$$

The $\breve{g}_i$ is employed to maintain the local geometric structure in both within-class and between-class scenarios through the appropriate use of the Laplacian matrix.

1.4.1.10 Extended LS (LSE)

The extended LS (LSE) employs a distance-based entropy approach to enhance its performance [55]. The LS employs k-means clustering for feature identification, whereas LSE utilizes a distance-based entropy approach. The LSE demonstrates enhanced stability and accuracy in the context of high-dimensional omics dataset [46].

The LS and LSE methodologies are grounded in the spectral FS. It identifies an accurate feature set by sample similarity measurement. The spectral FS is used to construct the unsupervised and supervised oriented FS for farmework. The spectral FS algorithms extract individual features and cannot identify redundant features.

1.4.1.11 Relief Method

The Relief method emphasizes the importance of the predictive features in relation to the metric. It exhibits a relationship with its values and distinguishes between examples of identical and different categories. It selects at random data from the training set and modifies the importance of each characteristic based on the disparities noted between the selected instance and its nearest counterparts. The most comparable instance is chosen from both the same category and the contrasting categories. It operates efficiently for datasets with a high number of dimensions and retains redundant features [46].

1.4.1.12 Correlation-based Filter Selection (CFS)

The selection process of CFS based on correlation is rooted in the principles of test theory. It utilizes a collection of heuristics to validate the sufficiency of feature subsets. This approach evaluates the effectiveness of distinct attributes in forecasting class labels and examines their interrelationships. This study focuses on the treatment of absent values within the omics data set [46, 56–58].

1.4.1.13 Fast Correlation-Based Filter (FCBF)

The Fast Correlation-Based Filter (FCBF) employs a relevance-redundancy methodology. It first computes each feature's symmetrical uncertainty (SU) and organizes these values into a list. It eliminates the superfluous elements from the compilation. The significance of feature relevance is considered insufficient for the process of FS in high-dimensional datasets. It prompts the introduction of a novel framework that decouples relevance from redundancy analysis [59–61]. The formulation of the FCBF FiR_i shown in Equation 1.11 [46]

$$\mathrm{FiR}_i = \frac{\left|\bar{X}_i^{(0)} - \bar{X}_i^{(1)}\right|}{\sqrt{\mathrm{var}(X_i)^{(0)} + \mathrm{var}(X_i)^{(1)}}}, \tag{1.11}$$

In this formulation, $\bar{X}_i^{(0)}$, $\bar{X}_i^{(1)}$, $\mathrm{var}(X_i)^{(0)}$, $\mathrm{var}(X_i)^{(1)}$ are the sample means and variances of feature i, for the patterns of each class.

1.4.1.14 Cooperative Feature Selection (CoFS)

The cooperative feature selection (CoFS) is based on the game-theoretic FS. The CoFS strategy mitigates the limitations of information-theoretic FS techniques. It tends to rank certain features with substantial discriminatory power poorly as a collective, notwithstanding their weakness [46, 62].

1.4.1.15 Analysis of Variance (ANOVA)

The ANOVA is an extremely widely used parametric statistical technique employed in biomedical, bioinformatics, machine learning, and statistics research. The basic objective of the ANOVA approach is to examine the disparities in groups by data means and variances. The ANOVA employs the F-statistic to illustrate the proportion of variances between groups compared to those within groups [63–65]. The F-statistic is formulated as shown in Equation 1.12 [39].

$$F_s = \frac{\textbf{\textit{Intergroup variance}}}{\textbf{\textit{Intragroup variance}}} = \frac{\sum_{i=1}^{P} q_i \left(\bar{X}_i - \bar{X}\right)^2 \Big/ (P-1)}{\sum_{ij=1}^{q} \left(X_{ij} - \bar{X}_i\right)^2 \Big/ (Q-P)} \tag{1.12}$$

The $\overline{X}_i$ represents the mean of the i group. The observation of the group i is $\boldsymbol{q}_i$. The overall mean is represented as $\overline{X}$. The jth experimental range of group *i* is X_{ij}. The number of groups is P, and Q is the number of observational groups.

1.4.1.16 Information Gain (IG)

The IG serves as a prominent univariate filter-based approach for FS, effectively pinpointing the highest-ranked features through established threshold criteria. The computation of information gain relies on the interdependence of two variables or attributes. The chosen attribute obtained through information gain methodologies demonstrated the greatest value in terms of information gain. The reduction in dimensionality of features is calculated through information gain and is based on the entropy of the omics dataset. The formulation of the IG is presented in Equations 1.13 and 1.14 [39, 66].

$$E_n(X) = -\sum_{n=1}^{L} A_n(Y)\log A_n^{(Y)} \tag{1.13}$$

$$EG = \text{Entropy} - \sum_{X} E_n(X) \tag{1.14}$$

The consistency-based feature selection (CBFS) [39], Pearson's correlation [67], Chi-square [68, 69], Markov Blanket Filter (MBF) [70], and MI [67] are more popular statistical-based filtering FS strategies [39]. The t-test serves as a univariate filter-based technique that assessing the statistical significance of features across different contexts. The t-test is a key method for detecting differentially expressed omics data in various experimental contexts [36, 39, 52].

Table 1.4 delineates the filter-based FS approach based on the sub category, implementation nature, basic idea, omics data type, merit, and demerit. The filter class uses a statistical approach to determine the accurate feature for the model implementation. The nature of the implementation and basic idea define its behaviour. The ANOVA and IG FS approaches are purely statistical filter methods. The Relief FS approach lies between filter and wrapper because it uses nearest neighbours and is known as data-driven. The MI-based methods are statistical filters. The VWmRMR is a filter-based FS method and also considered a hybrid based on its variance weights learned from the data in combination with MI.

1.4.2 Wrapper Class (Machine Learning-Based Feature Selection Techniques)

It employs a learning algorithm to select the features set during the FS phase. The wrapper is a closed-loop technique that employs the ML algorithm to assess the features [78]. It prioritizes features with the subset assessment method. Correlation and interdependence among the features are taken into account during FS. The inherent bias within the predictive algorithm contributes to enhancing its entire effectiveness [44].

The wrapper class evaluates the sufficiency of a feature subset by analysing the effectiveness of a machine learning algorithm. The wrappers necessitate a mechanism to explore

TABLE 1.4 Details of Filter-based Features Selection Approach

Category	Algorithm	Natures	Description	Omics data type	Advantages	Limitations	References
Filter (Univariate)	**Term Variance**	**Statistical**	Ranks features based on variance; assumes high variance equals higher importance.	Transcriptomics	• Simple, fast; works well with sparse and high-dimensional data.	• Ignores class label • It selects noisy features with high variance.	[46, 51]
	LS	**Statistical**	Preserves local data structure using graph Laplacians.	Transcriptomics, Proteomics	• Good for unsupervised learning; preserves manifold structure.	• Not suitable for nonlinear and supervised tasks • Sensitive to graph parameters.	[39, 46, 52]
	SCEFS	**Statistical**	Std. deviation + cosine similarity (exponentially scaled).	Transcriptomics	• Captures variance and feature direction.	• Sensitive to cosine similarity thresholding.	[39]
	SCRFS	**Statistical**	Std. deviation + reciprocal cosine similarity.	Transcriptomics	• Accentuates small-angle differences; easy to apply.	• Poor performance if data contains noise.	[39]
	SCAFS	**Statistical**	Anti-cosine similarity with standard deviation.	Transcriptomics	• Highlights anti-correlated features.	• May miss weak but relevant features.	[39]
	LSLS	**Statistical**	Laplacian Score with local sensitivity.	Transcriptomics	• Sensitive to neighbourhood structure.	• Depends on graph quality and K-nearest.	[39, 53]
	FPRS	**Statistical**	Uses fuzzy-rough logic to score features.	Genomics, Proteomics	• Robust to uncertain and imprecise data.	• Requires discretization or fuzzy membership tuning.	[39, 71,72]
	MIFS	**Statistical**	MI-based FS with redundancy control.	Transcriptomics	• Captures nonlinear relevance.	• May select redundant features in complex cases.	[32, 39, 73]
	CBFS	**Heuristic**	Consistency-based FS.	Transcriptomics	• Stable under noise.	• May overlook weak but consistent features.	[39, 72]
	ANOVA	**Statistical**	Uses F-tests to check if feature means differ significantly across classes.	Transcriptomics, Proteomics	• Simple, fast, interpretable • Effective for normally distributed data.	• Assumes normality and equal variance • Not ideal for multivariate interactions.	[39, 63]

	IG	**Statistical**	Measures entropy reduction after splitting; ranks features by their information value.	Transcriptomics, Epigenomics	• Captures nonlinear relations • Useful for categorical outputs.	• Biased towards features with more categories	[39, 66]
Filter (Multivariate)	**MIFS**	**Statistical**	Selects features with high mutual information and low redundancy.	Transcriptomics, Epigenomics	• Captures nonlinear dependencies • Effective for relevance scoring.	• It selects redundant features in high-dimensional data.	[40]
	MIFS-U	**Statistical**	MIFS with uniform prior to better balance redundancy and relevance.	Transcriptomics	• Better control over redundancy • Improved balance.	• Uniform prior may oversimplify true feature relationships.	[40, 47]
	mRMR	**Statistical**	Identifies attributes that exhibit the highest significance in relation to the objective while ensuring minimal overlap among them through the application of mutual information.	Genomics, Transcriptomics, Proteomics	• effective balance of relevance and redundancy.	• Computationally intensive for large feature spaces. • It ignores nonlinear interactions;	[40, 46, 48]
	NMIFS	**Statistical**	Scales mutual information for robustness and bias correction.	Transcriptomics, Epigenomics	• Improved robustness and comparability. • Balances MI contributions.	• It still suffers from feature redundancy.	[39,40, 49]
	INMIFS	**Statistical**	Refines NMIFS for better handling of redundancy and scaling.	Transcriptomics	• Better redundancy control • More stable selection.	• More parameters to tune • Increased complexity.	[39,40, 50]

(Continued)

TABLE 1.4 *(Continued)* Details of Filter-based Features Selection Approach

Category	Algorithm	Natures	Description	Omics data type	Advantages	Limitations	References
	VWmRMR	**Statistical + Heuristic**	mRMR extended with variance weighting; hybrid of filter and heuristic scoring.	Transcriptomics, Proteomics	• Adjusts for feature variance • Improves relevance.	• Sensitive to variance distribution • Heuristic may need tuning. • Heavily dependent on weight initialization	[39,40]
	LSE	**Statistical**	Combines local and global structures for better feature discrimination.	Transcriptomics, Epigenomics	• Improves over LS by capturing broader patterns.	• Computationally heavier than LS; sensitive to parameters.	[46]
	Relief	**ML-based Heuristic**	Assigns weights based on nearest-neighbour class distinction.	Genomics, Transcriptomics	• Captures feature interaction • Fast for small datasets.	• Less effective with noisy or irrelevant features • Struggles with large data.	[46]
	CFS	**Statistical**	Selects features with high correlation to class, low inter-correlation.	Transcriptomics, Epigenomics	• Simple and effective • Improves interpretability.	• It misses nonlinear relationships • Greedy selection be suboptimal.	[46, 56]
	FCBF	**Statistical**	Uses symmetrical uncertainty for fast correlation-based feature selection.	Transcriptomics, Epigenomics	• Faster than CFS • Suitable for high dimensional data.	• It misses subtle feature interactions • Threshold-sensitive.	[46, 59]
	CoFS	**Statistical + ML**	Combines outputs from multiple filter or ML-based methods to form a robust subset.	Multi-omics	• Increases robustness and stability	• Complexity increases with number of methods • Integration can be challenging.	[46, 62]
	REPES	**Statistical**	Clustering-based selection of representative features.	Transcriptomics, Proteomics	• Reduces redundancy and improves interpretability.	• Cluster selection may be unstable.	[39, 74]

LFSBSS	**Statistical**	Scatter separability in localized context.	Transcriptomics, Epigenomics	• Robust to imbalance and noise.	• Complex metric • Not scalable.	[39, 61]
SPFS	**Statistical**	Spectral graph-based FS.	Transcriptomics, Metabolomics	• Preserves data manifold • Good for unsupervised FS.	• It misses task-specific patterns.	[39, 75]
ECIA	**Statistical**	Improved initialization for clustering-based FS.	Transcriptomics	• Improves clustering-based selection reliability.	• Initialization may bias results.	[39, 76]
MRMR	**Statistical**	Balances high relevance and low redundancy.	Transcriptomics, Epigenomics	• Effective with gene expression data.	• Ignores feature interactions.	[39, 48, 77]
MBF	**Statistical**	Markov Blanket theory for conditional independence.	Transcriptomics	• Theoretically optimal • Minimal subset.	• Hard to estimate • Computationally complex.	[39]
MRmMC	**Statistical**	Ensures low multicollinearity in selected set.	Transcriptomics	• Improves feature independence.	• Ignores nonlinear dependencies.	[39, 54]
CGS	**Heuristic**	Constraint-guided sparse FS.	Transcriptomics	• Retains user-defined constraints.	• Manual constraint setting may limit generalization.	[39, 72]

the entirety of potential feature subsets. It evaluates the training process's efficacy and assesses the classifier by utilizing that specific subset of features. The wrappers determine the combinatorial problems by the goal function. The computation of the goal function is costly and insufficient for high-dimensional datasets. The floating search and branch and bound algorithms exemplify approaches employed for this search process [46].

The wrapper FS approaches are classified as randomized and deterministic classes [79]. The deterministic class used the deterministic framework to investigate disease diagnosis using omics data. The randomized wrappers improve FS through the application of the algorithm hypothesis during both the training and testing phases within the feature subset [79]. The wrapper technique faces computational efficiency-related issues compared to filter and embedding methods [79]. The wrapper-based FS strategies are utilized for the identification of accurate features in biological sequencing dataset [39]. The implementation process of the wrapper-based FS process is shown in Figure 1.7.

The genetic algorithm (GA) is a randomized wrapping approach and search-based FS approach. It employs the principles of natural selection as articulated by Charles Darwin's theory of evolution. The concept is grounded in the principles established by Charles Darwin regarding the process of natural selection. It provides a nearly optimal output and cannot guarantee strict overall minima [80].

The recursive feature elimination (RFE) identifies pertinent feature subsets from the wide range of the omics dataset [81]. This strategy enhances classification accuracy for diverse medical conditions [82]. The objective of the RFE method is to assess the ranking weights of feature subsets through classification. The features are organized based on their weights [83]. The RFE first applies the feature dataset to train the model and then computes the ranking weights for all features. It excludes the lowest weights associated with features to establish the final ranking of features [58, 84].

The support vector machine (SVM) RFE with correlation bias reduction (SVM-RFE-CBR) is the enhanced method of the support vector machine recursive feature elimination with correlation bias reduction (SVM-RFE-CBR). It combined the advantages of the correlation bias reduction (CBR) approach into the feature elimination procedure [85]. The SVM-RFE-CBR method employs a backward elimination approach. It has ability to manage feature sets with strongly correlated characteristics. The stability of SVM-RFE-CBR enhanced the accuracy of the FS when applying the various ensemble methods [32, 40].

The primary demerit of the wrapper strategy is its high computing cost, which arises from searching for the best set within a vast dimensional space. The wrapper approach carries a significant danger of overfitting. The sequential feature selection/sequential feature pruning (SFS/SPS) [39, 58] and SVM is a Boruta wrapper-oriented supervised FS technique that utilizes RF-SFS [80], SVM-RFE-GA, SVM-RFE-GS, SVM-RFE, and SVM-RFE-PSO [86].

Liang et al. [87] introduced a FS technique that mitigates the issue of elevated computing expenses and rectifies certain limitations of suboptimal methods. It identifies an effective subset of features without relying on exhaustive search. The greedy randomized adaptive search procedure (GRASP) is wrapper FS that uses the tabu search and has surpassed other methodologies. Esseghir et al. [88] proposed an alternative GRASP-based FS hybrid filter-wrapper approach.

The Flower Pollination Algorithm (FPA) [80], Artificial Bee Colony FS (ABCFS) [80, 89], Bayesian Network [84], Greedy Forward Selection (GFS) [90], Best Incremental Ranks Subset (BIRS) [84], and Slime Mold Algorithm (SMA) are some more popular wrapper methods for omics data analysis and modelling [39]. The AMOSA [91, 92] represents a dependable multi-objective simulated annealing optimization method designed for challenging scenarios.

Machine learning-based FS techniques are utilized for DNA methylation, quantifying gene expression, pathway analysis, exon expression, proteomics, miRNA expression, number variation, single-cell spatial transcriptomic data analysis, and other omics data applications [32, 93]. The machine learning FE methods use unsupervised, supervised, and semi-supervised FS strategies for dimensional reduction [39].

1.4.2.1 Filters-Based Supervised Learning FS

The predefined labelled data is utilized to discern the most significant features pertinent to the prediction task. This strategy improves the performance of the model by selecting positively impacting features. Supervised FS strategies are used by filter, wrapper, and embedded methods that are predominantly utilized for problems with classification [32, 39, 78].

Classification is a common supervised technique that analyses important biological information from the original dataset. The researchers mainly employ supervised FS techniques in NGS data analysis. It enables them to focus on the most informative characteristics. Regression analysis is another supervised learning method for forecasting continuous values.

The supervised FS method's first phase identifies the required feature subset from the labelled omics dataset. The selected features and associated labels are used to train the prediction model. This technique has challenges in the labelling process because the datasets are obtained from other sources, and it is costly and unreliable from a computational accuracy perspective. This situation increases the chances of overfitting during the FS process [78].

Table 1.5 outlines the aims, advantages, and limitations associated with supervised FS in the framework of omics data. The objective delineates the intention behind the supervised selection of features in omics data. The benefits demonstrate the efficacy of

TABLE 1.5 Objective, Advantages, and Limitations of Supervised Feature Selection on Omics Data

Objective	Advantages	Limitations
Identification of class-relevant features for disease classification, subtype prediction, or biomarker discovery	• Handles high-dimensional data efficiently. • Improves model interpretability and performance. • Reduces overfitting and computational cost. • Facilitates discovery of potential biomarkers. • Integrate prior biological knowledge.	• Requires the accurately labelled training data that is costly and unavailable. • Biased towards dominant classes in imbalanced datasets. • Risk of selecting features that perform well only for a specific model (model dependency). • It ignores interactions and multi-feature dependencies. • External validation is not used on limited generalization.

the supervised FS algorithm, whereas the drawbacks underscore its limitations in the context of omics data.

Numerous statistical methods utilizing filter strategies are employed in supervised learning to identify pertinent characteristics for various tasks [32]. The commonly utilized supervised FS approaches are applied in NGS and other omics data [39]. The pairwise constraint-guided sparse (CGS) method is grounded in discriminative regularization, which specifically targets the local discriminative structure inherent in the data [94]. The Fischer score [46], ReliefF, NMIFS, INMIFS, VWMRmR [32], CBFS [72], LSLS [53], Maximum Relevance Minimum Multicollinearity (MRmMC) [54], and MBF are supervised FS technique that uses scores to independently identify features and select the significant genes and other omics data, disease biomarkers, etc.

1.4.2.2 Filters-Based Unsupervised Learning FS

The unsupervised FS approaches effectively mitigate the curse of dimensionality in NGS datasets. It identifies the appropriate class of features and cell cluster identification [95]. The unsupervised FS approaches identify features of omics data without prior knowledge of the dataset's class labels [96]. It uses the distance metric to minimize the feature dimensionality and optimize the efficacy of subsequent analyses of high-dimensional omics data [93].

The clustering is an unsupervised FS approach that picks co-regulated omics data inside the same cluster [97]. It serves as a form of dimensionality reduction for the system in a variety of omics data such as scRNA-Seq, cell types, cellular variation, and so on [98]. The classical objective of the unsupervised FS approaches is to identify the optimal features without considering the feature class labels [39, 99].

Table 1.6 delineates the objectives, benefits, and constraints of unsupervised FS in omics data. The objective defines the purpose of unsupervised FS in omics data. The advantages highlight the benefits of the unsupervised FS technique, and the limitations delineate its drawbacks in the context of omics data.

The cosine similarity-oriented FS algorithm handles the outliner during the FS. The three proposed strategies utilize feature standard deviation and cosine similarity to analyse high-dimensional biological sequencing information. The more popular cosine

TABLE 1.6 Objective, Advantages, and Limitations of Unsupervised Feature Selection on Omics Data

Objective	Advantages	Limitations
Identify optimal features without using class labels.	• Class assumptions and label noise by unbiased • No need for labelled data • It reduces cost and time • Useful for exploratory analysis and dimensionality reduction • Helps in preprocessing for clustering or visualization tasks • Suitable for semi-supervised learning pipelines	• Requires a large volume of high-quality data to extract meaningful structure • Computationally intensive, especially with high-dimensional data • Difficult to validate results due to absence of labels • It selects irrelevant or redundant features on noise and complex correlations data • It less accurate than supervised methods in predictive tasks

similarity-oriented FS algorithm by Standard Deviation and Cosine Similarity with Reciprocal (SCRFS), Standard Deviation and Cosine Similarity with Exponent (SCEFS), and Standard Deviation and Anti-Cosine Similarity (SCAFS) [39].

The RE Presentative Feature Selection (REPFS) uses information entropy to identify the relevance features sets [74]. The UFS graph-based method was introduced for gene identification using multi-omics datasets [100]. Bhadra et al. [32] discuss Multivariate Localized Feature Selection Based on Scatter Separability (LFSBSS) [61]. The LS and Spectral Feature Selection (SPFS) [75] approach are more popular FS method of filters-based unsupervised learning. The Enhanced Center Initialization Algorithm (ECIA) [76] is a highly effective method for clustering a feature space, particularly when used with non-negative matrix factorization (NMF) [101].

1.4.2.3 Filters-Based Semi-supervised Learning FS

Semi-supervised FS techniques leverage the characteristics of supervised and unsupervised learning methods. It employs labelled and unlabelled data to derive the pertinent information. It predominantly depends on the comparison matrix and assesses aspects that exhibit the strongest matrix correlation [102]. The majority of semi-supervised approaches leverage clustering and manifold theory to derive more precise features. The labelled data is generally utilized to enhance the separation between data points belonging to different categories. The unlabelled data is employed to reveal the spatial geometric configuration [39].

Table 1.7 outlines the objectives, advantages, and limitations associated with semi-supervised FS in the framework of omics data. The objective delineates the intent behind the semi-supervised FS in omics data. The advantages illustrate the strengths of the semi-supervised FS algorithm, while the limitations highlight its shortcomings as applied to omics data.

The label propagation-based semi-supervised feature selection (LPFS) method identifies relevant omics data such as genes for diagnosing various illness stages. The LPFS employs a two-step technique. The initial step involves the implementation of a clustering technique based on label propagation. The application of matrix methods to identify the precise features. The computational implementation of LPFS is shown in Equation 1.15 [39, 102].

TABLE 1.7 Objective, Advantages and Limitations of Semi-Supervised Feature Selection on Omics Data

Objective	Advantages	Limitations
Identify relevant and discriminative features by leveraging both labelled and unlabelled data	• It reduces dependency on large labelled datasets and minimizing the annotation cost • It enhances generalization by exploiting intrinsic data structure • It suitable for high-dimensional omics data where labelled samples are scarce • Label propagation is adaptable to complex omics datasets	• It sensitive to quality and distribution of unlabelled data • It increased the model complexity and training time • It propagates incorrect labels and bias during self/co-training • Parameter tuning and model convergence can be challenging in heterogeneous biological data

$$\min(P,F)\sum_{nm=1}^{p} w_{nm} \left| \frac{1}{\sqrt{d_{nm}}} P_n - \frac{1}{\sqrt{d_{nm}}} P_m \right| + \mu \sum_{n=1}^{p} |P_n - Y_n|^2 + |XF - P|_F^2 + \beta |F|_{2,1} \tag{1.15}$$

This collection of multiple semi-supervised approaches is applicable to analysing biological sequencing datasets. The SemiFS semi-supervised methodology is based on the trace ratio that is resilient to noise [61]. The Semi-Supervised Maximum Discriminative Local Margin (semiMM) is used to identify the most discriminative features for classification purposes [103].

Table 1.8 explores the wrapper-based FS methodology, categorized by subcategory, implementation nature, fundamental concept, omics data type, advantages, and disadvantages. The wrapper class employs a machine learning-based statistical method to ascertain the optimal features for model implementation. Its behaviour is determined by the nature of the implementation and the fundamental concept. The GRASP FS method is search-based and frequently utilized in wrapper classes to optimize feature subsets. GAs are search-based frameworks that utilize population evolution to maximize subsets of features. The RFE is among the most utilized wrapper techniques. The SVM-RFE and its variations are traditional wrapper techniques reliant on model efficacy.

1.4.3 Embedded Class

The embedded approach is the integrated class of the wrapper and filter. The advantages of the FS procedure were incorporated into the learning algorithm. It is enhancing its ability to disregard certain subsets of features. The filters represent the most straightforward strategy and are anticipated to underperform compared to other methodologies. It frequently emerges as the sole viable alternative for high-dimensional datasets. The wrappers and embedded approaches are prohibitively costly compared to filter methods [46].

The embedded FS technique is utilized to diminish the by precise classifier-dependent approach. It outperforms the wrapper method in FS during model generation. The embedded techniques generally exhibit superior computational complexity compared to wrapper methods [67]. It does not utilize the complete feature subset to extract the feature and modelling [39, 105]. The implementation process of the embedded-based FS process is shown in Figure 1.7.

The construction of a classifier identifies the ideal feature subset. The appropriate feature subset selection is customized to align with the specific machine learning. The benefits of the embedded approach are equivalent to those of the wrapper method. It exhibits superior computing complexity compared to the wrapper technique [44, 106, 107].

The embedded method simultaneously trains the classifier and identifies a features subset. The FS of embedded is integral to the learning algorithm. These algorithms often operate by incorporating a sparsity-inducing regularize and prior into the objective function of the learning algorithm. It promotes the weights assigned to specific characteristics to attain a value of zero [46].

The sparse balanced iterative random forest (BIRF) [108], multinomial logistic regression (SMLR) algorithm [109], ridge regression [110], Bayesian logistic regression (BLogReg) [111],

TABLE 1.8 Details of Wrapper-based Features Selection Approach

Category	Algorithm	Nature	Description	Omics Data Type	Advantages	Limitations	References
Wrapper (Randomized)	**GRASP**	**Machine Learning**	Greedy randomized metaheuristic that iteratively searches feature subset space.	Genomics, Proteomics, Metabolomics	• Explores large search space • Avoids local optima • Flexible.	• High computational cost • sensitive to parameters.	[46, 88]
	SVM-RFE-PSO	**Machine Learning**	PSO-enhanced RFE with global search.	Genomics	• Better search than greedy RFE.	• Requires many iterations • Slower.	[39, 86]
	SVM-RFE-GA	**Machine Learning**	Genetic algorithm enhanced RFE.	Genomics	• Good at escaping local optima.	• Slow convergence • Parameter tuning.	[39, 86]
	AMOSA	**Metaheuristic**	Simulated annealing for multi-objective FS.	Genomics, Proteomics	• Avoids local optima • It works with complex data.	• Very slow • Difficult to tune objectives.	[39, 91,92]
	Genetic Algorithm (GA)	**Machine Learning**	Evolves feature subsets using crossover, mutation, and selection based on model fitness.	Genomics, Metabolomics, Proteomics	• Explores large feature spaces • It avoids local optima; flexible.	• Computationally expensive • Sensitive to parameter settings.	[39, 80]
Wrapper (Deterministic)	**SVM-RFE-CBR**	**Machine Learning**	It utilizing recursive feature elimination in conjunction with support vector machines, while addressing correlation bias through case-based reasoning.	Genomics, Transcriptomics	• Improves upon classic SVM-RFE • Reduces selection bias.	• Computationally expensive for high-dimensional data.	[32, 39,40, 85]
	SFS/SPS	**Machine Learning**	Greedy forward/backward subset search.	Any	• Easy to implement • It interpretable process.	• Computationally expensive • It selects suboptimal subsets.	[39, 104]
	SFS	**Machine Learning**	Greedy forward feature selection.	Any	• Simple and interpretable.	• Suboptimal subset • High computation.	[39]
	SVM-RFE	**Machine Learning**	Eliminates features based on SVM weights.	Genomics, Proteomics	• Great with high-dimensional low-sample-size data.	• Sensitive to irrelevant features and noise.	[39]
	SVM-RFE-GS	**Machine Learning**	Uses grid search for SVM-RFE tuning.	Genomics	• Performance-optimized RFE.	• Time-consuming • It requires large parameter search.	[39, 86]
	RFE	**Machine Learning**	It systematically eliminates the least significant features by utilizing model coefficients or importance metrics.	Genomics, Transcriptomics	• Produces compact and informative subsets • Model-integrated.	• It overfit on small datasets • expensive for large feature sets.	[39, 81]

SVM-embedded [112], and sparse logistic regression (SLogReg) [55] are more popular embedded FS approaches.

1.4.3.1 Discriminative Feature Selection (DFS)

The DFS is the filter-based FS. The DFS is integrated by the sparsity regularization and linear discriminant analysis (LDA) [113]. The method uses the row sparsity in the LDA transformation matrix by applying $I_{2,1}$-norm regularization for the purpose of FS. It optimizes the FS process by identifying the most discriminative features. It introduced an efficient approach to address the $I_{2,p}$-norm minimization problem. It minimizes the objective function in a monotonic fashion before convergence by $0 < p \leq 2$ [32, 40, 114].

1.4.3.2 Least Absolute Shrinkage and Selection Operator (LASSO)

The LASSO method identifies features by encouraging sparsity in the estimation of coefficients. It improves the precision of predictions and the clarity of model interpretation by integration of the beneficial aspects of ridge regression and subset selection. The LASSO equation is articulated in Equation 1.16 [39, 110, 115].

$$\hat{\beta} = \beta \arg\min \left\{ \frac{1}{2} \sum_{m=1}^{p} \left(a_m - \beta_0 - \sum_{n=1}^{p} b_{mn} \beta_n \right)^2 + \gamma \sum_{n=1}^{p} |\beta_n| \right\} \tag{1.16}$$

The Latent Kernel Feature Selection (LKFS) [116] is an unsupervised FS approach founded on the recently introduced shrinkage and expansion algorithm (SEA) [39]. The Unsupervised Discriminative Feature Selection (UDFS) [39], Nonnegative Discriminative Feature Selection (NDFS) [104], and CHNMF [101] are robust unsupervised embedded methods that outperform other clustering and FS techniques.

The Fuzzy Preference-based Roughest (FPRS) [71, 72], Local-Learning based Feature Selection (LLBFS) [78], nonlinear dynamic conditional relevance FS (NDCRFS) [86], and MIFS [32, 73] are more frequently used supervised embedded FS techniques.

Table 1.9 highlights the embedded-based FS methodology according to subcategory, implementation nature, fundamental concept, omics data type, advantages, and disadvantages. The embedded class leverages the benefits of machine learning and statistical methods to ascertain precise attributes. The SVM-embedded and BIRF techniques choose features during model training and classify them as embedded methods. The DFS can be integrated based on its compatibility with the classifier. The CoFS FS technique employs various filtering and hybrid approaches to identify a more robust subset called ensemble FS.

1.4.4 Hybrid Feature Selection Techniques

The hybrid method usually combines filter and wrapper strategies to reduce the computational load and select more relevant features. It helps to discard noisy features for improved classification accuracy. Hybrid approaches are progressively employed to diminish characteristics in omics data processing [39].

TABLE 1.9 Details of Embedded-based Features Selection Approach

Algorithm	Based on	Description	Omics Data Type	Advantages	Limitations	References
DFS	**Machine Learning**	Integrated with classifiers to maximize class separation.	Transcriptomics, Proteomics	• Incorporates class labels • Improves classifier performance. • Improves classification by weighting features	• Classifier-dependent • It overfits with small datasets • It requires label data • classifier-dependent.	[32, 39,40, 113,114]
LKFS	**Machine Learning**	Latent kernel space for scoring.	Multi-omics, Proteomics	• Captures complex nonlinear dependencies.	• Requires kernel and parameter tuning.	[39, 116]
NDFS	**Machine Learning**	Nonnegative, discriminative, sparse FS.	Transcriptomics	• Enhances class separability with sparsity.	• Sensitive to parameter selection.	[39, 104]
UDFS	**Machine Learning**	Sparse and discriminative FS without labels.	Transcriptomics	• No need for labels; robust under noise.	• It underperforms with noisy features.	[39, 75]
NMF	**Machine Learning**	Feature ranking via matrix factorization.	Transcriptomics, Metabolomics	• Interpretable latent factors; sparse result.	• Sensitive to noise and local minima.	[39, 101]
CHNMF	**Machine Learning**	Cluster-enhanced NMF for FS.	Multi-omics	• Combines grouping and decomposition.	• Requires careful tuning and validation.	[39, 101]
LASSO	**Machine Learning**	L1 penalty induces sparsity in models.	Any	• Combines FS and model building; fast.	• Unstable under correlated features.	[39, 110, 115]
LLBFS	**Machine Learning**	Local learning for adaptive feature ranking.	Transcriptomics	• Captures local feature behaviour.	• Overfit to local patterns.	[39, 78]
NDCRFS	**Machine Learning**	Selects features using dynamic relevance score.	Epigenomics	• Time-aware FS; good for longitudinal data.	• Complex scoring function.	[39, 86]
LPFS	**Semi-Supervised ML**	Uses label propagation for FS.	Transcriptomics	• Effective with few labelled samples.	• Sensitive to propagation structure.	[39, 102]
SemiFS	**Semi-Supervised ML**	Semi-supervised FS method.	Transcriptomics	• Leverages small labelled sets.	• Depends on label quality.	[39, 61]
Semi-Supervised MDLM	**Semi-Supervised ML**	Margin-based discriminative FS.	Transcriptomics	• Enhances class separation in semi-supervised setting.	• Poor generalization if margin is poorly set.	[39, 103]
BIRF	**Machine Learning**	Iterative Random Forests with balancing; useful in imbalanced datasets.	Epigenomics, Proteomics	• Handles class imbalance well; captures nonlinearity and interactions.	• Results depend on rebalancing strategy • interpretability may be low.	[39, 108]
SVM-embedded	**Machine Learning**	Uses trained SVM weights (e.g., linear) to select relevant features during training.	Genomics, Transcriptomics, Proteomics	• Fast and scalable; well-suited for high-dimensional datasets.	• It fails with nonlinear data unless kernelized • Sensitive to feature scaling.	[39, 112]

TABLE 1.10 Details of Hybrid-based Features Selection Approach

Category (Refined)	Algorithm	Based On	Description	Omics Data Type	Advantages	Limitations	References
Hybrid	**ECGUFS**	Ensemble + Statistical	Consensus-based ensemble unsupervised FS.	Multi-omics	• High robustness and feature stability.	• More complex and time-consuming.	[39, 75]
Hybrid/Wrapper (Randomized)	**IDGA**	Metaheuristic	Intelligent dynamic genetic algorithm.	Genomics	• Explores large search space • Adaptive.	• Expensive and time-consuming.	[39, 117]
	Genetic Bee Colony	Metaheuristic	Bee colony logic for global search.	Genomics	• Global exploration • Robust.	• Complex implementation • Slow for big data.	[39, 119]
Hybrid/Filter (Multivariate)	**MRMR**	Statistical	Classic MI-based feature ranking.	Transcriptomics	• Simple interpretable	• Ignores feature synergy.	[39]
Hybrid/Wrapper (Randomized)	**MIMAGA**	ML + MI	GA with MI-based objectives.	Genomics	• Balances mutual information and exploration.	• Requires GA parameter optimization.	[39, 119]

The Intelligent Dynamic Genetic Algorithm (IDGA) uses GAs, artificial intelligence, and reinforcement learning techniques to extract the features [117]. It utilized fitness function as RFR, and it is used to identify significant features from the omics data. The filter method generates a subset of features, and the IDGA method discerns the most informative subset [118]. The implementation process of the hybrid-based FS process is shown in Figure 1.7.

The Genetic Bee Colony Algorithm (GBC) represents a novel hybrid methodology for the selection of features in omics studies. This approach addresses the complexities involved in selecting omics features for both binary and multi-class cancer classification [119]. Ensemble Consensus-Guided Unsupervised Feature Selection (ECGUFS) has significantly enhanced the identification of crucial genes related to disease [75]. The RMR-BBHA is an additional hybrid methodology to analyse gene expression data [39].

The Minimum Redundancy and Maximum Relevance (MRMR) is a hybrid method that integrates wrapper strategies to address omics FS challenges. The research improved the MRMR approach by ABC, FPA, GAs, and so on [39].

The MIMAGA is an innovative hybrid FS technique. It integrates the advantages of the mutual information maximization (MIM) and adaptive genetic algorithm (AGA). The MIM identifies the omics data that significantly depends on other data for similar behaviour. The AGA achieves optimum results by identifying the best appropriate parameters for crossover and mutation probability [119].

Table 1.10 illustrates the hybrid-based FS methodology, categorizing it by subcategory, implementation nature, fundamental concept, omics data type, advantages, and disadvantages. The hybrid class leverages the benefits of the filter, wrapper, and embedding approaches to ascertain the precise attributes. The hybrid method first used the filter method to reduce the feature set. Then, the wrapper or embedded method was used to obtain the accurate feature subset for model implementation.

1.5 FEATURE EXTRACTION

The methodologies for FE derive quantitative information from biological sequences. The new feature space was transformed into a discrete domain, yielding multiple result sets. It transforms the biological sequencing data into a numerical input vector using computer programming languages. The primary aim of FE is to derive essential quantitative information and patterns from omics data, while also reducing its dimensionality and complexity [39]. The fundamental advantages of the FE approach are illustrated in Figure 1.8.

The FE approach generates a new feature set for modelling and data analysis based on numerical data. The modelling approach originally pre-processed the omics dataset by mitigating noise and outlier data, among other considerations. The pre-processing phase converts data from one format to another to clearly understand the relevant feature set. The FE employed the pre-processed omics data to identify the suitable pertinent characteristics from the high-dimensional omics dataset according to the application and direction. The conceptual framework of the FE process is shown in Figure 1.9.

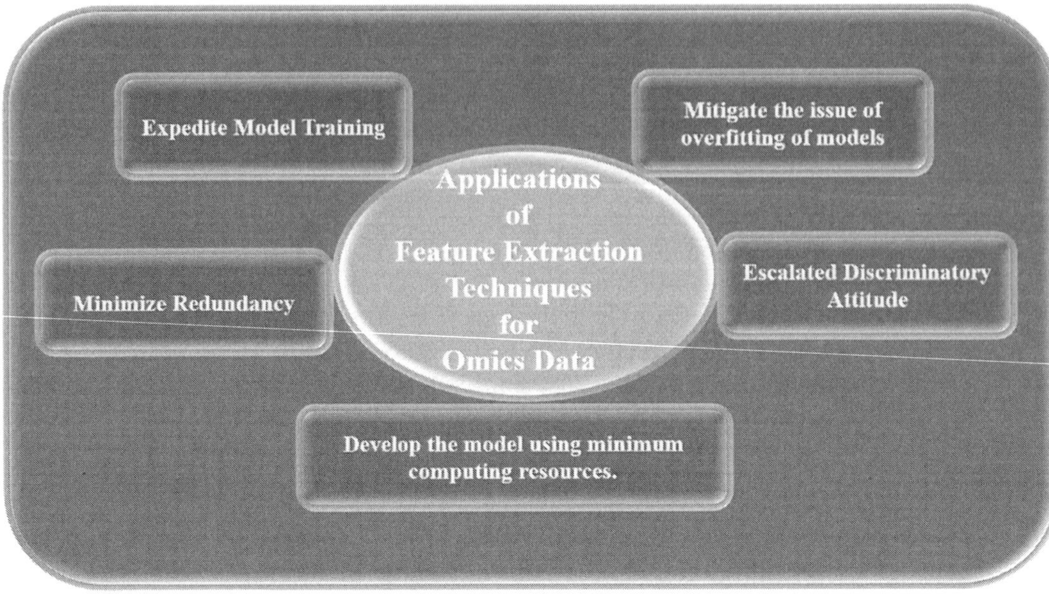

FIGURE 1.8 Fundamental advantages of the feature extraction approach.

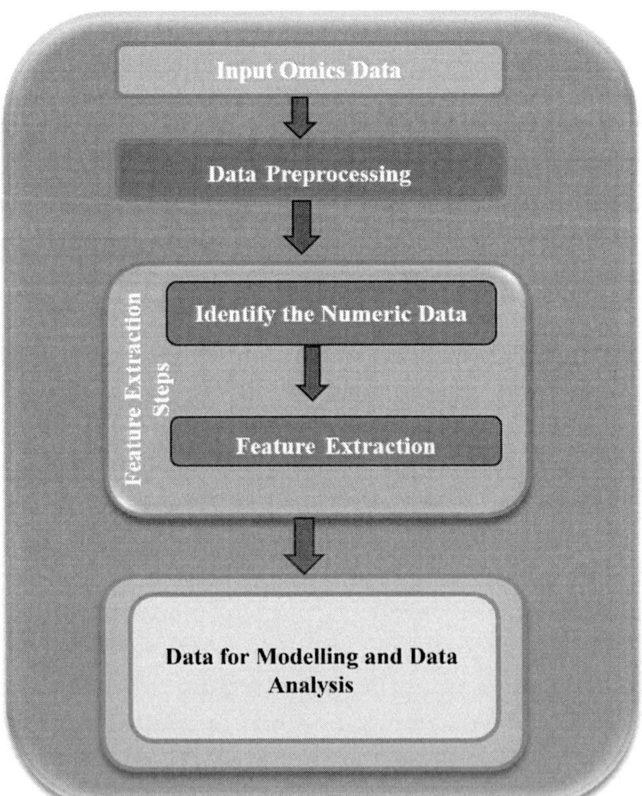

FIGURE 1.9 Conceptual framework of the feature extraction process.

The FE technique categories are linear and nonlinear. The linear FS method utilizes matrix calculations to ascertain the feature representations within a reduced-dimensional subspace. The more popular FE techniques are described as follows.

1.5.1 Principal Component Analysis (PCA)

The principal component analysis (PCA) is a frequently utilized and versatile multivariate unsupervised FE technique. It can reduce the dimensionality from high-dimensional omics data and NGS. The PCA helps to analysis the single-cell RNA-Seq data by reducing the dimensionality of the dataset [93]. It can reveal patterns and trends in omics data. It selects and extracts data by maximizing variance. The analysis employs the covariance matrix, along with eigenvalues and eigenvectors, to identify the principal components present in the dataset. The PCA is inadequate in accurately detecting nonlinear relationships within complex datasets, including miRNA expression, gene expression, similar variables, and exon expression [39, 120].

1.5.2 Supervised Principal Component Analysis (SPCA)

The supervised principal component analysis (SPCA) enhances the performance of the FE approach and focuses on the principal components derived from class variables. The SPCA approach is exceptionally proficient in retrieving significant genetic information. It is only outperformed in cross-validation assessments [41]. It ascertains a nonlinear relationship among the variables by mapping a low-dimensional manifold onto a high-dimensional space. It uses the kernel function in place of the lifting function. This approach suggests that the pertinent genetic data is located on a complex nonlinear structure. It is situated within the original data space and has a dimension that is less than that of the space itself [39].

1.5.3 Kernel Principal Component Analysis

The Kernel PCA represents a sophisticated approach to classification that utilizes a matrix-based framework for the purpose of reducing dimensionality in a nonlinear context. It retains information from the input data, and the dimensions of the matrix expand in a quadratic manner. Singular value decomposition (SVD) to achieve dimensionality reduction of microarray data based on the Kernel PCA [121]. It represents the input domain within a multi-dimensional feature landscape [39].

1.6 STABILITY MEASURES

The FS method uses the rank, weight, and subset approach to select the accurate features of the omics dataset. The weight methods first find the weight score for each feature and then extract the relevant features based on the maximum and minimum weight score-related criteria. The rank methods first determine the ranking of each feature and then extract the accurate features on the basis of the highest and lowest ranking score-related criteria. The subset methods first determine the index on each feature and then extract the suitable features on the basis of the minimum and maximum index score-related criteria. The three approaches are identified as the assessment standards for FS algorithm. This assessment

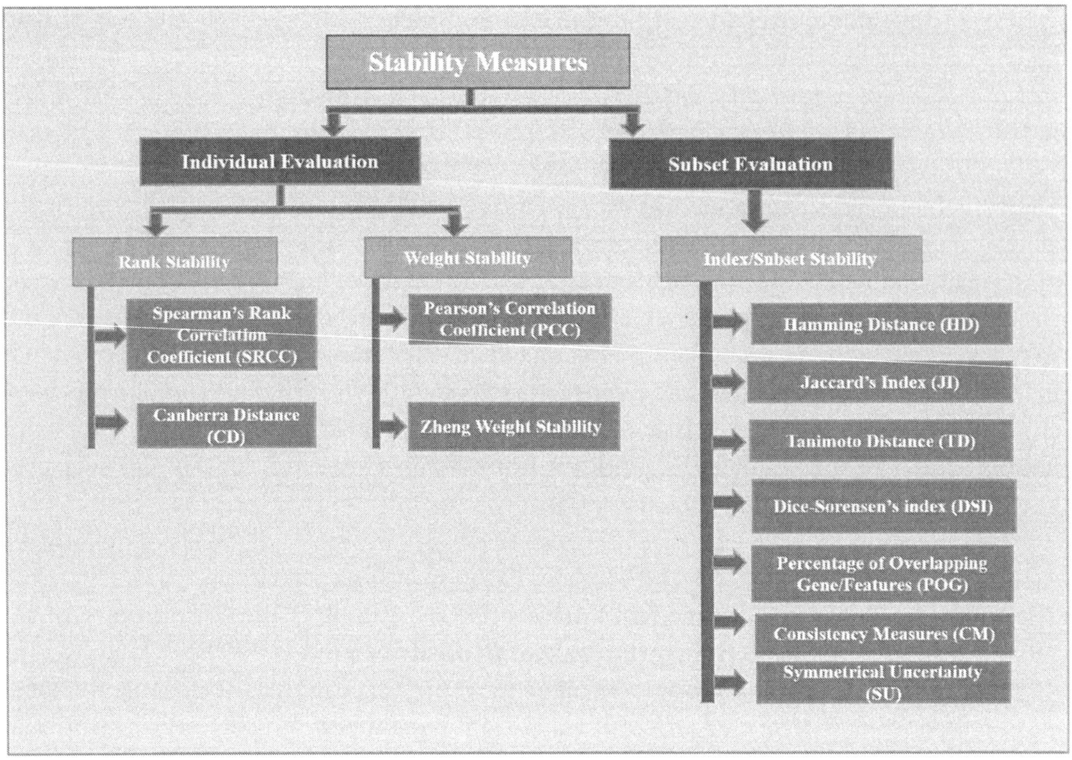

FIGURE 1.10 Classification of the stability measures.

approach is classified into two distinct components for evaluation [122]. The classification of the stability measures is shown in Figure 1.10.

1.6.1 Individual Evaluation

This evaluation technique uses the degree of relevance criteria to determine the accurate features on the basis of individual feature assessment. It assesses the performance of the rank and weight-related features selection algorithms. The subset evaluation measures the stability by rank and weight of FS algorithm

The weighted scoring $W = (\boldsymbol{w}_1, \boldsymbol{w}_2, \ldots\ldots\ldots, \boldsymbol{w}_m)$ is assigned to each feature of the dataset according to selection approach R^m and each weight is determined as $w \in W \in R^m$. The rank score $r = (\boldsymbol{r}_1, \boldsymbol{r}_2, \ldots\ldots\ldots, \boldsymbol{r}_m)$ is assigned by the FS algorithm, and it belongs under the $r \le r_i \le m$ [44].

1.6.1.1 Rank Stability

The relationship between features is assessed to ascertain the reliability of the FS methodology via feature ranking. The stability is determined by the degree of rank of features among the subset of selected features. The various measurements are used to assess the rank stability of the FS algorithm. The details are described as follows [44]:

1.6.1.1.1 Spearman's Rank Correlation Coefficient (SRCC) The SRCC measures the stability of (R_1, R_j) features ranked sets. The formulation of the SRCC is defined in Equation 1.17.

$$SRCC(R_1, R_j) = 1 - \frac{6\sum_{t=1}^{m}(R_{1t} - R_{jt})^2}{m(m^2 - 1)} \tag{1.17}$$

The stability results of the $SRCC(R_1, R_j)$ determine the range of [−1, 1]. The results −1 describes the stability as identical. The SRCC value −1 shows the inverse order stability, and the zero value shows no correlation between (R_1, R_j) features ranked sets. The overall stability of all feature subsets $ASRCC(R_i, R_j)$ shown in Equation 1.18.

$$ASRCC(R_i, R_j) = \frac{2}{|W|(|W| - 1)} \sum_{|W|-1}^{i=1} \sum_{|W|}^{j=i+1} SRCC(R_i, R_j) \tag{1.18}$$

1.6.1.1.2 Canberra Distance (CD) The Canberra distance measures the absolute difference between two rank sets [44]. The value of the CD is precisely proportional to its attributes. The formulation of the CD is shown in Equation 1.19. The results of the $CD(R_1, R_j)$ is bounded similarly $SRCC(R_1, R_j)$.

$$CD(R_i, R_j) = \sum_{m}^{t=1} \frac{|R_{it} - R_{jt}|}{R_{it} + R_{jt}} \tag{1.19}$$

1.6.1.2 Weight Stability

The weight stability criteria account for the significance of the feature set while assessing the robustness of the FS technique. The weights set (W_i, W_j) are utilized for the comprehensive feature set in omics data sets. It returns the correlation between these weights set as a measure of stability. The weight stability metrics cannot address subsets of features of varying sizes. The various measurements are used to assess the weight stability of the FS algorithm. The specifics are delineated as follows [44]:

1.6.1.2.1 Pearson's Correlation Coefficient (PCC) The PCC analyses the correlation among the weights of the selected feature subsets [123]. The PCC return derives within the range of [−1, 1]. The one value signifies that the weight vectors are perfectly correlated. The −1 value indicates that the weight vectors are anti-correlated. The zero value shows a blank correlation between weight vectors. The formulation of the PCC is shown in Equation 1.20.

$$PCC(W_i, W_j) = \frac{\Sigma (W_{it} - \mu_{W^1})(W_{jt} - \mu_{W_j})}{\sqrt{\Sigma (W_{it} - \mu_{W^1})^2 \Sigma (W_{jt} - \mu_{W_j})^2}} \tag{1.20}$$

1.6.1.2.2 Zheng Weight Stability This weight stability measures the stability of the features subset. It fulfills all the requisite characteristics of a stability metric. The formulation of the Zheng Weight stability is expressed in Equations 1.21 and 1.22 [90].

$$\Phi(z)=1-\frac{\frac{1}{m}\sum_{i=1}^{m} f_i^2}{\frac{k}{m}\left(1-\frac{k}{m}\right)} \tag{1.21}$$

$$f_i^2=\frac{N}{N-1}\cdot P_f\left(1-P_f\right) \tag{1.22}$$

The k ascertains the mean quantity of characteristics z selected in the feature collection F. The f_i^2 is the sample variance of Z_f.

1.6.2 Subset Evaluation

It uses the searching approach to determine the features for data analysis. The heuristic search generates the feature subsets. Each state of heuristic search specifies the feature subset for assessment inside the search space.

The feature subset is determined as $s = (s_1, s_2, \ldots\ldots, s_m)$. Each feature subset belongs $s \in \{0, 1\}$. The 0 value defined the absence and 1 determined the present of the feature in subset.

1.6.2.1 Index/Subset Stability

The subset evaluation measures the stability by index/subset of the features selection algorithm. The selected subset of characteristics is depicted as a binary vector of size 'm', with 0 indicating the absence and 1 indicating the presence of the characteristic. The stability is ascertained by the extent of intersection within the comprehensive collection of chosen attributes. The various measurements are used to assess the stability of the FS algorithm by index/subset. The details are described as follows [44]:

1.6.2.1.1 Hamming Distance (HD) The HD metric determines the degree of overlap between the two $(\boldsymbol{S_i}, \boldsymbol{S_j})$subgroups. It operates on the binary matrix of the specified feature subset. The size of the m is affecting the hamming distance H between $\left(S_i, S_j\right)$ groups. The formulation of the H shows the stable FS algorithm of omics data. The formulation of the $H\left(S_i, S_j\right)$ is shown as Equation 1.23.

$$H\left(S_i, S_j\right)=\sum_{k=1}^{m}\left|S_{ik}-S_{jk}\right| \tag{1.23}$$

The Hamming distance equally influences the stability of results regardless of whether the feature $\boldsymbol{S_i}$ is included or not during the cross-validation. It diminishes the stability of selected features relative to the overall complexity of the dataset. The overall Hamming distance H_t of all feature subsets is expressed in Equation 1.24.

$$H_t = \sum_{i=1}^{|W|-1} \sum_{j=i+1}^{|W|} H(S_i, S_j) \tag{1.24}$$

The overall stability of each pairwise feature subset in S is characterized by the Average Normalized Hamming Distance (ANHD). The W is characterized by the Average Normalized Hamming Index (ANHI). The NHI is shown in Equation 1.25, and ANHI in Equation 1.26.

$$\text{NHI}\left(S_1, S_1\right) = 1 - \frac{H(S_1, S_1)}{m,} \tag{1.25}$$

$$\text{ANHI}\left(S_1, S_j\right) = \frac{2\sum_{i=1}^{|W|-1} \sum_{j=i+1}^{|W|} \text{NHI}(S_i, S_j)}{|W|(|W|-1)} \tag{1.26}$$

The value of ANHI indicates the changes observed in one of the selected subsets. A higher value of ANHI provides greater insight into the variation within feature subsets. The Hamming distance is ineffective in quantifying the intersection between two subgroups.

1.6.2.1.2 Jaccard's Index (JI) The Jaccard index estimates the mean similarity among all pairwise selections of W feature subsets. The formulation Jaccard's index $J\left(S_i, S_j\right)$ is shown in Equation 1.27. The stability index (JS) of Jaccard's index is shown as Equation 1.28.

$$J\left(S_i, S_j\right) = \frac{|S_i \cap S_j|}{|S_i \cup S_j|} \tag{1.27}$$

$$JS = \frac{2}{|W|(|W|-1)} \sum_{i=1}^{|W|-1} \sum_{j=i+1}^{|W|} J(S_i, S_j) \tag{1.28}$$

The Jaccard's stability JS results come under the $JS = \{0, 1\}$ [124]. The 0 value signifies that the FS process is unstable, and 1 value suggests that the means algorithm is stable. The number of samples in the collection influences the Jaccard Index.

1.6.2.1.3 Tanimoto Distance (TD) The Tanimoto distance quantifies the degree of overlapping between two feature sets, and producing a number within a defined range [124]. The zero value implies that the two subsets are completely disjointed, and the one value signifies that the two subsets are identical. The generalized version of Jaccard's index $T\left(S_1, S_j\right)$ is shown in Equation 1.29 [44].

$$T\left(S_1, S_j\right) = 1 - \frac{|S_i| + |S_j| - 2|S_i \cap S_j|}{|S_i| + |S_j| - |S_i \cap S_j|} \tag{1.29}$$

1.6.2.1.4 Dice-Sorensen's Index (DSI) The DSI determines intersection between selected feature sets [44]. The formulation of the Dice (S_1, S_j) is shown in Equation 1.30.

$$\text{Dice}\left(S_1, S_j\right) = \frac{2\left|S_1 \cap S_j\right|}{\left|S_1\right| + \left|S_j\right|} \tag{1.30}$$

The results of Dice (S_1, S_j) imply under the 0 and 1. The 0 value implies that the two subsets are completely disjoint, while the value of 1 signifies that the two subsets are identical. The DSI occasionally provides marginally superior and more significant stability results due to the absence of division by the union of subsets.

1.6.2.1.5 Percentage of Overlapping Gene/Features (POG) This assesses the reliability of chosen feature through the calculation of their degree of overlap. The POG is determined by non-symmetric as $POG(S_i, S_j) \neq POG(S_j, S_i)$. The POG is determined by symmetric as $S_i = S_j$ [44]. The formulation of POG is shown in Equation 1.31.

$$POG\left(S_i, S_j\right) = \frac{\left|S_i \cap S_j\right|}{\left|S_i\right|} \tag{1.31}$$

The POG matrix assesses the coherence among genes that are expressed differentially. The limitation of POG lies in its failure to consider the interrelation among features. This demerit is resolved by POGR, and its formulation is Equation 1.32.

$$POGR\left(S_i, S_j\right) = \frac{\left|S_i \cap S_j\right| + Z}{\left|S_i\right|} \tag{1.32}$$

The value Z determines the number of genes and other omics data. It is missing in S_j but significantly positively associated with at least one gene in S_i. It captures the correlation between features and singular characteristics. The normalized form of POG and POGR eliminates the reliance between the results. The normalized POG and POGR are shown in Equations 1.33 and 1.34. It is bounded by the interval of [−1, 1].

$$nPOG\left(S_1, S_j\right) = \frac{\left|S_1 \cap S_j\right| - E\left(\left|S_1 \cap S_j\right|\right)}{\left|S_1\right| - E\left(\left|S_1 \cap S_j\right|\right)} \tag{1.33}$$

$$nPOGR\left(S_1, S_j\right) = \frac{\left|S_1 \cap S_j\right| + Z - E\left(\left|S_1 \cap S_j\right|\right) + E(Z)}{\left|S_1\right| - E\left(\left|S_1 \cap S_j\right|\right) - E(Z)} \tag{1.34}$$

The E determines the expected value intersection between (S_i, S_j) features subset. The E(Z) defines the features set of S_i are not shared in other features set and positivity correlate to S_j.

1.6.2.1.6 Kuncheva Index (KI) or Consistency Index (IC) The considerable magnitude of the chosen feature subsets leads to stability estimators suggesting an increased overlap among the feature subsets. This elimination is that KI uses corrections to discard the intersection between the selected feature subsets by chance [125]. The formulation of the Kuncheva index-based consistency index is shown in Equation 1.35.

$$IC(S_i, S_j) = \frac{|S_i \cap S_j| * m - k^2}{k * (m - k)} \tag{1.35}$$

The formulation result of KI is showing the range of [−1, 1]. The one value shows the (S_i, S_j) features subset is identical, and −1 shows the subsets do not overlap. The zero shows the independent relationship between the drawn list features subset. The average of the pairwise consistency index shows the consistency of the (S_i, S_j) features subset. The formulation of Average pairwise consistency A_{Ic} index is shown in Equation 1.36.

$$A_{Ic} = \frac{2}{|W|(|W| - 1)} \sum_{i=1}^{|W|-1} \sum_{j=i+1}^{|W|} I_c(S_i, S_j) \tag{1.36}$$

1.6.2.1.7 Consistency Measures (CM) The values of stability produced by various estimators for the identical system are limited to specific ranges. The available metrics apply solely to the FS issue with a predetermined subset size (k). The stability metric of value 1 shows the robust FS technique, while value 0 signifies the poorest stability level of FS algorithms. This metric assessment of stability is based on the regularity of feature appearances [126].

Suppose X is the subset of Y and all features are represented in S. The F denotes the presence of feature f in S. The N represents the quantity of feature occurrences in S. The minimal value F_{min} is related to value 1 and it determines the feature occurrence f. The maximum value F_{max} determines the maximum feature value of m. The stability value is bounded by ranges. The 0 value shows the feature is present in only one feature subset, and 1 indicates that the feature exists in every subset. The formulation of the consistency $C(f)$ of the feature is shown in Equation 1.37, and consistency of the entire system $C(S)$ is shown in Equation 1.38.

$$C(f) = \frac{Ff - F_{min}}{F_{max} - F_{min}} \tag{1.37}$$

$$C(S) = \frac{1}{|X|} \sum_{f \in X} C(f) \tag{1.38}$$

1.6.2.1.8 Symmetrical Uncertainty (SU) This feature stability measure is based on the Suisan entropy and determines the feature correlation by a nonlinear approach. It considers feature values by stability and avoids the features. The SU defines the correlation of associated features in the subset. It uses the information gain metric for symmetric measurement. The formulation of the $SU(S_1, S_j)$ and information gain IG are shown in Equations 1.39 and 1.40. The property of the IG is demonstrated in Equations 1.41 and 1.42

$$SU(S_1, S_j) = 2\left[\frac{IG(S_1|S_j)}{H(S_1) + H(S_j)}\right] \tag{1.39}$$

$$IG = H(S_1) - H(S_1|S_j) \tag{1.40}$$

Where

$$H(S_1) = \sum_{xes} P(x) \cdot \log_2(P(x)) \tag{1.41}$$

$$H(S_1|S_j) = \sum_{yes} P(y) \sum_{xes} P(xy) \cdot \log_2(P(x|y)) \tag{1.42}$$

1.7 STABILITY MEASURES PROPERTIES FOR FEATURE SELECTION

Stability refers to the responsiveness of the selected feature set to variances in the offered extracted data for the data analysis and modelling process. FS addresses the 'curse of dimensionality' as a significant strategy for dimensionality reduction. The feature stability evaluates the susceptibility of FS methods. It defined the minor variations in the omics dataset. Numerous stability metrics have been employed to assess the stability of FS to identify pertinent characteristics.

The stability method evaluates the reliability of FS algorithms in response to variations in input data by determining the mean similarity across all combinations of chosen subsets. The central issue in assessing stability lies in determining the efficacy of the stability metric for the process of FS. This section elucidates the diverse characteristics of stability metrics that assist in the precise selection of features from the omics dataset [44, 127].

1.7.1 Iterative Process

The FS technique produces different sizes of feature subsets each time. The excellence FS algorithm requires selecting an optimal feature subset by executing the algorithm multiple times. The robust stability measure should consistently account for this characteristic.

1.7.2 Bounds

The bounding determines the upper and lower bound of the FS approach. The bounding property focuses on the stability value. The stability value is generally restricted by ranges such as [0, 1], [−1, 1], etc. The accuracy of the results falls within this range to determine stability.

1.7.3 Maximum Stability

The deterministic section of the feature determines the maximum stability. The deterministic selection leads to maximum stability, while the maximum stability leads to deterministic selection.

1.7.3.1 Deterministic Selection to Maximum Stability

The FS algorithm must use the deterministic selection process to obtain the maximum stability of the feature subset. The deterministic nature of the selection process identifies the maximum stability, which determines the important feature based on the data analysis results [127]. This property is based on the deterministic approach.

1.7.3.2 Maximum Stability to Deterministic Selection

Maximum stability to deterministic selection determines the consistent value of maximum stability regardless of the number of selected features. This property first identifies the important features for analysis and then selects all these features by deterministic and FS approaches. This property relies on predefined features for data modelling using a deterministic approach.

1.7.4 Correction

It guarantees that the procedure for selecting features identifies a random quantity of features while maintaining consistent estimated stability values. This property determines the estimated stability value of each FS algorithm, which should be equal. The correction determines the accuracy level of the FS approach. It determines if one feature is independent of another feature.

1.7.5 Monotonicity

The monotonicity determines the stronger stability of feature subsets by utilizing the cross-section of two feature subsets. It implicitly delineates the concept of stability by utilizing the stable FS technique that produces feature sets with a significant intersection. It provides that the similarity value escalates with the magnitude of the intersection of the feature sets.

The comparative analysis of stability measures properties to stability measures for the accurate FS approach is shown in Table 1.11.

TABLE 1.11 Stability Measures and Their Properties for Optimal Feature Selection

Stability Measures	Iterative Process	Bounds	Maximum Stability	Correction	Monotonicity
Jaccard	Yes	Yes	Yes	No	Yes
Hamming	Yes	Yes	Yes	No	Yes
Dice	Yes	Yes	Yes	No	Yes
POG	Yes	Yes	Yes	No	No
Kuncheva	No	Yes	Yes	Yes	No
nPOG	Yes	No	Yes	Yes	No
Wald	Yes	No	No	Yes	Yes
CW_{rel}	Yes	Yes	No	Yes	Yes
Tanimoto	Yes	Yes	No	Yes	Yes
Symmetrical	Yes	Yes	No	Yes	Yes
Canberra	Yes	Yes	No	Yes	No
Spearman	Yes	Yes	No	Yes	Yes
Pearson	Yes	Yes	No	Yes	Yes

1.8 CONCLUSION

The rapid progression of technology has generated intricate and high-dimensional datasets in the realm of medical research and diverse scientific disciplines. The research in bioinformatics, biostatistics, and other medical disciplines utilizes omics data for modeling and data analysis. The omics data encompass one or more types of information derived from genomics, transcriptomics, proteomics, metabolomics, epigenomics, microbiomics, and lipidomics. The omics data-related research faces problems related to large dimensionality and small sample sizes. The large dimensional determines the small sample size, dataset shift, data complexity, outliers, and class imbalance-related challenges in omics data. The FE and selection are well-known approaches to resolving considerable dimensionality-related challenges. The FS approach selects the most relevant features, and the FE approach extracts the feature subset for modeling and data analysis. The modelling and data analysis help in disease diagnosis and subtyping, drug discovery, personalized medicine, biomarker discovery, etc.

This chapter initially outlines the various difficulties associated with the foundational aspects of bioinformatics and biostatistics regarding the structure of omics data. The following section thoroughly analyses the FS and FS methodology. The precise FS method enhances model effectiveness while reducing overfitting and computational expenses in high-dimensional datasets such as omics. The FS techniques are characterized as filter, wrapper, embedding, and hybrid. The filter class applies statistical parameters, whereas the wrapper employs a machine learning prediction model to choose the feature set. The embedded class utilizes the advantages of both the filter and wrapper classes. The hybrid class employs many approaches to precisely ascertain specific characteristics. The examination of each FS class reveals that each method possesses distinct advantages and disadvantages regarding the challenges presented.

This chapter intends to enhance the understanding of FS, FE, feature validation methods, stability measures, and stability measures properties about omics data and its practical applications. The examination of the FS methodology facilitates the uncovering of concealed molecular insights pertinent to particular human disease therapies and diagnostic processes.

REFERENCES

1. E. S. Boja, C. R. Kinsinger, H. Rodriguez, and P. Srinivas, "Integration of omics sciences to advance biology and medicine," *Clin Proteomics*, vols. 11, no. 1, p. 45, 2014. doi: 10.1186/1559-0275-11-45.
2. A. Khoshnam, D. Chafamo, and N. Tokcan, "Tensor-based approaches for omics data analysis: Applications, challenges, and future directions," *La Matematica*, vol. 4, no. 2, pp. 486–520, 2025. doi: 10.1007/s44007-025-00149-x.
3. K. J. Karczewski and M. P. Snyder, "Integrative omics for health and disease," *Nat Rev Genet*, vol. 19, no. 5, pp. 299–310, 2018. doi: 10.1038/nrg.2018.4.
4. R. Margolis *et al*, "The national institutes of Health's big data to knowledge (BD2K) initiative: Capitalizing on biomedical big data," *J Am Med Inf Assoc*, vol. 21, no. 6, pp. 957–958, 2014. doi: 10.1136/amiajnl-2014-002974.

5. Z. Yao, G. Zhu, J. Too, M. Duan, and Z. Wang, "Feature selection of OMIC data by ensemble swarm intelligence based approaches," *Front Genet*, vol. 12, Mar. 2022, doi: 10.3389/fgene.2021.793629.
6. X. Yu, F. Abbas-Aghababazadeh, Y. A. Chen, and B. L. Fridley, "Statistical and Bioinformatics Analysis of Data from Bulk and Single-Cell RNA Sequencing Experiments," 2021, pp. 143–175. doi: 10.1007/978-1-0716-0849-4_9.
7. S. Aldridge and S. A. Teichmann, "Single cell transcriptomics comes of age," *Nat Commun*, vol. 11, no. 1, p. 4307, 2020. doi: 10.1038/s41467-020-18158-5.
8. Y. Hasin, M. Seldin, and A. Lusis, "Multi-omics approaches to disease," *Genome Biol*, vol. 18, no. 1, p. 83, 2017. doi: 10.1186/s13059-017-1215-1.
9. B. B. Misra, C. Langefeld, M. Olivier, and L. A. Cox, "Integrated omics: Tools, advances and future approaches," *J Mol Endocrinol*, vol. 62, no. 1, pp. R21–R45, 2019. doi: 10.1530/JME-18-0055.
10. D. Chicco, F. Cumbo, and C. Angione, "Ten quick tips for avoiding pitfalls in multi-omics data integration analyses," *PLoS Comput Biol*, vol. 19, no. 7, p. e1011224, 2023. doi: 10.1371/journal.pcbi.1011224.
11. J. G. Liao and K.-V. Chin, "Logistic regression for disease classification using microarray data: Model selection in a large *p* and small *n* case," *Bioinformatics*, vol. 23, no. 15, pp. 1945–1951, 2007. doi: 10.1093/bioinformatics/btm287.
12. X. Han *et al*, "Mapping the mouse cell atlas by microwell-seq," *Cell*, vol. 172, no. 5, pp. 1091–1107.e17, 2018. doi: 10.1016/j.cell.2018.02.001.
13. D. Lähnemann *et al*, "Eleven grand challenges in single-cell data science," *Genome Biol*, vol. 21, no. 1, p. 31, 2020. doi: 10.1186/s13059-020-1926-6.
14. E. López de Maturana *et al*, "Challenges in the integration of omics and non-omics data," *Genes*, vol. 10, no. 3, p. 238, 2019. doi: 10.3390/genes10030238.
15. L. Zhao *et al*, "DeepOmix: A scalable and interpretable multi-omics deep learning framework and application in cancer survival analysis," *Comput Struct Biotechnol J*, vol. 19, pp. 2719–2725, 2021. doi: 10.1016/j.csbj.2021.04.067.
16. G. L. Libralon, A. C. P. L. F. de Carvalho, and A. C. Lorena, "Pre-processing for noise detection in gene expression classification data," *Journal of the Brazilian Computer Society*, vol. 15, no. 1, pp. 3–11, 2009. doi: 10.1007/BF03192573.
17. Y. Li, U. Mansmann, S. Du, and R. Hornung, "Benchmark study of feature selection strategies for multi-omics data," *BMC Bioinformatics*, vol. 23, no. 1, p. 412, 2022. doi: 10.1186/s12859-022-04962-x.
18. L. Gidskehaug, E. Anderssen, A. Flatberg, and B. K. Alsberg, "A framework for significance analysis of gene expression data using dimension reduction methods," *BMC Bioinformatics*, vol. 8, no. 1, p. 346, 2007. doi: 10.1186/1471-2105-8-346.
19. J. Li *et al*, "Feature selection," *ACM Comput Surv*, vol. 50, no. 6, pp. 1–45, 2018. doi: 10.1145/3136625.
20. Y. Han, L. Huang, and F. Zhou, "A dynamic recursive feature elimination framework (dRFE) to further refine a set of OMIC biomarkers," *Bioinformatics*, vol. 37, no. 15, pp. 2183–2189, 2021. doi: 10.1093/bioinformatics/btab055.
21. M. A. Hambali, T. O. Oladele, and K. S. Adewole, "Microarray cancer feature selection: Review, challenges and research directions," *Int J Cognit Comput Eng*, vol. 1, pp. 78–97, 2020. doi: 10.1016/j.ijcce.2020.11.001.
22. E. R. Dougherty, "Small sample issues for microarray-based classification," *Comp Funct Genomics*, vol. 2, no. 1, pp. 28–34, 2001. doi: 10.1002/cfg.62.
23. V. López, A. Fernández, S. García, V. Palade, and F. Herrera, "An insight into classification with imbalanced data: Empirical results and current trends on using data intrinsic characteristics," *Inf Sci*, vol. 250, pp. 113–141, 2013. doi: 10.1016/j.ins.2013.07.007.

24. M. Galar, A. Fernández, E. Barrenechea, and F. Herrera, "EUSBoost: Enhancing ensembles for highly imbalanced data-sets by evolutionary undersampling," *Pattern Recognit*, vol. 46, no. 12, pp. 3460–3471, 2013. doi: 10.1016/j.patcog.2013.05.006.
25. J. A. Sáez, J. Luengo, and F. Herrera, "Predicting noise filtering efficacy with data complexity measures for nearest neighbor classification," *Pattern Recognit*, vol. 46, no. 1, pp. 355–364, 2013. doi: 10.1016/j.patcog.2012.07.009.
26. J. G. Moreno-Torres, J. A. Saez, and F. Herrera, "Study on the impact of partition-induced dataset shift on k-fold cross-validation," *IEEE Trans Neural Netw Learn Syst*, vol. 23, no. 8, pp. 1304–1312, 2012. doi: 10.1109/TNNLS.2012.2199516.
27. V. Bolon-Canedo, N. Sanchez-Marono, and A. Alonso-Betanzos, "On the Effectiveness of Discretization on Gene Selection of Microarray Data," in *The 2010 International Joint Conference on Neural Networks (IJCNN)*, IEEE, Jul. 2010, pp. 1–8. doi: 10.1109/IJCNN.2010.5596825.
28. J. G. Moreno-Torres, J. A. Saez, and F. Herrera, "Study on the impact of partition-induced dataset shift on k-Fold cross-validation," *IEEE Trans Neural Netw Learn Syst*, vol. 23, no. 8, pp. 1304–1312, 2012. doi: 10.1109/TNNLS.2012.2199516.
29. J. Hardin and D. M. Rocke, "Outlier detection in the multiple cluster setting using the minimum covariance determinant estimator," *Comput Stat Data Anal*, vol. 44, no. 4, pp. 625–638, 2004. doi: 10.1016/S0167-9473(02)00280-3.
30. F. F. G. Navarro and L. A. B. Muñoz, "Gene Subset selection in microarray data using entropic filtering for cancer classification," *Expert Syst*, vol. 26, no. 1, pp. 113–124, 2009. doi: 10.1111/j.1468-0394.2008.00489.x.
31. A. Bommert, T. Welchowski, M. Schmid, and J. Rahnenführer, "Benchmark of filter methods for feature selection in high-dimensional gene expression survival data," *Brief Bioinform*, vol. 23, no. 1, p. bbab354, 2022.
32. T. Bhadra, S. Mallik, N. Hasan, and Z. Zhao, "Comparison of five supervised feature selection algorithms leading to top features and gene signatures from multi-omics data in cancer," *BMC Bioinform*, vol. 23, no. S3, p. 153, 2022. doi: 10.1186/s12859-022-04678-y.
33. R. Zebari, A. Abdulazeez, D. Zeebaree, D. Zebari, and J. Saeed, "A comprehensive review of dimensionality reduction techniques for feature selection and feature extraction," *J Appl SciTechnol Trends*, vol. 1, no. 1, pp. 56–70, 2020. doi: 10.38094/jastt1224.
34. J. Sheng, J. Amankwah-Amoah, Z. Khan, and X. Wang, "COVID-19 pandemic in the new era of big data analytics: Methodological innovations and future research directions," *Br J Manag*, vol. 32, no. 4, pp. 1164–1183, 2021. doi: 10.1111/1467-8551.12441.
35. F. Qiu *et al*, "Mutational slime mould algorithm for gene selection," *Biomedicines*, vol. 10, no. 8, p. 2052, 2022. doi: 10.3390/biomedicines10082052.
36. S. Mallik, T. Bhadra, and U. Maulik, "Identifying epigenetic biomarkers using maximal relevance and minimal redundancy based feature selection for multi-omics data," *IEEE Trans Nanobioscience*, vol. 16, no. 1, pp. 3–10, 2017. doi: 10.1109/TNB.2017.2650217.
37. T. Bhadra and U. Maulik, "Unsupervised feature selection using iterative shrinking and expansion algorithm," *IEEE Trans Emerg Top Comput Intell*, vol. 6, no. 6, pp. 1453–1462, 2022. doi: 10.1109/TETCI.2022.3199704.
38. A. Bommert, T. Welchowski, M. Schmid, and J. Rahnenführer, "Benchmark of filter methods for feature selection in high-dimensional gene expression survival data," *Brief Bioinform*, vol. 23, no. 1, 2022. doi: 10.1093/bib/bbab354.
39. K. Borah, H. S. Das, S. Seth, K. Mallick, Z. Rahaman, and S. Mallik, "A review on advancements in feature selection and feature extraction for high-dimensional NGS data analysis," *Funct Integr Genom*, vol. 24, no. 5, p. 139, 2024. doi: 10.1007/s10142-024-01415-x.
40. T. Bhadra, S. Mallik, N. Hasan, and Z. Zhao, "Comparison of five supervised feature selection algorithms leading to top features and gene signatures from multi-omics data in cancer," *BMC Bioinform*, vol. 23, Mar. 2022, doi: 10.1186/s12859-022-04678-y.

41. E. Barshan, A. Ghodsi, Z. Azimifar, and M. Zolghadri Jahromi, "Supervised principal component analysis: Visualization, classification and regression on subspaces and submanifolds," *Pattern Recognit*, vol. 44, no. 7, pp. 1357–1371, 2011. doi: 10.1016/j.patcog.2010.12.015.
42. T. M. Khoshgoftaar, A. Fazelpour, H. Wang, and R. Wald, "A Survey of Stability Analysis of Feature Subset Selection Techniques," in *2013 IEEE 14th International Conference on Information Reuse & Integration (IRI)*, IEEE, Aug. 2013, pp. 424–431. doi: 10.1109/IRI.2013.6642502.
43. J. Miao and L. Niu, "A survey on feature selection," *Procedia Comput Sci*, vol. 91, pp. 919–926, 2016. doi: 10.1016/j.procs.2016.07.111.
44. U. M. Khaire and R. Dhanalakshmi, "Stability of feature selection algorithm: A review," *J King Saud Univ – Comput Inf Sci*, vol. 34, no. 4, pp. 1060–1073, 2022. doi: 10.1016/j.jksuci.2019.06.012.
45. F. Yang and K. Z. Mao, "Robust feature selection for microarray data based on multicriterion fusion," *IEEE/ACM Trans Comput Biol Bioinform*, vol. 8, no. 4, pp. 1080–1092, 2011. doi: 10.1109/TCBB.2010.103.
46. A. J. Ferreira and M. A. T. Figueiredo, "Efficient feature selection filters for high-dimensional data," *Pattern Recognit Lett*, vol. 33, no. 13, pp. 1794–1804, 2012. doi: 10.1016/j.patrec.2012.05.019.
47. N. Kwak and C.-H. Choi, "Input feature selection by mutual information based on Parzen window," *IEEE Trans Pattern Anal Mach Intell*, vol. 24, no. 12, pp. 1667–1671, 2002. doi: 10.1109/TPAMI.2002.1114861.
48. H. Peng, F. Long, and C. Ding, "Feature selection based on mutual information criteria of max-dependency, max-relevance, and min-redundancy," *IEEE Trans Pattern Anal Mach Intell*, vol. 27, no. 8, pp. 1226–1238, 2005. doi: 10.1109/TPAMI.2005.159.
49. P. A. Estevez, M. Tesmer, C. A. Perez, and J. M. Zurada, "Normalized mutual information feature selection," *IEEE Trans Neural Netw*, vol. 20, no. 2, pp. 189–201, 2009. doi: 10.1109/TNN.2008.2005601.
50. L. T. Vinh, N. D. Thang, and Y.-K. Lee, "An Improved Maximum Relevance and Minimum Redundancy Feature Selection Algorithm Based on Normalized Mutual Information," in *2010 10th IEEE/IPSJ International Symposium on Applications and the Internet*, IEEE, Jul. 2010, pp. 395–398. doi: 10.1109/SAINT.2010.50.
51. L. Liu, J. Kang, J. Yu, and Z. Wang, "A Comparative Study on Unsupervised Feature Selection Methods for Text Clustering," in *2005 International Conference on Natural Language Processing and Knowledge Engineering*, IEEE, pp. 597–601. doi: 10.1109/NLPKE.2005.1598807.
52. S. Bandyopadhyay, S. Mallik, and A. Mukhopadhyay, "A survey and comparative study of statistical tests for identifying differential expression from microarray data," *IEEE/ACM Trans Comput Biol Bioinform*, vol. 11, no. 1, pp. 95–115, 2014. doi: 10.1109/TCBB.2013.147.
53. B. Liao, Y. Jiang, W. Liang, W. Zhu, L. Cai, and Z. Cao, "Gene selection using locality sensitive laplacian score," *IEEE/ACM Trans Comput Biol Bioinform*, vol. 11, no. 6, pp. 1146–1156, 2014. doi: 10.1109/TCBB.2014.2328334.
54. G. Manikandan, E. Susi, and S. Abirami, "Flexible-Fuzzy Mutual Information based Feature Selection on High Dimensional Data," in *2018 Tenth International Conference on Advanced Computing (ICoAC)*, IEEE, Dec. 2018, pp. 237–243. doi: 10.1109/ICoAC44903.2018.8939115.
55. S. K. Shevade, and S. S. Keerthi, "A simple and efficient algorithm for gene selection using sparse logistic regression," *Bioinformatics*, vol. 19, no. 17, pp. 2246–2253, 2003. doi: 10.1093/bioinformatics/btg308.
56. M. A. Hall, "Correlation-based feature selection of discrete and numeric class machine learning," 2000. (Working paper 00/08). Hamilton, New Zealand: University of Waikato, Department of Computer Science.
57. B. Sahu, S. Dehuri, and A. Jagadev, "A study on the relevance of feature selection methods in microarray data," *Open Bioinforma J*, vol. 11, no. 1, pp. 117–139, 2018. doi: 10.2174/1875036201811010117.

58. Z. M. Hira, and D. F. Gillies, "A review of feature selection and feature extraction methods applied on microarray data," *Adv Bioinform*, vol. 2015, pp. 1–13, 2015. doi: 10.1155/2015/198363.
59. L. Yu, and H. Liu, "Efficient feature selection via analysis of relevance and redundancy," *J Mach Learn Res*, vol. 5, pp. 1205–1224, 2004.
60. Y. Khourdifi, and M. Bahaj, "Feature Selection with Fast Correlation-Based Filter for Breast Cancer Prediction and Classification Using Machine Learning Algorithms," in *2018 International Symposium on Advanced Electrical and Communication Technologies (ISAECT)*, IEEE, Nov. 2018, pp. 1–6. doi: 10.1109/ISAECT.2018.8618688.
61. S. Vanjimalar, D. Ramyachitra, and P. Manikandan, "A Review on Feature Selection Techniques for Gene Expression Data," in *2018 IEEE International Conference on Computational Intelligence and Computing Research (ICCIC)*, IEEE, Dec. 2018, pp. 1–4. doi: 10.1109/ICCIC.2018.8782294.
62. X. Sun, Y. Liu, J. Li, J. Zhu, H. Chen, and X. Liu, "Feature evaluation and selection with cooperative game theory," *Pattern Recognit*, vol. 45, no. 8, pp. 2992–3002, 2012. doi: 10.1016/j.patcog.2012.02.001.
63. T. K. Kim, "Understanding one-way ANOVA using conceptual figures," *Korean J Anesthesiol*, vol. 70, no. 1, p. 22, 2017. doi: 10.4097/kjae.2017.70.1.22.
64. S. Solorio-Fernandez, J. F. Martinez-Trinidad, J. A. Carrasco-Ochoa, and Y.-Q. Zhang, "Hybrid Feature Selection Method for Biomedical Datasets," in *2012 IEEE Symposium on Computational Intelligence in Bioinformatics and Computational Biology (CIBCB)*, IEEE, May 2012, pp. 150–155. doi: 10.1109/CIBCB.2012.6217224.
65. U. Moorthy, and U. D. Gandhi, "RETRACTED ARTICLE: A novel optimal feature selection technique for medical data classification using ANOVA based whale optimization," *J Ambient Intell Humaniz Comput*, vol. 12, no. 3, pp. 3527–3538, 2021. doi: 10.1007/s12652-020-02592-w.
66. X. Xu, H. Gu, Y. Wang, J. Wang, and P. Qin, "Autoencoder based feature selection method for classification of anticancer drug response," *Front Genet*, vol. 10, Mar. 2019, doi: 10.3389/fgene.2019.00233.
67. B. Venkatesh, and J. Anuradha, "A review of feature selection and its methods," *Cybern Inform Technol*, vol. 19, no. 1, pp. 3–26, 2019. doi: 10.2478/cait-2019-0001.
68. D. Dittman, T. Khoshgoftaar, R. Wald, and A. Napolitano, "Similarity Analysis of Feature Ranking Techniques on Imbalanced DNA Microarray Datasets," in *2012 IEEE International Conference on Bioinformatics and Biomedicine*, IEEE, Oct. 2012, pp. 1–5. doi: 10.1109/BIBM.2012.6392708.
69. S. Rathore, M. A. Iftikhar, and M. Hussain, "A Novel Approach for Automatic Gene Selection and Classification of Gene Based Colon Cancer Datasets," in *2014 International Conference on Emerging Technologies (ICET)*, IEEE, Dec. 2014, pp. 42–47. doi: 10.1109/ICET.2014.7021014.
70. J. Tang *et al*, "Computational advances of tumor marker selection and sample classification in cancer proteomics," *Comput Struct Biotechnol J*, vol. 18, pp. 2012–2025, 2020. doi: 10.1016/j.csbj.2020.07.009.
71. U. Maulik, and D. Chakraborty, "Fuzzy preference based feature selection and semisupervised SVM for cancer classification," *IEEE Trans Nanobiosci*, vol. 13, no. 2, pp. 152–160, 2014. doi: 10.1109/TNB.2014.2312132.
72. D. Chakraborty, and U. Maulik, "Identifying cancer biomarkers from microarray data using feature selection and semisupervised learning," *IEEE J Transl Eng Health Med*, vol. 2, pp. 1–11, 2014. doi: 10.1109/JTEHM.2014.2375820.
73. R. Battiti, "Using mutual information for selecting features in supervised neural net learning," *IEEE Trans Neural Netw*, vol. 5, no. 4, pp. 537–550, 1994. doi: 10.1109/72.298224.
74. Y. Wang, "Unsupervised representative feature selection algorithm based on information entropy and relevance analysis," *IEEE Access*, vol. 6, pp. 45317–45324, 2018. doi: 10.1109/ACCESS.2018.2863752.

75. X. Guo, X. Jiang, J. Xu, X. Quan, M. Wu, and H. Zhang, "Ensemble consensus-guided unsupervised feature selection to identify Huntington's disease-associated genes," *Genes*, vol. 9, no. 7, p. 350, 2018. doi: 10.3390/genes9070350.
76. T. Chandrasekhar, K. Thangavel, E. Elayaraja, and E. N. Sathishkumar, "Unsupervised Gene Expression Data Using Enhanced Clustering Method," in *2013 IEEE International Conference on Emerging Trends in Computing, Communication and Nanotechnology (ICECCN)*, IEEE, Mar. 2013, pp. 518–522. doi: 10.1109/ICE-CCN.2013.6528554.
77. M. R. Afrash, E. Mirbagheri, M. Mashoufi, and H. Kazemi-Arpanahi, "Optimizing prognostic factors of five-year survival in gastric cancer patients using feature selection techniques with machine learning algorithms: A comparative study," *BMC Med Inform Decis Mak*, vol. 23, no. 1, p. 54, 2023. doi: 10.1186/s12911-023-02154-y.
78. J. C. Ang, A. Mirzal, H. Haron, and H. N. A. Hamed, "Supervised, unsupervised, and semi-supervised feature selection: A review on gene selection," *IEEE/ACM Trans Comput Biol Bioinform*, vol. 13, no. 5, pp. 971–989, 2016. doi: 10.1109/TCBB.2015.2478454.
79. N. Pudjihartono, T. Fadason, A. W. Kempa-Liehr, and J. M. O'Sullivan, "A review of feature selection methods for machine learning-based disease risk prediction," *Front Bioinform*, vol. 2, Jun. 2022, doi: 10.3389/fbinf.2022.927312.
80. H. Almazrua, and H. Alshamlan, "A comprehensive survey of recent hybrid feature selection methods in cancer microarray Gene expression data," *IEEE Access*, vol. 10, pp. 71427–71449, 2022. doi: 10.1109/ACCESS.2022.3185226.
81. I. Guyon, J. Weston, S. Barnhill, and V. Vapnik, "Gene selection for cancer classification using support vector machines," *Mach Learn*, vol. 46, no. 1/3, pp. 389–422, 2002. doi: 10.1023/A:1012487302797.
82. H. Sanz, C. Valim, E. Vegas, J. M. Oller, and F. Reverter, "SVM-RFE: Selection and visualization of the most relevant features through non-linear kernels," *BMC Bioinform*, vol. 19, no. 1, p. 432, 2018. doi: 10.1186/s12859-018-2451-4.
83. M.-L. Huang, Y.-H. Hung, W. M. Lee, R. K. Li, and B.-R. Jiang, "SVM-RFE based feature selection and Taguchi parameters optimization for multiclass SVM classifier," *Sci World J*, vol. 2014, pp. 1–10, 2014. doi: 10.1155/2014/795624.
84. Y. Saeys, I. Inza, and P. Larrañaga, "A review of feature selection techniques in bioinformatics," *Bioinformatics*, vol. 23, no. 19, pp. 2507–2517, 2007. doi: 10.1093/bioinformatics/btm344.
85. K. Yan, and D. Zhang, "Feature selection and analysis on correlated gas sensor data with recursive feature elimination," *Sens Actuators B Chem*, vol. 212, pp. 353–363, 2015. doi: 10.1016/j.snb.2015.02.025.
86. L. Zhang, "A feature selection algorithm integrating maximum classification information and minimum interaction feature dependency information," *Comput Intell Neurosci*, vol. 2021, no. 1, 2021. doi: 10.1155/2021/3569632.
87. J. Liang, S. Yang, and A. Winstanley, "Invariant optimal feature selection: A distance discriminant and feature ranking based solution," *Pattern Recognit*, vol. 41, no. 5, pp. 1429–1439, 2008. doi: 10.1016/j.patcog.2007.10.018.
88. M. A. Esseghir, "Effective Wrapper-Filter Hybridization through GRASP Schemata," in *Proceedings of the Fourth International Workshop on Feature Selection in Data Mining*, H. Liu, H. Motoda, R. Setiono, and Z. Zhao, Eds., in Proceedings of Machine Learning Research, vol. 10. Hyderabad, India: PMLR, Jul. 2010, pp. 45–54. [Online]. Available: https://proceedings.mlr.press/v10/esseghir10a.html
89. A. M. Ismail *et al*, "Artificial bee colony algorithm in estimating kinetic parameters for yeast fermentation pathway," *J Integr Bioinform*, vol. 20, no. 2, 2023. doi: 10.1515/jib-2022-0051.
90. W. Zheng *et al*, "Multi-feature based network revealing the structural abnormalities in autism spectrum disorder," *IEEE Trans Affect Comput*, vol. 12, no. 3, pp. 732–742, 2021. doi: 10.1109/TAFFC.2018.2890597.

91. A. K. Alok, S. Saha, and A. Ekbal, "Semi-supervised clustering for gene-expression data in multiobjective optimization framework," *International Journal of Machine Learning and Cybernetics*, vol. 8, no. 2, pp. 421–439, 2017. doi: 10.1007/s13042-015-0335-8.
92. S. Mallik, "A multi-molecular fusion to detect transcriptomic signature in tissue-specific cancer," *Eurasian J Med Oncol*, 2022. doi: 10.14744/ejmo.2022.53376.
93. S. Seth, S. Mallik, T. Bhadra, and Z. Zhao, "Dimensionality reduction and Louvain agglomerative hierarchical clustering for cluster-specified frequent biomarker discovery in single-cell sequencing data," *Front Genet*, vol. 13, Feb. 2022, doi: 10.3389/fgene.2022.828479.
94. M. Liu and D. Zhang, "Pairwise constraint-guided sparse learning for feature selection," *IEEE Trans Cybern*, vol. 46, no. 1, pp. 298–310, 2016. doi: 10.1109/TCYB.2015.2401733.
95. K. K. Pandey and D. Shukla, "Min max kurtosis distance based improved initial centroid selection approach of k-means clustering for big data mining on Gene expression data," *Evol Systems*, vol. 14, no. 2, pp. 207–244, 2023. doi: 10.1007/s12530-022-09447-z.
96. K. K. Pandey and D. Shukla, "Cancer tissue detection using improved k-means initialization method for multi-dimensional microarray big data clustering," *J Ambient Intell Humaniz Comput*, vol. 14, no. 7, pp. 9277–9303, 2023. doi: 10.1007/s12652-022-04428-1.
97. K. K. Pandey and D. Shukla, "Min-max kurtosis stratum mean: An improved k-means cluster initialization approach for microarray Gene clustering on multidimensional big data," *Concurr Comput*, vol. 34, no. 23, 2022. doi: 10.1002/cpe.7185.
98. K. K. Pandey and D. Shukla, "Maxmin distance sort heuristic-based initial centroid method of partitional clustering for big data mining," *Pattern Anal Appl*, vol. 25, no. 1, pp. 139–156, 2022. doi: 10.1007/s10044-021-01045-0.
99. K. K. Pandey and D. Shukla, "NDPD: An improved initial centroid method of partitional clustering for big data mining," *J Adv Manag Res*, vol. 20, no. 1, pp. 1–34, 2023. doi: 10.1108/JAMR-07-2021-0242.
100. T. Swarnkar and P. Mitra, "Graph-based unsupervised feature selection and multiview clustering for microarray data," *J Biosci*, vol. 40, no. 4, pp. 755–767, 2015. doi: 10.1007/s12038-015-9559-8.
101. N. Yu, M.-J. Wu, J.-X. Liu, C.-H. Zheng, and Y. Xu, "Correntropy-based hypergraph regularized NMF for clustering and feature selection on multi-cancer integrated data," *IEEE Trans Cybern*, vol. 51, no. 8, pp. 3952–3963, 2021. doi: 10.1109/TCYB.2020.3000799.
102. X. Jiang, M. Chen, W. Wang, W. Song, and G. N. Lin, "Label Propagation Based Semi-supervised Feature Selection to Decode Clinical Phenotype of Huntington's Disease," 2019, pp. 529–542. doi: 10.1007/978-3-030-26763-6_51.
103. Z. Li, B. Liao, L. Cai, M. Chen, and W. Liu, "Semi-supervised maximum discriminative local margin for Gene selection," *Sci Rep*, vol. 8, no. 1, p. 8619, 2018. doi: 10.1038/s41598-018-26806-6.
104. S. Solorio-Fernández, J. A. Carrasco-Ochoa, and J. F. Martínez-Trinidad, "A review of unsupervised feature selection methods," *Artif Intell Rev*, vol. 53, no. 2, pp. 907–948, 2020. doi: 10.1007/s10462-019-09682-y.
105. G. Liu, J. Ma, T. Hu, and X. Gao, "A feature selection method with feature ranking using genetic programming," *Conn Sci*, vol. 34, no. 1, pp. 1146–1168, 2022. doi: 10.1080/09540091.2022.2049702.
106. C. Khanji, L. Lalonde, C. Bareil, M.-T. Lussier, S. Perreault, and M. E. Schnitzer, "Lasso regression for the prediction of intermediate outcomes related to cardiovascular disease prevention using the TRANSIT quality indicators," *Med Care*, vol. 57, no. 1, pp. 63–72, 2019. doi: 10.1097/MLR.0000000000001014.
107. C. Kang, Y. Huo, L. Xin, B. Tian, and B. Yu, "Feature selection and tumor classification for microarray data using relaxed lasso and generalized multi-class support vector machine," *J Theor Biol*, vol. 463, pp. 77–91, 2019. doi: 10.1016/j.jtbi.2018.12.010.

108. A. Anaissi, P. J. Kennedy, M. Goyal, and D. R. Catchpoole, "A balanced iterative random forest for gene selection from microarray data," *BMC Bioinform*, vol. 14, no. 1, p. 261, 2013. doi: 10.1186/1471-2105-14-261.
109. B. Krishnapuram, L. Carin, M. A. T. Figueiredo, and A. J. Hartemink, "Sparse multinomial logistic regression: Fast algorithms and generalization bounds," *IEEE Trans Pattern Anal Mach Intell*, vol. 27, no. 6, pp. 957–968, 2005. doi: 10.1109/TPAMI.2005.127.
110. R. Muthukrishnan and R. Rohini, "LASSO: A Feature Selection Technique in Predictive Modeling for Machine Learning," in *2016 IEEE International Conference on Advances in Computer Applications (ICACA)*, IEEE, Oct. 2016, pp. 18–20. doi: 10.1109/ICACA.2016.7887916.
111. G. C. Cawley and N. L. C. Talbot, "Gene Selection in cancer classification using sparse logistic regression with bayesian regularization," *Bioinformatics*, vol. 22, no. 19, pp. 2348–2355, 2006. doi: 10.1093/bioinformatics/btl386.
112. A. Jovic, K. Brkic, and N. Bogunovic, "A Review of Feature Selection Methods with Applications," in *2015 38th International Convention on Information and Communication Technology, Electronics and Microelectronics (MIPRO)*, IEEE, May 2015, pp. 1200–1205. doi: 10.1109/MIPRO.2015.7160458.
113. H. Tao, C. Hou, F. Nie, Y. Jiao, and D. Yi, "Effective discriminative feature selection with nontrivial solution," *IEEE Trans Neural Netw Learn Syst*, vol. 27, no. 4, pp. 796–808, 2016. doi: 10.1109/TNNLS.2015.2424721.
114. S. P. Potharaju and M. Sreedevi, "Distributed feature selection (DFS) strategy for microarray gene expression data to improve the classification performance," *Clin Epidemiol Glob Health*, vol. 7, no. 2, pp. 171–176, 2019. doi: 10.1016/j.cegh.2018.04.001.
115. M. Gupta and B. Gupta, "A novel gene expression test method of minimizing breast cancer risk in reduced cost and time by improving SVM-RFE gene selection method combined with LASSO," *J Integr Bioinform*, vol. 18, no. 2, pp. 139–153, 2021. doi: 10.1515/jib-2019-0110.
116. M. Palazzo, P. Beauseroy, and P. Yankilevich, "Unsupervised Feature Selection for Tumor Profiles Using Autoencoders and Kernel Methods," in *2020 IEEE Conference on Computational Intelligence in Bioinformatics and Computational Biology (CIBCB)*, IEEE, Oct. 2020, pp. 1–8. doi: 10.1109/CIBCB48159.2020.9277699.
117. M. Dashtban and M. Balafar, "Gene Selection for microarray cancer classification using a new evolutionary method employing artificial intelligence concepts," *Genomics*, vol. 109, no. 2, pp. 91–107, 2017. doi: 10.1016/j.ygeno.2017.01.004.
118. E. Pashaei and E. Pashaei, "Gene Selection for Cancer Classification using a New Hybrid of Binary Black Hole Algorithm," in *2020 28th Signal Processing and Communications Applications Conference (SIU)*, IEEE, Oct. 2020, pp. 1–4. doi: 10.1109/SIU49456.2020.9302351.
119. H. M. Alshamlan, G. H. Badr, and Y. A. Alohali, "Genetic bee colony (GBC) algorithm: A new gene selection method for microarray cancer classification," *Comput Biol Chem*, vol. 56, pp. 49–60, 2015. doi: 10.1016/j.compbiolchem.2015.03.001.
120. J. F. Pinto da Costa, H. Alonso, and L. Roque, "A weighted principal component analysis and its application to gene expression data," *IEEE/ACM Trans Comput Biol Bioinform*, vol. 8, no. 1, pp. 246–252, 2011. doi: 10.1109/TCBB.2009.61.
121. V.-S. Ha and H.-N. Nguyen, "C-KPCA: Custom Kernel PCA for Cancer Classification," 2016, pp. 459–467. doi: 10.1007/978-3-319-41920-6_36.
122. S. A. Mostafa *et al*, "Examining multiple feature evaluation and classification methods for improving the diagnosis of Parkinson's disease," *Cogn Syst Res*, vol. 54, pp. 90–99, 2019. doi: 10.1016/j.cogsys.2018.12.004.
123. S. Geman, E. Bienenstock, and R. Doursat, "Neural networks and the Bias/Variance dilemma," *Neural Comput*, vol. 4, no. 1, pp. 1–58, 1992. doi: 10.1162/neco.1992.4.1.1.

124. M. Mohammadi, H. Sharifi Noghabi, G. Abed Hodtani, and H. Rajabi Mashhadi, "Robust and stable gene selection via maximum–minimum correntropy criterion," *Genomics*, vol. 107, no. 2–3, pp. 83–87, 2016. doi: 10.1016/j.ygeno.2015.12.006.
125. L. I. Kuncheva, "A stability index for feature selection," in *Artificial intelligence and applications*, 2007, pp. 421–427.
126. P. Somol and J. Novovičová, "Evaluating stability and comparing output of feature selectors that optimize feature subset cardinality," *IEEE Trans Pattern Anal Mach Intell*, vol. 32, no. 11, pp. 1921–1939, 2010. doi: 10.1109/TPAMI.2010.34.
127. S. Nogueira and G. Brown, "Measuring the Stability of Feature Selection," 2016, pp. 442–457. doi: 10.1007/978-3-319-46227-1_28.

## CHAPTER 2

Advanced Feature Selection and Extraction Techniques for Omics Data Analysis

Applications in Multi-Omics Integration

M. M. Mohamed Mufassirin and A. Alan Steve Amath

2.1 INTRODUCTION

Technological advancements have revolutionized biomedical research, enabling the comprehensive study of various biological layers, collectively known as omics. The term "omics" refers to large-scale biological studies with fields such as genomics, transcriptomics, proteomics, and metabolomics [1]. The suffix "-ome" represents the complete set of molecules studied within each domain, including the genome (DNA), transcriptome (RNA), proteome (proteins), and metabolome (metabolites). Traditionally, researchers relied on single-omics approaches, which analyze only one biological layer at a time. While this has provided valuable insights, it often fails to capture the complex interactions and regulatory mechanisms within biological systems. Genomics focuses on an organism's DNA, studying genetic variations that influence hereditary traits and diseases. Transcriptomics examines RNA molecules, providing insights into gene expression and regulation. Proteomics explores the structure and function of proteins, which are essential for cellular activities [2]. Metabolomics investigates small molecules involved in metabolic processes, offering a snapshot of an organism's biochemical state [3].

To address the limitations associated with single-omics approaches, multi-omics data integration techniques have emerged as powerful tools in biological and medical research. By combining information from multiple omics layers, researchers can gain a comprehensive systems-level understanding of biological processes, which uncover complicated molecular pathways, and identify disease biomarkers more accurately. This approach is particularly valuable in personalized medicine, where treatment plans can be tailored based on an individual's genetic, transcriptomic, and metabolic profiles. However, it also presents challenges, including data heterogeneity, high dimensionality, and missing values,

DOI: 10.1201/9781003591658-2

which require advanced computational techniques for effective analysis. Among these, feature selection and feature extraction methods help reduce data complexity while preserving biologically meaningful information, improving the interpretability and predictive performance of computational models.

Each omics layer, genomics, transcriptomics, proteomics, and metabolomics have separate data characteristics, making direct comparisons across layers difficult. Variations in data generation technologies and analytical platforms introduce biases and inconsistencies and complicating integration efforts. Additionally, normalization techniques must account for technical variations while ensuring biological relevance, yet no universal normalization method exists for multi-omics datasets. Another significant challenge is missing data, as not all samples contain information for every omics layer [4]. This issue requires sophisticated imputation methods to avoid bias while maintaining data integrity. Computationally, multi-omics data are inherently high-dimensional, requiring advanced statistical and machine-learning (ML) techniques to extract meaningful patterns. Moreover, many ML models used in multi-omics integration function as "black boxes," posing challenges in model interpretability [1]. Understanding the interactions and regulatory networks between different omics layers remains crucial, which require robust bioinformatics pipelines and domain expertise to derive meaningful biological conclusions.

Feature selection is a crucial step in multi-omics data analysis. This aimed at identifying the most informative variables while eliminating irrelevant or redundant features. This process enhances computational efficiency, improves model interpretability, and mitigates the curse of dimensionality. Various techniques have been developed to optimize feature selection in omics research. In this chapter, we have broadly categorized statistical, bioinformatics, and ML-based approaches. Also, feature extraction complements feature selection by transforming raw high-dimensional data into a more informative, lower-dimensional representation. Classical methods such as PCA and t-SNE provide effective ways to visualize complex multi-omics datasets, while deep learning-based techniques, including auto-encoders and convolutional neural networks (CNNs), offer automated solutions for feature extraction [1, 3, 5–7]. Hybrid approaches combining feature selection and extraction are gaining traction and provide enhanced accuracy and interpretability for multi-omics integration.

In recent years, omics data have revolutionized healthcare and biomedical research, offering insights into disease mechanisms, biomarker discovery, and therapeutic development. Researchers can uncover complex biological interactions by analyzing multi-omics datasets, leading to advances in precision medicine and early disease detection. A significant application is in neurodegenerative diseases such as Amyotrophic Lateral Sclerosis (ALS), where multi-omics integration has provided valuable insights into genetic risk factors, protein misfolding, and metabolic dysregulation [4, 8]. Recent advancements in ALS research have leveraged multi-omics approaches to identify novel biomarkers and therapeutic targets, which certainly accelerate drug discovery and improve patient outcomes [8]. Furthermore, omics data play a crucial role in cancer research, infectious

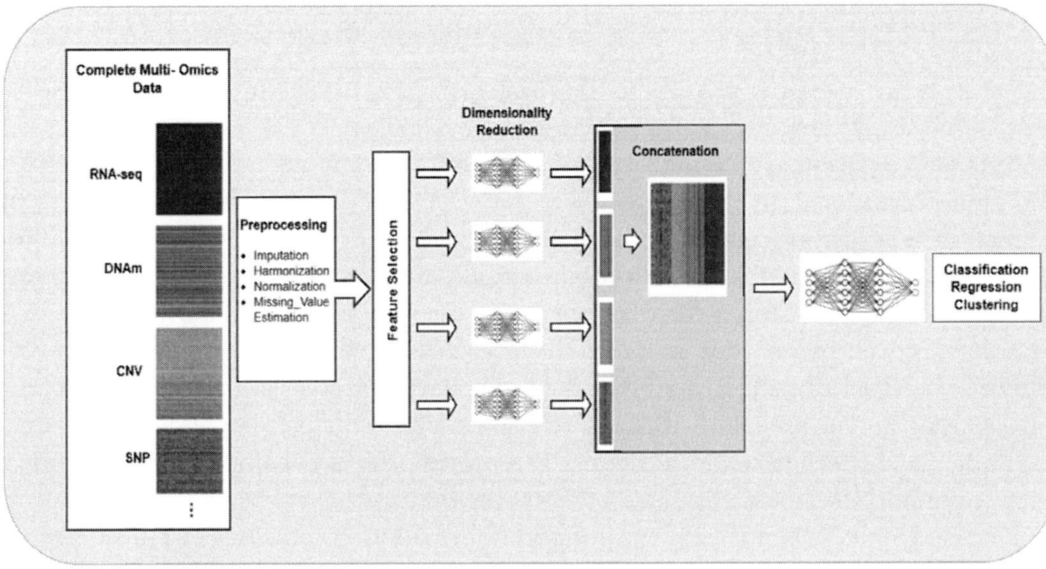

FIGURE 2.1 Pipeline of multi-omics data integration analyses.

disease studies, and personalized treatment strategies, demonstrating their widespread applicability in modern medicine.

In fact, most ML models for multi-omics data analyses follow a common pipeline. It consists of several key steps to ensure effective analysis. Initially, the raw multi-omics data undergoes pre-processing for cleaning and standardization. Feature selection or dimensionality reduction techniques are then applied to minimize the number of multi-omics variables. An autoencoder, a deep learning model, is used to extract low-dimensional latent variables through its bottleneck layer. The refined multi-omics variables are subsequently concatenated into a unified dataset for integration. Additional feature selection or reduction techniques may be applied to enhance data quality. Finally, the integrated dataset is analyzed for various tasks, including classification, regression, and clustering, enabling meaningful biological insights [3]. Figure 2.1. illustrates a common pipeline for multi-omics data integration.

Overall, this chapter provides an in-depth exploration of feature selection and extraction techniques that are tailored to multi-omics data integration. It covers statistical, bioinformatics, and ML-based approaches for feature selection, alongside classical and deep learning-driven feature extraction methods. Additionally, the chapter discusses the applications of multi-omics integration in biomarker discovery, drug development, and precision medicine, with a particular focus on recent advancements in ALS research. Key challenges in omics data analysis, including data heterogeneity, scalability, and ethical considerations, are also addressed. Finally, the chapter highlights emerging trends and their potential impact on future multi-omics research. By bridging computational techniques with explainable AI (XAI) in biomedical applications, this chapter aims to provide a comprehensive overview of the state-of-the-art methodologies driving multi-omics integration forward in biomedical research.

2.2 FEATURE SELECTION TECHNIQUES IN OMICS DATA

Omics data contains vast amounts of biological information, making it highly complex and high-dimensional. In this context, features refer to measurable variables within the dataset, such as genes, proteins, or metabolites, that provide insights into biological processes. However, not all features contribute equally to the study's objective, such as disease classification, biomarker discovery, or drug response prediction. Many features may be redundant, noisy, or irrelevant, leading to increased computational complexity and reduced model performance. Notably, the feature selection technique is a key step in omics data analysis, as it helps identify the most relevant and informative features while eliminating those that do not add significant value. This process improves computational efficiency, enhances model interpretability, and increases predictive accuracy by reducing overfitting and focusing only on biologically meaningful information. By selecting the most relevant features, researchers can uncover critical molecular markers associated with diseases, improve diagnostic models, and develop targeted therapies.

On the other hand, the feature selection is used in the feature dimensionality reduction process. It happens from high-dimensional data. High-dimensional data refers to datasets with a large number of features or attributes [9]. These datasets are increasingly common due to advancements in data collection techniques across various fields, including biomedical research, computational biology, and artificial intelligence (AI). While high-dimensional data contains valuable insights, it also presents challenges due to the curse of dimensionality, a phenomenon where an increase in the number of features leads to higher computational complexity, increased storage requirements, and difficulties in extracting meaningful patterns.

As dimensionality increases, traditional analytical methods become inefficient, often leading to overfitting and reduced model performance. To overcome these challenges, feature dimensionality reduction is applied. This process maps high-dimensional data into a lower-dimensional space while retaining its most informative characteristics. The goal is to eliminate redundant or irrelevant features, thereby improving computational efficiency, reducing noise, and enhancing the accuracy of ML models. Feature dimensionality reduction is widely used in pattern recognition, ML, and data mining, helping to simplify complex problems and improve model interpretability.

There are two primary approaches to dimensionality reduction: feature selection and extraction. The feature selection process typically involves three primary approaches: statistical methods for dimensionality reduction, bioinformatics tools for omics-specific feature selection, and ML approaches for high-dimensional datasets.

2.2.1 Statistical Methods for Dimensionality Reduction

Statistical methods for dimensionality reduction are essential in handling high-dimensional datasets, particularly in omics research, where vast amounts of biological data need to be analyzed efficiently. These techniques strengthen statistical principles to identify relationships between features and reduce dimensionality while preserving essential biological information. The core assumption behind these methods is that features exhibit specific structures or correlations that can be exploited to retain the most informative aspects of the dataset. By focusing on key variations, statistical dimensionality reduction enhances computational efficiency, prevents overfitting, and improves model interpretability.

One widely used statistical approach is univariate feature selection, which evaluates each feature independently using statistical tests such as t-tests, ANOVA, or chi-squared tests [3]. In this practice, features are ranked based on their statistical significance, and only the most relevant ones are selected for further analysis. While this method is simple and effective in identifying strong individual associations between features and the target variable, it does not account for interactions between multiple features. In contrast, multivariate feature selection methods consider the relationships between multiple features and select subsets that are collectively informative. Multivariate techniques such as linear regression, logistic regression, and Cox proportional hazards models are commonly employed to analyze feature interactions, ensuring that selected features contribute meaningfully to prediction models [5].

Another powerful statistical approach for feature selection is regularization. This technique involves adding a penalty term to ML models to shrink the coefficients of less important features. LASSO (L1 regularization) is particularly effective in selecting a subset of features by forcing some coefficients to zero, thereby eliminating irrelevant variables. On the other hand, ridge regression (L2 regularization) reduces the impact of less significant features by shrinking their coefficients without eliminating them entirely [5]. These regularization techniques help manage complex datasets with numerous correlated features, which improves both predictive performance and model generalization. By applying these statistical methods, researchers can efficiently reduce dimensionality, extract meaningful biological insights, and enhance the accuracy of ML models in omics studies.

2.2.2 Bioinformatics Tools for Omics-Specific Feature Selection

In the early years, researchers started to analyze single omics datasets such as genomics, transcriptomics, proteomics, and metabolomics. Bioinformatics tools in omics data refer to specialized software, algorithms, and databases used to analyze and interpret large-scale biological data derived from different omics fields. Mainly, these tools help to filter out redundant and non-informative variables while preserving essential biological signals [10], improving the accuracy and interpretability of disease classification, biomarker discovery, and functional analysis.

In past research, bioinformatics tools were primarily used to analyze single-omics data, which focuses on individual layers such as genomics, transcriptomics, proteomics, or metabolomics. However, recent advancements in the field have shifted toward integrating multi-omics data to gain a more comprehensive understanding of complex biological systems. Today, numerous tools are available for integrating and analyzing multi-omics data, enabling the identification of key biomarkers and molecular interactions across different omics layers. Additionally, network-based feature selection methods are now widely utilized to identify important features by considering the relationships and interactions within biological networks. In the following subsections, we will explore these approaches in more detail.

2.2.2.1 Genomics and Transcriptomics

In past research, tools like DESeq2 and edgeR have been commonly used for differential gene expression analysis in RNA sequencing data in genomics and transcriptomics. These tools apply statistical models to identify genes with significant expression differences, aiding in disease association studies [3]. Both tools employ statistical models to detect significant gene expression differences while accounting for variability in sequencing depth and biological replicates. DESeq2 uses a negative binomial distribution with shrinkage estimation for dispersion and fold change. This can improve the statistical power for small sample sizes. edgeR, on the other hand, applies an empirical Bayes approach to estimate dispersion, making it efficient for datasets with complex experimental designs. In omics-specific feature selection, these tools filter out genes with minimal or insignificant expression variation, focusing only on biologically relevant markers.

Another tool is GSEA. Gene Set Enrichment Analysis (GSEA) is a computational method used to determine whether a predefined set of genes shows statistically significant differences in expression between two biological conditions [11]. Unlike traditional differential expression analysis, which focuses on individual genes, GSEA evaluates gene sets that share common biological functions, pathways, or regulatory mechanisms. This approach helps identify biologically meaningful patterns that might be overlooked when analyzing genes individually. GSEA plays a crucial role in omics-specific feature selection by prioritizing gene sets relevant to diseases or biological processes, reducing dimensionality while retaining functionally significant information.

Additionally, Weighted Gene Co-Expression Network Analysis (WGCNA) is a network-based method that clusters genes with similar expression patterns across multiple samples, forming co-expression modules [11]. These modules often correspond to biological pathways or cellular functions, making WGCNA a powerful tool for identifying key regulatory genes (hub genes) that are central to a given process. In feature selection and dimensionality reduction, WGCNA filters out less relevant genes by focusing on biologically meaningful clusters rather than individual genes, thereby simplifying complex omics datasets while preserving essential information for further analysis.

2.2.2.2 Proteomics and Metabolomics

In metabolomics and proteomics, past records mention that MetaboAnalyst and Perseus have been utilized in many research studies. MetaboAnalyst works by pre-processing metabolomics data through normalization, scaling, and transformation to ensure comparability across samples [12]. It employs various statistical techniques such as principal component analysis (PCA), partial least squares discriminant analysis (PLS-DA), and hierarchical clustering to reduce dimensionality and highlight significant metabolic variations. Also, it integrates pathway enrichment analysis and ML methods to refine feature selection, allowing researchers to identify disease-associated metabolites while filtering out irrelevant or redundant data.

On the other hand, Perseus specializes in proteomics data analysis, particularly for mass spectrometry-based studies. It processes raw proteomics data by handling missing values, normalizing expression levels, and applying statistical tests such as t-tests and ANOVA to detect significantly different proteins between experimental conditions. Furthermore, Perseus incorporates clustering techniques, correlation networks, and enrichment analysis to identify biologically relevant protein groups, helping researchers uncover meaningful patterns in high-dimensional datasets. Both tools play a crucial role in dimensionality reduction by eliminating noise and irrelevant features, ensuring that only the most informative biomarkers are retained for further biological interpretation and disease research.

2.2.2.3 Multi-Omics Integration

Beyond single omics data nowadays, multi-omics data analysis shows more accurate results. Multi-omics factor analysis (MOFA) and pathway/network-based approaches, such as those utilizing KEGG (Kyoto Encyclopedia of Genes and Genomes) and Reactome, play a critical role in integrating diverse omics datasets while reducing dimensionality. MOFA is a statistical framework that identifies shared and unique features across multi-omics data by decomposing high-dimensional datasets into a lower-dimensional latent space, capturing key biological variations [13]. This helps in uncovering hidden patterns across genomics, transcriptomics, proteomics, and metabolomics data.

Additionally, network-based feature selection methods utilize biological networks, such as protein–protein interaction networks, to identify important nodes or subnetworks associated with the phenotype of interest [14]. Here, mainly, techniques like KEGG help visualize molecular interactions, metabolic networks, and signaling cascades, enabling researchers to focus on functionally relevant biomarkers rather than statistically dominant but biologically irrelevant features. These approaches collectively enhance the interpretability of complex omics data, improve feature selection, and streamline computational analysis, ultimately leading to more accurate disease classification, biomarker discovery, and precision medicine applications. Ontology-based feature selection methods also leverage ontologies like Gene Ontology to select features that are functionally related to the biological process or disease being studied [5]. These approaches provide a more structured and interpretable selection of features compared to purely statistical methods. However, with advancements in AI, ML, and cloud computing, bioinformatics tools are becoming more scalable and efficient [2], enabling automated feature selection from large omics datasets.

2.2.3 Machine Learning Approaches for High-Dimensional Datasets

ML approaches play a crucial role in analyzing high-dimensional datasets, which are common in omics studies where the number of features significantly exceeds the number of samples. These techniques help identify relevant features, reduce computational complexity, and improve predictive accuracy. Various ML techniques are employed based on the nature of the data and the research objective. Supervised learning and Unsupervised learning methods are predominantly applied to the feature selection and feature extractions in omics dataset. Supervised learning methods, such as support vector machines (SVM),

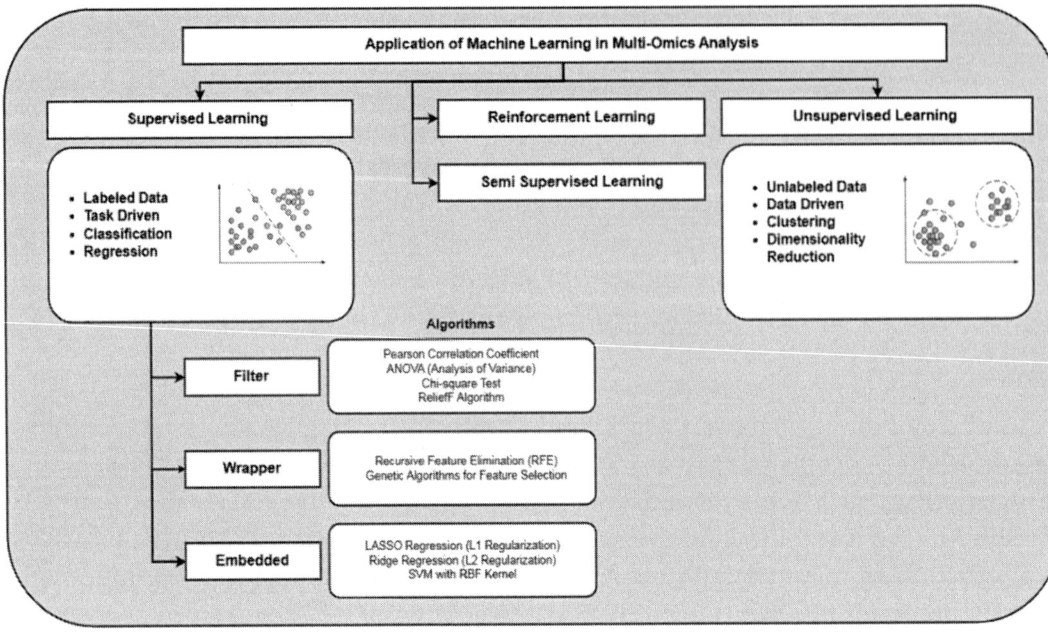

FIGURE 2.2 Types of machine learning techniques applied in omics data analysis.

random forest (RF), and neural networks, are widely used for classification and regression tasks in omics research. Unsupervised learning techniques, including K-means clustering and PCA, assist in discovering hidden patterns without requiring labeled data. In addition to these two ML techniques, Semi-supervised learning and Reinforcement learning are also applied in omics studies. Semi-supervised learning combines labeled and unlabeled data to enhance learning efficiency, making it particularly useful for biological datasets with limited annotations. Additionally, reinforcement learning, though less frequently applied in omics studies, provides a decision-making framework to optimize predictive models. Integrating these ML techniques enables the extraction of meaningful insights from complex biological data, facilitating advancements in precision medicine and disease diagnosis. Figure 2.2. illustrates the types of ML techniques applied in omics data analysis.

2.2.3.1 Supervised Feature Selection Methods

Supervised learning involves training a model on labeled data, where each input is associated with a corresponding output [4]. In supervised learning, feature selection is critical for improving model performance and reducing complexity by identifying the most relevant features for the task at hand. Supervised learning-based feature selection techniques are broadly categorized into filter, wrapper, and embedded methods, each offering unique advantages in selecting informative features [15, 16].

Embedded Methods perform feature selection during the model training process. Embedded methods incorporate feature selection as part of the model's learning algorithm [11]. For instance, LASSO (Least Absolute Shrinkage and Selection Operator) is an embedded method that applies regularization to the regression model, which penalizes irrelevant features and shrinks their coefficients to zero. In this approach, feature selection occurs while training the model, making it more efficient than using separate selection steps. Embedded methods are beneficial for handling high-dimensional datasets while ensuring that only the most relevant features contribute to the model's predictions.

The next method wrapper evaluates subsets of features by training a model on different combinations of features and measuring performance using a predefined criterion, such as accuracy or cross-validation [10]. Recursive feature elimination (RFE) is a classic example of a wrapper method [17]. RFE iteratively removes the least important features, retrains the model, and selects the best subset of features based on model performance. Wrapper methods tend to be computationally expensive, as they require training multiple models for different feature subsets, but they are often highly accurate for identifying the optimal feature set.

Furthermore, filter methods are independent of the learning algorithm and focus on assessing the relationship between individual features and the target variable. Mutual information-based selection, for example, measures the dependency between features and labels, selecting those that offer the most predictive power. Filter methods are typically faster than wrapper methods because they do not involve model training and instead rely on statistical tests to rank features. However, they may not always capture interactions between features as effectively as wrapper or embedded methods. These three approaches, embedded, wrapper, and filter, each have advantages and limitations in dimensionality reduction and feature selection. Embedded methods like LASSO integrate feature selection into the model training process, wrapper methods like RFE evaluate subsets of features through model performance, and filter methods like mutual information focus on statistical relationships. Together, they offer different strategies for reducing dimensionality, improving model interpretability, and enhancing predictive accuracy in supervised learning.

2.2.3.2 Unsupervised Feature Selection Methods

Unsupervised feature selection techniques are designed to identify relevant features from data without relying on labeled outcomes. Two prominent unsupervised methods are autoencoders and clustering-based feature selection [18]. Autoencoders are a type of deep learning model used for dimensionality reduction. They consist of an encoder that compresses high-dimensional data into a lower-dimensional representation, followed by a decoder that reconstructs the original input. The model learns to retain the most important features during the compression process, effectively reducing the number of features while preserving essential information. Autoencoders are particularly useful when dealing with complex datasets like images or gene expression data, where traditional linear methods might fail to capture intricate relationships.

Clustering-based feature selection groups similar features into clusters using techniques like k-means or hierarchical clustering. The method then selects representative features from each cluster, reducing redundancy and ensuring that only the most informative features are retained. By identifying and removing highly correlated features, clustering-based selection simplifies the dataset, improves model performance, and reduces overfitting, making it a valuable approach for high-dimensional and noisy data. Both of these unsupervised methods contribute to feature dimensionality reduction by identifying and retaining the most relevant and informative features, helping to improve the efficiency and interpretability of models while reducing computational costs.

2.2.3.3 Ensemble-Based Feature Selection

Ensemble-based feature selection is a technique that combines multiple feature selection methods to improve the robustness, stability, and accuracy of selected features in ML models. Instead of relying on a single feature selection algorithm, ensemble methods aggregate the results from different techniques to obtain a more reliable subset of features. Ensemble-based feature selection enhances feature relevance by integrating multiple selection algorithms through strategies like voting, averaging importance scores, or meta-learning. It includes bagging-based methods (e.g., RF) that aggregate feature importance from bootstrapped subsets, boosting-based methods (e.g., Gradient Boosting) that refine selection iteratively, and hybrid approaches that combine filter, wrapper, and embedded techniques for optimal feature ranking.

For example, RF and gradient boosting machines (GBM) use the concept of multiple models working together to improve feature selection and predictive performance. RF is a powerful ensemble method that builds multiple decision trees. It assigns feature importance scores based on how much each feature contributes to reducing classification errors across all the trees. Features with higher importance scores are considered more relevant and are selected for further analysis, making it a useful technique for handling high-dimensional datasets. GBM, on the other hand, also build an ensemble of decision trees but work by sequentially combining multiple weak learners to create a stronger, more accurate predictive model. Each subsequent model corrects the errors made by the previous one, resulting in a robust approach to feature selection. Both RF and GBM effectively identify the most significant features while reducing the dimensionality of the dataset. Their ensemble-based approach helps capture complex interactions among features, which traditional methods may overlook, improving model accuracy and efficiency.

2.3 FEATURE EXTRACTION METHODS IN OMICS DATA

Feature extraction is a fundamental step in multi-omics research that transforms high-dimensional biological data into lower-dimensional representations while preserving essential biological information [19]. This process is beneficial for integrating diverse omics datasets, as different layers, such as genomics, transcriptomics, and proteomics, often have distinct feature spaces and scales. By reducing dimensionality, feature extraction enhances model performance, uncovers meaningful biological patterns, and lowers computational costs. Feature extraction techniques can be broadly categorized into classical statistical methods, deep learning-based approaches, and hybrid strategies. Classical methods include PCA and independent component analysis (ICA), which transform data into fewer, more informative features while minimizing redundancy [5]. Deep learning-based approaches, such as autoencoders and CNNs, learn hierarchical feature representations directly from raw omics data. Hybrid methods combine statistical and ML techniques to maximize interpretability and accuracy. In multi-omics integration, feature extraction helps unify disparate datasets by condensing complex biological information into a standardized format [5]. For instance, gene activity levels and metabolite concentrations can be transformed into a smaller set of biologically relevant features, ensuring that downstream analyses capture the most significant variations. Effective feature extraction enhances predictive modelling and facilitates the discovery of key biomarkers and disease mechanisms.

2.3.1 Classical Methods: Principal Component Analysis (PCA), t-SNE, and Others

These methods help researchers uncover meaningful patterns, improve visualization, and enhance computational efficiency in multi-omics data analysis. PCA is a widely used linear dimensionality reduction technique that transforms the original dataset into a set of orthogonal principal components, which capture the maximum variance in the data [20]. PCA is particularly effective in gene expression analysis, where it helps identify disease subtypes, group samples based on genetic similarity, and improve data visualization. However, PCA assumes that the relationships in the data are linear, which may not always be the case in biological systems [5]. To address the limitations of linear methods like PCA, t-distributed Stochastic Neighbor Embedding (t-SNE) is often used as a nonlinear dimensionality reduction technique. t-SNE is particularly effective for visualizing high-dimensional omics data in two or three dimensions while preserving the local structure of the data [20]. This method has been successfully applied in clustering cancer subtypes based on transcriptomic and proteomic data, helping researchers identify biologically significant patterns.

Several other classical techniques are commonly used in multi-omics research for feature extraction and dimensionality reduction: **ICA** is used to separate independent biological signals from mixed omics datasets, such as identifying individual genetic components in complex gene expression data. **Linear discriminant analysis (LDA)** is a supervised learning method that maximizes the separation between predefined classes, making it useful for disease classification tasks. **Non-negative matrix factorization (NMF)** decomposes complex datasets into biologically meaningful components, which is particularly useful for analyzing transcriptomic and metabolomic data. **Multidimensional scaling (MDS)** is a technique used for visualizing similarities or dissimilarities in high-dimensional omics datasets. These classical methods remain fundamental in omics research, often complementing deep learning-based feature extraction techniques to enhance interpretability and improve data-driven discoveries.

2.3.2 Deep Learning-Based Feature Extraction Techniques

With the increasing complexity of multi-omics data, deep learning-based feature extraction techniques have gained popularity due to their ability to automatically learn hierarchical representations of biological information. These methods enhance the identification of meaningful patterns, improve classification and clustering tasks, and aid in biomarker discovery. Autoencoders are neural networks trained to reconstruct their input, with the hidden layer providing a compressed representation of the data that serves as extracted features [21]. Different types of autoencoders, such as variational autoencoders (VAEs) and denoising autoencoders (DAEs), can be used to extract diverse feature types. VAEs generate meaningful latent representations by learning probabilistic distributions, making them useful for biological data modeling, while DAEs improve feature robustness by reconstructing noisy input data [21]. Although CNNs are primarily used for image analysis, they have been successfully applied to non-image omics data, such as genomics and transcriptomics [22].

CNNs extract spatial and hierarchical patterns from biological sequences, aiding in disease diagnosis and prognosis prediction. For example, CNNs have been used to predict gene regulatory interactions and identify functional genomic regions by learning meaningful patterns from DNA sequences.

RNNs and LSTMs are particularly effective for analyzing sequential omics data, such as time-series gene expression profiles. These models capture temporal dependencies and dynamic changes in biological systems, making them valuable for understanding disease progression and treatment responses. LSTMs have been applied in transcriptomics to predict gene expression patterns across different developmental stages or disease states. Recent advancements in deep learning have led to the use of transformer-based architectures, such as BERT-like models adapted for genomics and proteomics. These models leverage attention mechanisms to capture long-range dependencies between genetic elements, improving the prediction of protein–protein interactions and functional gene annotations. Transformers enhance the interpretability of multi-omics data and provide robust feature extraction for complex biological networks. Cross-modal neural networks integrate multiple omics layers and extract features that capture relationships between different types of biological data [23]. These models are instrumental in multi-omics research, where combining genomics, transcriptomics, and proteomics data improves understanding complex biological interactions. Deep learning-based feature extraction techniques continue to evolve, offering powerful tools for analyzing high-dimensional biological data, improving predictive models, and uncovering novel insights in multi-omics research.

2.3.3 Hybrid Approaches Combining Feature Selection and Extraction

While feature extraction focuses on transforming data into lower-dimensional representations, feature selection techniques aim to identify the most relevant features without altering their original structure. Hybrid approaches combine both techniques to enhance interpretability, improve model performance, and reduce overfitting. One common hybrid method involves using PCA for dimensionality reduction followed by ML-based feature selection. For instance, PCA can be applied to transcriptomic data to extract significant components, followed by RF or RFE to select the most informative genes associated with a particular disease. Another effective hybrid approach is autoencoder-based feature extraction combined with statistical feature selection. Autoencoders can reduce omics data dimensionality by capturing latent representations, and statistical methods like t-tests, ANOVA, or mutual information-based selection can refine these features further. This approach has been successfully applied in cancer biomarker discovery [10], where it helps identify key genetic and proteomic signatures. In some cases, multi-view learning approaches are used to integrate multiple omics datasets, where feature extraction techniques like canonical correlation analysis (CCA) or deep multi-view networks help capture meaningful relationships between genomics, proteomics, and metabolomics data. These hybrid strategies improve multi-omics integration and enhance the ability to discover novel biomarkers and therapeutic targets.

Table 2.1 provides an overview of recent multi-omics data integration studies, summarizing key details such as study references, publication year, ML models used, datasets

TABLE 2.1 List of Multi-Omics Data Integration Techniques in the Recent Publications

No	Study References	Year	ML Model	Dataset	Findings
1	Zeng et al. [24]	2020	Multi-layer convolutional neural network (CNN)	TCGA	The proposed method achieved 90.02% accuracy in identifying PAM50 subtypes of breast cancer patients.
2	Wang et al. [25]	2020	Multi-omics graph convolutional networks. (MORONET)	ROSMAP for Alzheimer's, disease (AD) classification LGG for grade classification in low-grade glioma (LGG) KIPAN for kidney cancer classification BRCA for breast invasive carcinoma (BRCA)	MORONET outperformed other state-of-the-art supervised multi-omics integrative analysis methods across various biomedical classification tasks.
3	Park et al. [15]	2021	Supervised feature extraction learning using triplet loss-super. FELT	GDSC, CCLE, CTRP, PDXEncyclopedia, TCGA	Super.FELT outperformed other existing methods in predicting drug response (sensitive or resistant) using multi-omics data. Effective dimensionality reduction
4	Pierre-Jean et al. [26]	2022	Penalized integrative matrix factorization-PInMF	OmicsSIMLA	Applied to diet and cancer datasets, where it discovered interpretable clusters linked to clinical data.
5	Taguchi et al. [27]	2022	Advanced KTD-based unsupervised FE	TCGA,GSE1644117	Required less memory and CPU time, making it more computationally efficient compared to existing approaches.
6	Azman et al. [28]	2023	SVM-RFE, stacked denoising autoencoder (SDAE) Variational autoencoder (VAE)	Lung Squamous Cell Carcinoma (LUSC) multi-omics dataset	All 5 VAE models achieved 100% accuracy and an AUC score of 1.000. Feature subsets of size 1000 and 4000 provided the best classification results.
7	Srivastava et al. [29]	2023	Recursive feature elimination (RFE)	Clinical and genetic data from ovarian cancer patients	Ensemble learning models achieved approximately 99% accuracy.
8	Rad, H.N et al. [8]	2024	Fuzzy k-means clustering and deep learning	MOALS: New York Genome Center	Accuracy for gene Expression: 84.85% and Gene Expression + SNV: 92.06%.
9	Dhinakaran et al. [16]	2025	Synergistic Kruskal-RFE selector (SKR)	Four broad medical datasets	Feature reduction Ratio: 89%.

analyzed, and significant findings. By organizing these aspects in a structured format, the table facilitates a clear comparison of different methodologies and their effectiveness in multi-omics analysis, highlighting advancements in predictive modeling and biological interpretation.

2.4 APPLICATIONS OF MULTI-OMICS INTEGRATION

Multi-omics integration has revolutionized biological research by combining genomics, proteomics, metabolomics, and other omics layers to gain a holistic understanding of complex diseases. By integrating these diverse datasets, researchers can uncover hidden relationships between molecular pathways, enabling breakthroughs in disease diagnostics, biomarker discovery, and personalized medicine. This approach is particularly valuable for studying complex diseases like ALS.

2.4.1 Integrating Genomics, Proteomics, and Metabolomics Data

Integrating genomics, proteomics, and metabolomics data enhances biological insight by providing a deeper understanding of disease mechanisms and identifying potential therapeutic targets. Genomics reveals genetic variations and mutations that contribute to disease susceptibility, while proteomics helps decipher protein expression patterns and molecular interactions [2]. Metabolomics, on the other hand, captures biochemical changes within cells, reflecting metabolic disruptions associated with disease. By combining these omics layers, researchers gain a comprehensive view of the underlying biological processes, offering a more complete perspective compared to single-omics approaches [30]. This multi-omics integration enables the construction of disease-associated networks, where genomic, epigenomic, and proteomic findings are interconnected to map key regulatory pathways [3]. Identifying overlapping patterns across omics levels allows researchers to pinpoint crucial biomarkers and molecular signatures linked to specific diseases. Such integration has been instrumental in fields like cancer research, neurodegenerative diseases, and metabolic disorders, enabling more precise disease classification and personalized therapeutic strategies. By leveraging multi-omics data, scientists can move beyond isolated molecular snapshots to a holistic systems biology approach, ultimately improving disease diagnosis, prognosis, and treatment development.

2.4.2 Case Studies in Biomarker Discovery for Diseases

The case study examines the disease known as ALS to understand the impact and implications of application of multi-omics data. ALS is a progressive neurodegenerative disease with no known cure, but multi-omics approaches have significantly improved our understanding of its complex molecular architecture [31]. By integrating genomics, transcriptomics, proteomics, and metabolomics data, researchers have identified key genetic mutations such as SOD1 and C9orf72, as well as protein misfolding patterns and metabolic disruptions associated with ALS. These findings provide crucial insights into the disease's pathogenesis, helping to uncover novel diagnostic, prognostic, and therapeutic biomarkers. Multi-omics analyses have also revealed molecularly distinct subtypes of ALS, identifying unique drivers and pathways that differentiate patient groups at a mechanistic level [31]. This stratification enables precision medicine approaches tailored to specific disease subtypes. Moreover, ML models utilizing multi-omics data have been successfully applied for ALS diagnosis and survival prediction, improving early detection and patient outcome assessments.

Recent studies have utilized multi-omics data for ALS and introduced the MOALS platform for comprehensive data processing and analysis [8, 31]. The methodology of MOALS consists of three key stages. First, pre-processing and feature selection, where transcriptome data undergoes initial processing, and relevant genes associated with ALS pathways are identified through clustering of gene expression profiles while reducing noise in the dataset. After pre-processing, the second stage is variant extraction, in which single-nucleotide variants (SNVs) and small insertions and deletions (indels) are detected using sequence alignment and variant calling tools. The raw sequencing data is aligned to the human reference genome (GRCh38) using bwa-mem2, followed by variant calling through GATK (version 4.2.5.0) and annotation with VEP (Version 104.2). Variants are then filtered and prioritized using biologically driven custom scripts. Finally, multi-omics integration and classification, where class probability estimates from individual omics layers are combined using a Variational Auto Encoder (VAE) to improve predictive performance. This systematic approach enhances the accuracy of feature selection, variant prioritization, and classification, providing a robust framework for ALS research using multi-omics data. A schematic representation outlining each step is illustrated in Figure 2.3.

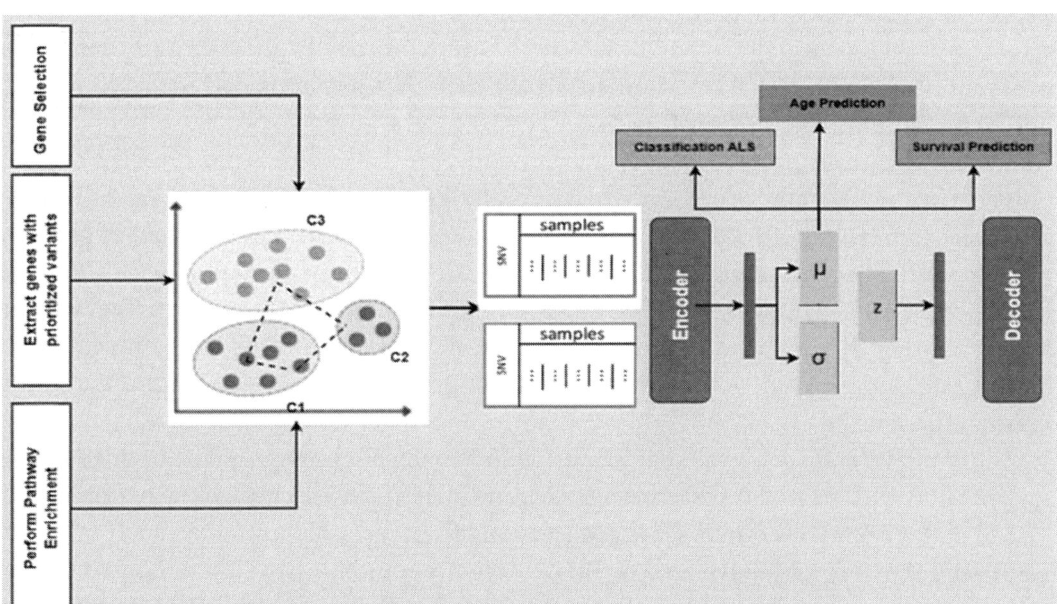

FIGURE 2.3 The overview of the implemented method for ALS diagnosis.

2.4.3 Role in Drug Discovery and Precision Medicine

Multi-omics data integration plays a crucial role in drug discovery and precision medicine by enabling the identification of novel therapeutic targets and patient-specific treatment strategies. By analyzing genomic mutations, protein interactions, and metabolic shifts, researchers can design targeted therapies that align with an individual's molecular profile. This approach enhances treatment efficacy and minimizes adverse effects. For instance, in ALS research, multi-omics methods have facilitated drug repurposing, leading to the use of existing drugs like Riluzole and the development of new therapeutic candidates targeting RNA metabolism and neuroinflammation.

One of the key applications of multi-omics in precision medicine is therapeutic target identification, where integrated analysis uncovers new molecular targets for drug development [4]. Additionally, personalized treatment plans can be designed based on a patient's unique molecular characteristics, optimizing therapy selection and improving patient outcomes [4]. A notable example is SELECT, a precision oncology framework that leverages tumor transcriptome analysis to predict patient responses to cancer therapies, ensuring more effective and targeted treatment strategies. Moreover, network medicine-informed multi-omics integration has been instrumental in drug repurposing for ALS, identifying potential drug targets and previously approved medicines that can be redirected for ALS treatment [32]. These advancements highlight the transformative impact of multi-omics in modern drug discovery and personalized medicine.

2.4.4 Recent Advancements in Multi-Omics-Based ALS Research

Recent advancements in multi-omics research have significantly enhanced our understanding of ALS pathology, paving the way for more precise diagnostics and therapeutic interventions. Single-cell RNA sequencing (scRNA-seq) has provided valuable insights into neuronal and glial cell dysfunction, shedding light on how specific cell populations contribute to disease progression [8]. Additionally, proteogenomics has revealed connections between abnormal protein aggregation and genetic mutations, helping to identify key molecular drivers of ALS. These discoveries contribute to the development of genome-guided personalized diagnosis and treatment strategies, offering a roadmap for improved patient care [31]. The integration of AI and ML in multi-omics research has further accelerated biomarker discovery and treatment predictions. ML algorithms are being employed to integrate complex multi-omics datasets, enabling more accurate ALS diagnosis and survival predictions [8].

Additionally, pathway-level analyses using clustering algorithms have provided deeper insights into ALS-related transcriptomic changes, helping researchers identify critical biological pathways involved in disease progression [8]. Large-scale initiatives such as Project MinE and Answer ALS have also facilitated cross-population studies, uncovering novel genetic risk factors and potential therapeutic targets. These advancements highlight the transformative impact of multi-omics approaches in ALS research, bringing us closer to effective precision medicine strategies for this devastating disease.

2.4.5 Success Stories of Biomarker Discovery and Therapeutic Development in ALS

Multi-omics approaches have played a crucial role in advancing biomarker discovery and therapeutic development for ALS, leading to significant breakthroughs in disease diagnosis, prognosis, and treatment. One of the most notable success stories is the identification of the neurofilament light (NFL) chain as a robust biomarker for ALS progression. Elevated NFL levels in cerebrospinal fluid and blood have been associated with disease severity, making it a valuable tool for monitoring disease progression and evaluating treatment efficacy. Additionally, metabolomics studies have revealed disruptions in lipid metabolism among ALS patients, providing further insights into disease mechanisms and potential therapeutic targets. Integrative multi-omics research has also contributed to the development of cutting-edge therapies, such as antisense oligonucleotide (ASO) treatments.

A significant milestone in ALS therapy is the approval of Tofersen, an ASO drug designed to target SOD1 mutations, one of the genetic causes of ALS. This breakthrough underscores the power of multi-omics in identifying precise molecular targets and developing personalized interventions. Furthermore, transcriptome-wide analyses have helped identify potential genetic markers and drug candidates, paving the way for novel therapeutic strategies [33]. Advancements in multi-omics-driven diagnostic models have further improved ALS classification and prediction [8]. The MOALS model, an integrated multi-omics framework, has demonstrated superior accuracy in diagnosing ALS compared to traditional models. Additionally, multimodal MRI techniques have enhanced diagnostic accuracy and sensitivity to longitudinal changes in ALS patients, making early detection more reliable [34]. These success stories illustrate the transformative impact of multi-omics in ALS research, driving innovation in disease understanding, biomarker discovery, and targeted therapies. As technology continues to evolve, multi-omics integration will remain a cornerstone in developing effective, personalized treatments for ALS and other complex neurodegenerative disorders.

2.5 CHALLENGES AND OPPORTUNITIES IN MULTI-OMICS RESEARCH

Multi-omics research holds immense potential for understanding biological systems, disease mechanisms, and personalized medicine. However, several challenges hinder the effective integration, analysis, and application of multi-omics data. These challenges primarily revolve around data heterogeneity, computational scalability, and ethical concerns. Overcoming these obstacles requires advanced bioinformatics tools, AI techniques, and ethical frameworks to ensure responsible data usage. The emergence of XAI provides an opportunity to enhance AI-driven multi-omics research's transparency, interpretability, and reliability.

2.5.1 Overcoming Data Heterogeneity and Scalability Issues

Multi-omics research faces significant challenges related to data heterogeneity and scalability, which must be addressed to enable seamless integration and efficient analysis. Data heterogeneity arises due to differences in the nature, scale, and measurement techniques of various omics layers, such as genomics, transcriptomics, proteomics, and metabolomics.

For instance, while genomic data consists of discrete nucleotide sequences, proteomic and metabolomic data are continuous variables obtained through mass spectrometry. These fundamental differences make it challenging to unify data within a single analytical framework. To overcome this, researchers use advanced data harmonization techniques, including batch effect correction and statistical transformations like Z-score normalization. Additionally, graph-based approaches and network biology facilitate the identification of meaningful biological relationships across different omics layers, improving integration and interpretation [4, 35].

Scalability is another critical issue, as high-throughput omics technologies generate vast amounts of data that demand substantial storage and computational resources. Large-scale studies, particularly those involving thousands of samples, require sophisticated infrastructure for efficient data processing. Traditional computational methods struggle to handle such large datasets, leading to high processing times and increased computational costs [35]. To address this, researchers leverage cloud computing, high-performance computing (HPC), and distributed computing frameworks like Apache Spark and Tensor Flow, enabling efficient data management and analysis. ML-based dimensionality reduction techniques, such as PCA and VAEs, further enhance scalability by reducing computational complexity while retaining critical biological information. These advancements help overcome data integration challenges, making multi-omics research more effective in uncovering novel insights into complex diseases like ALS.

2.5.2 Ethical Considerations in Omics Data Analysis

As multi-omics research expands, ethical considerations play a vital role in ensuring responsible data collection, analysis, and sharing. Given that omics data contains highly sensitive genetic and molecular information, researchers must prioritize privacy, security, and informed consent. For example, genomic data can reveal information about inherited diseases, ancestry, and predispositions to certain conditions, raising concerns about potential misuse and genetic discrimination [4]. To mitigate these risks, international regulations such as the General Data Protection Regulation (GDPR) and the Health Insurance Portability and Accountability Act (HIPAA) enforce guidelines on obtaining informed consent, anonymizing data, and implementing secure storage mechanisms.

Another ethical challenge in multi-omics research is ensuring equitable access to data and benefits. Many large-scale studies are conducted in high-income countries, potentially limiting the generalizability of findings to diverse populations. To address this, researchers are developing open-access multi-omics repositories that promote global collaboration while maintaining ethical data-sharing standards. The adoption of the FAIR (Findable, Accessible, Interoperable, Reusable) principles is essential for ensuring transparency, reproducibility, and equitable research participation [35]. Algorithmic bias in AI-driven multi-omics research is another growing concern. ML models trained on non-representative datasets can lead to biased findings and disparities in healthcare recommendations. To counter this, researchers must use diverse and inclusive datasets, apply unbiased feature selection techniques, and ensure transparency in AI-driven decision-making processes [36]. By addressing these ethical considerations, multi-omics research can advance in a responsible and equitable manner, ultimately benefiting global healthcare and precision medicine.

2.5.3 Future Trends: Explainable AI in Omics Research

AI has revolutionized multi-omics research by enabling high-throughput data analysis, pattern recognition, and predictive modeling [3]. However, traditional deep learning models function as "black boxes," making it difficult for researchers and clinicians to interpret their decision-making processes [1]. This lack of transparency poses challenges in clinical applications, where trust and justification for AI-driven predictions are crucial. To address this issue, XAI techniques are being integrated into multi-omics research, ensuring that AI models provide interpretable and trustworthy insights.

Several XAI methods are gaining traction in omics research. SHAP (Shapley Additive Explanations) assigns importance values to features, helping researchers understand how specific genes, proteins, or metabolites contribute to a model's prediction. LIME (Local Interpretable Model-Agnostic Explanations) simplifies complex models locally, making it easier to interpret individual predictions without analyzing the entire model. Additionally, attention mechanisms in deep learning highlight key genomic or proteomic regions influencing predictions, improving the identification of critical biomarkers. The integration of XAI in multi-omics research enhances biomarker discovery, drug development, and precision medicine by making AI-driven insights more interpretable and actionable. For example, in ALS research, XAI can help identify the most influential genetic and molecular markers, leading to improved early diagnosis and personalized treatment strategies. Furthermore, combining multi-omics data with clinical and environmental exposure data can provide a more comprehensive understanding of diseases, leading to more accurate predictions and targeted interventions [3]. By increasing transparency and trust, XAI will play a crucial role in advancing AI-driven multi-omics research, ensuring that predictive models are accurate and clinically meaningful.

However, multi-omics research presents both challenges and opportunities in advancing biological and medical sciences. Overcoming data heterogeneity and scalability issues requires innovative bioinformatics techniques and high-performance computing solutions. Addressing ethical considerations ensures responsible data collection, privacy protection, and equitable healthcare advancements. The integration of XAI in multi-omics analysis represents a promising future direction, allowing researchers and clinicians to gain clearer insights into complex biological interactions. As computational and ethical challenges are tackled, multi-omics research will continue to drive groundbreaking discoveries in disease mechanisms, biomarker identification, and personalized medicine.

2.6 CONCLUSION

Feature selection and extraction techniques play a fundamental role in the analysis of high-dimensional omics data, enabling researchers to identify biologically significant variables while improving computational efficiency and model interpretability. Feature selection methods, including statistical approaches, bioinformatics-driven tools, and ML-based techniques, help refine datasets by retaining only the most informative features. Feature extraction methods, such as PCA, t-SNE, deep learning-based encoders, and hybrid approaches, transform complex data into lower-dimensional representations while preserving essential biological information. The integration of these methods has significantly advanced biomarker discovery, disease classification, and personalized medicine. Looking ahead, multi-omics integration is poised to revolutionize biomedical research by providing a comprehensive understanding of biological systems. The incorporation of AI-driven approaches, such as explainable deep learning models, will enhance the interpretability of omics data, leading to more precise and reliable insights. Furthermore, advancements in cloud computing, federated learning, and blockchain technology may help address challenges related to data heterogeneity, scalability, and security. As multi-omics research continues to evolve, it holds great promise for personalized healthcare, precision medicine, and novel therapeutic development, ultimately improving patient outcomes and our understanding of complex diseases.

REFERENCES

1. Mohammed, M. A., Abdulkareem, K. H., Dinar, A. M., & Zapirain, B. G. (2023). Rise of deep learning clinical applications and challenges in omics data: A systematic review. Diagnostics, 13(4), 664.
2. Mufassirin, M. M., Newton, M. H., & Sattar, A. (2023). Artificial intelligence for template-free protein structure prediction: a comprehensive review. Artificial Intelligence Review, 56(8), 7665–7732.
3. Kang, M., Ko, E., & Mersha, T. B. (2022). A roadmap for multi-omics data integration using deep learning. Briefings in Bioinformatics, 23(1), Bbab454. https://academic.oup.com/bib/article/23/1/bbab454/6425807
4. Acharya, D., & Mukhopadhyay, A. (2024). A comprehensive review of machine learning techniques for multi-omics data integration: Challenges and applications in precision oncology. Briefings in Functional Genomics, 23(5), 549–560. https://academic.oup.com/bfg/article/23/5/549/7643360
5. Feldner-Busztin, D., Firbas Nisantzis, P., Edmunds, S. J., Boza, G., Racimo, F., Gopalakrishnan, S., … & De Polavieja, G. G. (2023). Dealing with dimensionality: The application of machine learning to multi-omics data. Bioinformatics, 39(2), Btad021. https://academic.oup.com/bioinformatics/article/39/2/btad021/6986971
6. Subramanian, I., Verma, S., Kumar, S., Jere, A., & Anamika, K. (2020). Multi-omics data integration, interpretation, and its application. Bioinformatics and Biology Insights, 14, 1177932219899051. https://journals.sagepub.com/doi/full/10.1177/1177932219899051
7. Li, H., Shi, M., Ren, K., Zhang, L., Ye, W., Zhang, W., … & Xia, X. Q. (2023). Visual omics: A web-based platform for omics data analysis and visualization with rich graph-tuning capabilities. Bioinformatics, 39(1), Btac777. https://academic.oup.com/bioinformatics/article/39/1/btac777/6865031
8. Rad, H. N., Su, Z., Trinh, A., Newton, M. H., Shamsani, J., & Karim, A.Nygc Als Consortium (2024). Amyotrophic lateral sclerosis diagnosis using machine learning and multi-omic data integration. Heliyon, 10(20). https://doi.org/10.1016/j.heliyon.2024.e38583
9. Jia, W., Sun, M., Lian, J., & Hou, S. (2022). Feature dimensionality reduction: A review. Complex & Intelligent Systems, 8(3), 2663–2693. https://link.springer.com/article/10.1007/S40747-021-00637-X
10. Mufassirin, M. M., & Ragel, R. G. (2018, December). A Novel Filter-Wrapper Based Feature Selection Approach for Cancer Data Classification. In 2018 IEEE International Conference on Information and Automation for Sustainability (ICIAfS) (pp. 1–6). IEEE.

11. Taylor, P., Griffiths, N., Hall, V., Xu, Z., & Mouzakitis, A. (2022). Feature selection for supervised learning and compression. Applied Artificial Intelligence, 36(1), 2034293. https://www.tandfonline.com/doi/full/10.1080/08839514.2022.2034293
12. Nguyen, Q. H., Nguyen, H., Oh, E. C., & Nguyen, T. (2024). Current approaches and outstanding challenges of functional annotation of metabolites: A comprehensive review. Briefings in Bioinformatics, 25(6), Bbae498. https://academic.oup.com/bib/article/25/6/bbae498/7820077
13. Hayes, C. N., Nakahara, H., Ono, A., Tsuge, M., & Oka, S. (2024). From omics to multi-omics: A review of advantages and tradeoffs. Genes, 15(12), 1551. https://www.mdpi.com/2073-4425/15/12/1551
14. Cheng, F., Yu, M., Xu, J., Dutta, R., Trapp, B., & Pieper, A. (2024). Network Medicine Informed Multi-Omics Integration Identifies Drug Targets and Repurposable Medicines for Amyotrophic Lateral Sclerosis. https://www.researchsquare.com/article/rs-4271755/v1
15. Park, S., Soh, J., & Lee, H. (2021). Super FELT: Supervised feature extraction learning using triplet loss for drug response prediction with multi-omics data. BMC bioinformatics, 22(1), 269. https://link.springer.com/article/10.1186/s12859-021-04146-z
16. Dhinakaran, D., Srinivasan, L., Raja, S. E., Valarmathi, K., & Nayagam, M. G. (2025). Synergistic feature selection and distributed classification framework for high-dimensional medical data analysis. MethodsX, 103219. https://www.sciencedirect.com/science/article/pii/S2215016125000664
17. Abualigah, L., Aldulaimi, A., Al, S., Alabool, H., Diabat, M., & Shehab, M. (2019). Optimization algorithms to solve feature selection problem: A review. International Journal of Science Applied Information Technology, 8(6), 66–72.
18. Nicora, G., Vitali, F., Dagliati, A., Geifman, N., & Bellazzi, R. (2020). Integrated multi-omics analyses in oncology: A review of machine learning methods and tools. Frontiers in Oncology, 10, 1030. https://www.frontiersin.org/journals/oncology/articles/10.3389/fonc.2020.01030/full
19. Meng, C., Zeleznik, O. A., Thallinger, G. G., Kuster, B., Gholami, A. M., & Culhane, A. C. (2016). Dimension reduction techniques for the integrative analysis of multi-omics data. Briefings in Bioinformatics, 17(4), 628–641. https://academic.oup.com/bib/article/17/4/628/2240645
20. Mehtonen, J., Pölönen, P., Häyrynen, S., Dufva, O., Lin, J., Liuksiala, T., … & Heinäniemi, M. (2019). Data-driven characterization of molecular phenotypes across heterogeneous sample collections. Nucleic Acids Research, 47(13), E76–E76. https://academic.oup.com/nar/article/47/13/e76/5477460
21. Zhang, Y. (2018, March). A Better Autoencoder for Image: Convolutional Autoencoder. In ICONIP17-DCEC. Available Online: Http://Users. Cecs. Anu. Edu. Au/Tom. Gedeon/Conf/Abcs2018/Paper/Abcs2018_Paper_58. Pdf (Accessed On 23 March 2017). https://theory.sinp.msu.ru/lib/exe/fetch.php/archive/dlcp/kryukov22/biblio/abcs2018_paper_58-conv_ae.pdf
22. Sharma, A., Lysenko, A., Boroevich, K. A., Vans, E., & Tsunoda, T. (2021). Deepfeature: Feature selection in nonimage data using convolutional neural network. Briefings in Bioinformatics, 22(6), Bbab297. https://academic.oup.com/bib/article/22/6/bbab297/6343526
23. Bica, I., Velickovic, P., Xiao, H., & Li, P. (2018, April). Multi-Omics Data Integration Using Cross-Modal Neural Networks. in ESANN. https://www.esann.org/sites/default/files/proceedings/legacy/es2018-93.pdf
24. Zeng, J., Cai, H., & Akutsu, T. (2020, January). Breast Cancer Subtype by Imbalanced Omics Data through a Deep Learning Fusion Model. In Proceedings of the 2020 10th International Conference on Bioscience, Biochemistry and Bioinformatics (pp. 78–83). https://dl.acm.org/doi/abs/10.1145/3386052.3386063
25. Wang, T., Shao, W., Huang, Z., Tang, H., Zhang, J., Ding, Z., & Huang, K. (2020). Moronet: multi-omics integration via graph convolutional networks for biomedical data classification. BioRxiv, 2020-07. https://www.biorxiv.org/content/10.1101/2020.07.02.184705v1.abstract
26. Pierre-Jean, M., Mauger, F., Deleuze, J. F., & Le Floch, E. (2022). PIntMF: Penalized integrative matrix factorization method for multi-omics data. Bioinformatics, 38(4), 900–907. https://academic.oup.com/bioinformatics/article/38/4/900/6443074

27. Taguchi, Y. H., & Turki, T. (2022). Novel feature selection method via kernel tensor decomposition for improved multi-omics data analysis. BMC Medical Genomics, 15(1), 37. https://link.springer.com/article/10.1186/s12920-022-01181-4
28. Azman, N. S., Samah, A. A., Lin, J. T., Majid, H. A., Shah, Z. A., Wen, N. H., & Howe, C. W. (2023). Support vector machine–Recursive feature elimination for feature selection on multi-omics lung cancer data. Progress in Microbes & Molecular Biology, 6(1). https://journals.hh-publisher.com/index.php/pmmb/article/view/790
29. Srivastava, J., & Rathi, M. (2024). Biomarkers to Predictive Modeling: Leveraging Feature Selection and Ensemble Learning for Ovarian Cancer Prognosis. In Proceedings of the 2024 Sixteenth International Conference on Contemporary Computing (pp. 51–58). https://dl.acm.org/doi/abs/10.1145/3675888.3676030
30. Wang, Z., & He, Y. (2021). Precision omics data integration and analysis with interoperable ontologies and their application for COVID-19 research. Briefings in Functional Genomics, 20(4), 235–248. https://academic.oup.com/bfg/article/20/4/235/6307952
31. Morello, G., Salomone, S., D'Agata, V., Conforti, F. L., & Cavallaro, S. (2020). From multi-omics approaches to precision medicine in amyotrophic lateral sclerosis. Frontiers in Neuroscience, 14, 577755. https://www.frontiersin.org/journals/neuroscience/articles/10.3389/fnins.2020.577755/full
32. Yu, W., Yu, F., Li, M., Yang, F., Wang, H., Song, H., & Huang, X. (2024). Quantitative association between lead exposure and amyotrophic lateral sclerosis: A Bayesian network-based predictive study. Environmental Health, 23(1), 2. https://link.springer.com/article/10.1186/s12940-023-01041-3
33. Park, H. R., & Yang, E. J. (2021). Oxidative stress as a therapeutic target in amyotrophic lateral sclerosis: Opportunities and limitations. Diagnostics, 11(9), 1546. https://www.mdpi.com/2075-4418/11/9/1546
34. Pisharady, P. K., Eberly, L. E., Adanyeguh, I. M., Manousakis, G., Guliani, G., Walk, D., & Lenglet, C. (2023). Multimodal MRI improves diagnostic accuracy and sensitivity to longitudinal change in amyotrophic lateral sclerosis. Communications Medicine, 3(1), 84. https://www.nature.com/articles/s43856-023-00318-5
35. Krassowski, M., Das, V., Sahu, S. K., & Misra, B. B. (2020). State of the field in multi-omics research: from computational needs to data mining and sharing. Frontiers in Genetics, 11, 610798. https://www.frontiersin.org/journals/genetics/articles/10.3389/fgene.2020.610798/full
36. Tarazona, S., Arzalluz-Luque, A., & Conesa, A. (2021). Undisclosed, unmet and neglected challenges in multi-omics studies. Nature Computational Science, 1(6), 395–402. https://www.nature.com/articles/s43588-021-00086-z

CHAPTER 3

Role of Bioinformatics and Feature Selection Approaches in Analyzing Metagenomics Data

Anita Kachari and Deepranjan Pathak

3.1 INTRODUCTION

3.1.1 Metagenomics

Metagenomics studies all the genetic material (DNA) we obtain from environmental samples, e.g., soil, water, or the human gut. In metagenomics, we do not need to grow the microorganisms in culture, and we can study a whole community of microbes, including bacteria, viruses, fungi, and others, most of which cannot be grown using traditional culturing techniques. Often, we extract DNA from a sample that contains many organisms; sequence that DNA using high-throughput technology; and then analyze it using bioinformatics tools to determine which microbes are present and potentially what functions their genes may serve. The metagenomic approach can lead to discoveries of new microorganisms, understanding microbial ecosystems, and utilizing that knowledge in medicine, agriculture, environmental science, and biotechnology (Duan et al., 2024; Yen & Johnson, 2021).

3.1.2 Importance in Microbiome Research, Ecology, and Clinical Applications

Metagenomics is a powerful tool that drives both microbiome research and informs applications related to ecology and clinical practice. In terms of microbiome research, metagenomics assists researchers in exploring the complexity of the entire community of microorganisms that are part of "us". The microorganisms reside in the human gut, skin, or mouth and demonstrate how these microbes can affect our health, our digestion, our immunity, and even contribute to our disease (Yen & Johnson, 2021). From an ecological context, metagenomics helps researchers better define the diversity and functional

DOI: 10.1201/9781003591658-3

attributes of communities of microorganisms that inhabit soil, oceans, or extreme environments to provide researchers a better understanding of how microorganisms cycle nutrients, how the environment is changing because of climate change, and how the cycling of microbial communities can stabilize ecosystems (Nguyen et al., 2023). From a clinical perspective, metagenomics facilitates the identification of pathogens directly from patient samples without diagnostic culture, which provides faster and more accurate diagnoses of infections, including infections caused by rare or unknown organisms (Chiu & Miller, 2019). Metagenomics promotes the identification of antibiotic resistance genes and supports personalized medicine using an individual's microbial structural profile (de Abreu et al., 2021).

3.1.3 Role of Bioinformatics in Analyzing Metagenomic Data

Bioinformatics is crucial in working with metagenomic data and developing biological meaning from raw DNA sequences. When a metagenomic sample is sequenced, it generates millions or billions of short DNA sequences derived from many organisms, some of which are uncultivated and/or described, and many of which are unknown (Almeida & De Martinis, 2019). The large and complex nature of metagenomic data requires sophisticated computational tools and systems, which is why bioinformatics is important. The initial stage in bioinformatics is quality control and filtering to remove errors and contaminants from the raw sequence data (Xiong, 2006). Second, bioinformatics tools assemble the short DNA reads into long sequences or contigs, reconstructing portions of microbial genomes (Wick et al., 2017). Then these sequences are annotated, meaning the software predicts the genes, proteins, and possible functions that might be associated with them using databases like NCBI, KEGG, or IMG for comparison. Bioinformatics also allows for taxonomic profiling, which identifies the present microbes and estimates their abundance in a sample. This is valuable to researchers because it can provide insight into microbial life's diversity and community structure in different habitats. Functional profiling indicates what biological functions the microbes may occupy (e.g., production of vitamins, degradation of contaminants, or causing disease).

Bioinformatics will be used to:

- Assemble short reads into more extended sequences (contigs).
- Identify which organisms are present (taxonomy analysis).
- Predict genes and protein functions (functional annotation).
- Compare microbial communities in different environments or under different conditions.
- Detect pathogens or antibiotic resistance genes in clinical samples.

Bioinformatics makes it possible to manage and interpret the enormous amounts of data generated in metagenomics. Bioinformatics turns its raw identifiers into biological significance.

3.2 TYPES OF METAGENOMIC SEQUENCING

3.2.1 Shotgun Sequencing

Shotgun metagenomic sequencing is a modern method for examining the total genes of all organisms in a complicated environmental sample. Shotgun sequencing has one clear advantage over targeted methods that examine only specific genes, like the 16S rRNA gene, in that it is non-targeted and processes all organisms and their associated DNA. Rather than using targeted methods, shotgun metagenomic sequencing randomly cuts all of the extracted DNA into small fragments, sequences those fragments, and uses computational methods to assemble and analyze the data from the fragments (Breitwieser et al., 2018). It documents the DNA from all organisms in a sample, including bacteria, archaea, fungi, viruses, and microorganisms that may not have been previously described, with no culture required.

One advantage of shotgun metagenomics is its ability to identify taxonomic (what organisms exist) and functional (what they can do) data. Shotgun metagenomics can identify metabolic pathways, virulent factors, antibiotic resistance genes, and other genes affecting functional outcomes. Researchers can also find low-abundance species and evaluate their roles in the community (this could be relevant in studies of the human microbiome, human health, soil health, marine ecosystems, and disease diagnostics) (Quince et al., 2017). Because this is not possible with targeted sequencing, it will become helpful as the field progresses.

On the other hand, shotgun sequencing requires large amounts of DNA and increased computational power, and the cost will be much greater than that of targeted sequences. Notwithstanding these difficulties, it is regarded as one of the most effective instruments for comprehending microbial communities, composition, dynamics, and role in clinical and environmental contexts.

3.2.2 Amplicon Sequencing (16S/18S rRNA, ITS)

Amplicon sequencing is a type of targeted metagenomics that involves sequencing specific genetic markers for identifying and classifying microorganisms in a sample. Markers that are most commonly used include the 16S rRNA gene for bacteria and archaea, the 18S rRNA gene for eukaryotes, and the ITS (Internal Transcribed Spacer) for fungi (Caporaso et al., 2012). In this technique, marker-specific regions are amplified using PCR, and sequencing is conducted, allowing researchers to define the taxonomic diversity of a microbial community. Amplicon sequencing is a popular method because it is relatively low-cost, rapid, and straightforward, particularly when researchers only require overall microbial identification or want to determine community diversity. However, amplicon sequencing only provides limited functional information because only small genome regions are targeted, meaning gene function cannot be assessed, and viruses and unknown organisms will not be reported (Boers et al., 2019; Navgire et al., 2022).

3.2.3 Long-read Sequencing (Nanopore, PacBio)

Long-read sequencing is a cutting-edge DNA sequencing technology that produces longer reads than traditional short-read sequencing. Oxford Nanopore Technologies (ONT) and Pacific Biosciences (PacBio) are the primary platforms. Long-read sequencing can sequence entire genes or whole genomes in single reads, with the spans of single reads often tens of thousands or millions of base pairs in length. Generating long reads from a single reaction helps scientists resolve complex regions that may include repeats, structural variants, or multiple genetically similar species located in metagenomic samples. Improvements in genome assembly, gene annotation, and strain-level identification are possible through long-read sequencing (ONS and PacBio) and are beneficial for metagenomics. Long-read sequencing typically has a higher error rate than short-read sequencing. However, improvements and hybrid approaches (the use of both long and short reads) have dramatically improved the accuracy of long-read sequencing, which has also increased the usability of long-read sequencing technology in research and clinical applications (Olivucci et al., 2024).

3.3 METAGENOMIC DATA PROCESSING PIPELINE

The sequence processing pipeline in metagenomics refers to the workflows used to clean, analyze, and assemble the raw DNA sequence data generated from various environmental or clinical samples. Because metagenomic reads originate from complex, multi-species samples and contain sequencing noise, contaminants, and low-quality fragments, a rigorous processing pipeline is essential. It ensures that errors are removed before downstream steps such as assembly into contigs or direct read-based analysis. Once the cleaned reads have been prepared, a taxonomic classification can be done on the microbial data, either using databases and tools to identify which microbes are present in the sample (e.g., Kraken, MetaPhlAn, or Kaiju) (Menzel et al., 2016; Truong et al., 2015; Wood & Salzberg, 2014). Moreover, assigning taxonomic identity/resolution. At the same time, functional annotation of DNA, protein, and metabolic pathways can also be performed using tools such as HUMAnN or PROKKA (Blanco-Míguez et al., 2024). Once the cleaned reads have been established, taxonomic classification and functional annotation of the samples, either in clinical or environmental contexts, we may conduct statistical and/or comparative analysis to compare samples (e.g., to understand differences between samples), visualize the microbial diversity, and potentially detect relationships to environmental or clinical factors. From the original sample to raw sequencing data to cleaned DNA sequence data of microbial diversity and functions, a well-established pipeline brings a new understanding to microbial communities and their capacities in clinical settings or environmental systems.

3.3.1 Preprocessing: Quality Control: FastQC, Trimmomatic

Preprocessing, which includes quality control, is an important step in metagenomic studies. It allows researchers to ensure that their raw sequencing data is reliable, consistent, and safe for accurate downstream analysis. Environmental or clinical samples typically contain a mixture of microbial DNA, host DNA, and contaminants. If these different sources of DNA are not handled correctly, errors in sequencing or biased results are likely. Programs such as FastQC provide a dashboard of the raw reads to visualize, or at least have an overview of their structural quality as it relates to various issues such as per-base sequence quality, GC content, sequence duplication (multiple copies of a sequence in that read), adapter sequences, and overrepresented sequences. This diagnostic step is important to assess problematic issues that need rectifying (Bolger et al., 2014). Once the potential quality issues have been identified, you can use any number of applications to address the problems. Many researchers would probably turn to applications such as Trimmomatic, Cutadapt, or fastp to trim low-quality bases at the ends of reads, remove adapter contamination, and filter low-quality or too-short sequences, or a combination thereof. These cleaning steps almost universally improve the dataset's quality by reducing noise, removing most errors, and allowing for more accurate read alignment, assembly, and annotation, if anyone was comparing against a database or reference sequence (Bolger et al., 2014). By retaining high-quality reads, important advantages of preprocessing lie in its sensitivity and specificity in reliably identifying microbes while reducing false positives across the dataset and are related to assigning a more accurate taxonomic classification, functional profiles, and comparative analysis. In conclusion, optimizing the dependability and biological significance of the outcomes derived from intricate microbial communities requires quality control in metagenomic preprocessing.

3.3.2 Host Read Removal: Bowtie2, BWA

Host read removal is a critical quality control and preprocessing step in metagenomic analyses. The step is essential when samples are taken from host-associated environments (e.g., human gut, skin, blood, and animal tissues). In these environments, much of the sequenced DNA can be derived from the host (e.g., human DNA) rather than the microbes under investigation. Host-derived reads must be identified and filtered from the sequencing dataset to assess only microbial data. There are multiple tools available to do this, such as Bowtie2 and BWA, that align the sequencing reads to the reference genome of the host (e.g., the reference genome of the human genome) (Langmead & Salzberg, 2012; Li & Durbin, 2009). Suppose a read aligns to the host genome reference. In that case, it is labelled a host read and filtered out, leaving only non-host reads (predominantly microbial) to be used for further downstream analyses. Removal of host reads is therefore an important step because it not only improves accuracy and efficiency of downstream processes, such as taxonomic classification, functional annotation, and genome assembly, by removing unnecessary data and computation costs, but also maintains patient privacy in clinical cases, as human genetic information does not exist in the final analyses. Finally, host removal is critical to ensure that metagenomic results indicate the microbial community instead of the host background.

3.3.3 Assembly: MEGAHIT, metaSPAdes

In metagenomics, assembly is the process of merging short DNA sequencing reads into longer, continuous sequences called contigs, representing parts of microbial genomes. Because metagenomic samples contain DNA from various organisms, the assembly process is more complicated than assembling a single genome.

MEGAHIT and metaSPAdes are assemblers designed specifically for metagenomic data.

- MEGAHIT is a fast, memory-efficient tool that can handle relatively large and complex datasets, making it ideal when merging reads from highly diverse microbial communities (Li et al., 2015).
- metaSPAdes is another familiar assembler with the advantage of producing high-quality assemblies regardless of uneven coverage and genetic variability (Greenwald et al., 2017).

These types of tools allow researchers to reconstruct partial or near-complete microbial genomes from mixed samples, and this is an important first step in identifying microbes, studying their functions and genes, and finding new species in environmental or clinical studies.

3.3.4 Binning: MetaBAT, MaxBin, and CONCOCT

Binning in metagenomics is the assignment of assembled DNA sequences (contigs) to groups (bins). Ideally, each bin would represent a single microbial genome. Binning is an important step, often necessary because metagenomic data commonly arise from mixed microbial communities. Binning allows users to reconstruct individual genomes from that mixture.

Binning tools used in metagenomics include, but are not limited to, the following:

- **MetaBAT** directly utilizes sequence composition (e.g., GC content) and coverage information (the number of times each sequence appears) across the user-defined samples to group contigs together (Kang et al., 2015).
- **MaxBin** generates probabilistic models that rely on the presence of specific single-copy marker genes, sequence composition, and abundance for binning contigs based on what genes they carry (Wu et al., 2014).
- **CONCOCT** uses the sequence composition and coverage information over multiple samples to utilize clustering algorithms to separate genomes based on more closely related contigs (Alneberg et al., 2014).

Binning helps recover metagenome-assembled genomes (MAGs), which can provide important information about the biology, evolution, and function of uncultured or novel microbes. Binning represents an important step in which assembled sequences are molded into biological units to understand individual organisms in complex communities better.

3.3.5 Taxonomic Classification: Kraken2, MetaPhlAn, Kaiju

Taxonomic classification in metagenomics assigns DNA sequences to the respective taxa (i.e., what type of organisms), including species, genus, or phylum. Taxonomic classification is important for knowing what microorganisms are present in a sample and their prevalence. As metagenomic data contain sequences from a highly diverse variety of organisms, various tools and software are available to assist with correct classification. The most common analysis tools are as follows:

- Kraken2 is a fast and effective taxonomic classifier that identifies short sequenced DNA (k-mers) from the sample and links it back to a reference database of known microbial genomes. Kraken2 classifies at the taxonomic hierarchy level based on the matching sequences' lowest common ancestor. Kraken2 is designed to handle large datasets and can classify at the species level (Menzel et al., 2016; Truong et al., 2015; Wood & Salzberg, 2014).
- MetaPhlAn (Metagenomic Phylogenetic Analysis) uses a curated database of clade-specific marker genes. MetaPhlAn does not analyze all the DNA from the sample; it uses unique markers exclusively found in certain microbial groups. This approach is quicker and more precise for profiling microbial communities, especially with dominant species (Truong et al., 2015).
- Kaiju is a classifier that operates at the protein level, which is an important distinction when aligning DNA sequences against databases with annotated protein sequences, and being able to identify microbes that may be conspicuously missed through DNA-based tools, particularly ones with highly divergent genomes, separating them into individual cells. Kaiju is particularly potent and efficient in taxonomically classifying unknown or novel organisms (Menzel et al., 2016).

3.3.6 Functional Annotation: HUMAnN, eggNOG, KEGG, COGs

Functional annotation in metagenomics is the linking of biological function, pathways, or gene families, typically by means of a DNA or protein sequence, to identify what microbial genes do. Functional annotation goes beyond merely identifying what microbes are present to understanding their metabolic actions, ecological roles, and possible consequences for health or the environment. Many tools and databases, like The HMP Unified Metabolic Analysis Network (HUMAnN), eggNOG, Kyoto Encyclopedia of Genes and Genomes (KEGG), and Clusters of Orthologous Groups (COGs), are suitable for functional annotation. HUMAnN links metagenomic "sequences" to metabolic pathways in an ecosystem and quantifies their activity, indicating a community's metabolic activity (Franzosa et al., 2018). eggNOG (the evolutionary genealogy of genes) and COGs classify genes into functional categories based on the evolutionary relationship of taxonomically referenced genes (Cantalapiedra et al., 2021). KEGG links genes with biological pathways, such as carbohydrate metabolism, drug resistance, and signal transduction. Together, these tools help researchers explore the functional prospects of microbial communities, for example, in nutrient cycling, promoting disease, or detoxifying the environment (Kanehisa et al., 2023).

3.4 BIOINFORMATICS TOOLS AND WORKFLOWS

Bioinformatic tools and workflows are software programs and pipelines allowing users to process, analyze, and/or interpret large biological datasets originating from DNA, RNA, or protein sequences (Di Tommaso et al., 2017). In research areas such as metagenomics, bioinformatics tools and workflows are imperative to manage the large amounts of read data produced by high-throughput sequencing technologies (Quince et al., 2017). Bioinformatics workflows, which combine tools offered for general bioinformatics analysis while preserving subjectivity, can automate and mitigate the need for users to organize steps such as quality control, assembly, sequence alignment, taxonomic classification, functional annotation, and subsequent statistical analyses (Uritskiy et al., 2018). It serves an important role in microbiome studies, ecological work, clinical diagnostics, drug discovery, and biotechnology, as the analysis of metagenomes allows researchers to elucidate microbial diversity, conduct pathogen detection, catalog unique genes, and study biological functionality in complex communities. However, a wide array of tools, programs, and pipelines can be utilized for established metagenomic workflows and quick-starting programs, such as QIIME2, Galaxy, MG-RAST, and Nextflow. From this perspective, bioinformatics tools and workflows allow biological data to be quickly developed from raw data into meaningful discoveries and insights, and increase the ROI in the public return of research by enhancing pace, credibility, and reproducibility. Bioinformatics can also enable the direct detection of pathogens and antibiotic resistance genes, exploit clinical samples, and potentially facilitate quicker and better diagnoses. Second, bioinformatics and metagenomics can be applied to environmental science and ecology, which allows research on a variety of microbial diversities from microbial communities associated with sediments, soils, deep ocean environments, and extreme environments, as well as understanding microbial roles in important processes like carbon cycling, pollution remediation, and alleviation of climate change. Third, bioinformatics and metagenomics can also be transferred into agricultural and plant science to study plant-associated microbiomes, and demonstrate improved crop yield, soil fertility, and resiliency against disease and pests; and beneficial microbes that can be isolated and developed into bio-fertilizers or biopesticides; and finally, this technology can be used in biotechnology and drug discovery, where the delivered protein sequences from entrapment of uncultured microbes using bioinformatics can discover novel enzymes, novel antibiotics, and bioactive compounds that can potentially lead to a human pharmacopoeia. In biotechnology, examples also include the recreation of higher oligosaccharides and new bioactive compounds, and the advancement of synthetic biology will also benefit from basic gene cluster and pathway discovery. Bioinformatics is essential throughout the continuum of discovery and innovation in life sciences, which ultimately can turn a large and complex data set into meaningful information for humans.

3.4.1 Comparative Tools (MG-RAST, QIIME2, Mothur)

Comparative methods such as MG-RAST, QIIME2, and mothur are practical in metagenomics, as they allow for analyzing, comparing, and visualizing microbial communities across different samples in different environments or conditions. Comparative methods in metagenomics research allow researchers to see how microbial diversity and composition change depending on disease, environment, diet, treatment, and many other factors.

Comparative tools are important in metagenomics science for analyzing and comparing the microbial communities across different samples or conditions, and applications such as MG-RAST (Metagenomics Rapid Annotation using Subsystem Technology), QIIME2 (Quantitative Insights Into Microbial Ecology 2), and mothur are often utilized. MG-RAST is a web-based database and analysis platform capable of automatic quality control, annotation, taxonomic classification, and functional analysis of raw metagenomic data. MG-RAST allows users to compare datasets based on their microbial composition or gene function and is helpful for large-scale environmental or clinical metagenomic studies. QIIME2 is a robust open-source analysis pipeline mainly used for amplicon sequencing-based data, such as 16S rRNA, so it is more helpful in sequencing data than metagenomic sequencing. It allows for the analysis of microbial diversity, taxonomic classification, visualization, and statistical testing, and it is commonly used for microbiome research to compare microbial communities established under different conditions. Mothur is also a widely used tool for analyzing amplicon-based data that supports functionality, such as sequence alignment, OTU (operational taxonomic unit) clustering, calculation of diversity metrics, and comparison of communities. Both of these tools allow researchers to measure alpha diversity (diversity within a sample) and beta diversity (diversity between samples) and to compare community composition in different environments (i.e., healthy vs diseased), conduct statistical analyses (i.e., PERMANOVA), and create visualizations such as heatmaps, PCA plots, and phylogenetic trees. In the end, these tools have allowed researchers to understand the influence of various ecological factors, diseases, or treatments on the composition and function of microbial communities, making them incredibly useful in microbiome, ecological, agricultural, and clinical research.

3.4.2 Visualization Tools (Krona, Pavian, Phinch)

Visualization tools such as Krona, Pavian, and Phinch are specialized software used in metagenomics for visual interaction with and interpreting complex microbial data straightforwardly and intuitively. Each visualization tool takes potentially large, dimension-rich datasets (e.g., taxonomic classifications or functional annotations) and creates an easy-to-understand visual.

- Krona allows users to access interactive, multi-level pie charts (sunburst charts), drill down through taxonomic hierarchies, and check the relative abundance of different microbial groups within a sample.
- Pavian is a visualization tool designed specifically to visualize Kraken2 classification results. It offers interactive bar plots, Sankey diagrams, and tables. Each visualization example has proper functionality, such as identifying and comparing microbial contents across multiple samples. For example, visualization of multiple samples makes it suitable for clinical studies or investigations into pathogen detection.
- Phinch is an open-source tool for visualizing and analyzing microbiome data in BIOM format (as can be exported from software such as QIIME or similar tools). It has interactive visualizations, such as bar charts, bubble plots, and heatmaps for the studied samples, and allows the user to visually explore diversity patterns, taxonomic composition, and sample metadata.

These tools are crucial to metagenomics because they enable researchers to communicate findings more efficiently, produce hypotheses for additional study, and swiftly spot trends, patterns, and outliers in microbial communities.

3.4.3 Workflow Management (Snakemake, Nextflow)

Workflow management tools help metagenomics researchers generate, organize, and automate complex data analysis pipelines. Metagenomic studies are laborious endeavors that comprise many steps including– quality control, assembly, and annotation are a few that utilize multiple software programs. Performing these tasks in sequential manner, manually it can be laborious and uncertain. In contrast, workflow managers allow a lab organization or programmatic combination of each step in a pipeline that automates, replicates, and scales the entire process.

- Snakemake is a workflow management system written in Python that allows users to define the pipeline analysis steps as a "rule" declared in a Snakefile. The Snakemake system performs efficient task execution on a local computer or high-performance/compute cluster and understands which files depend on one another by establishing a definable linear order (Köster & Rahmann, 2012).
- Nextflow is a flexible workflow management tool built on its scripting language. It offers multiple platforms, such as cloud services, and specific integration to scalable containers like Docker or Singularity with a consistent computing environment. Where Nextflow shines in pipelines is its capacity for parallel execution (Zhang et al., 2021).

These tools are significant in a metagenomics project to structure data analysis, allow high-throughput workflow pipelines with few errors from manually entering code, saving time, and allow workflows to remain consistent, replicable, or transferable between projects or labs. In all cases, it is important to effectively structure large datasets and complex pipelines to maintain organizational integrity.

3.5 CHALLENGES IN METAGENOMIC BIOINFORMATICS

Metagenomic bioinformatics entails the analysis of large and complex sequence datasets to investigate microbial communities from a wide range of environments. While this science is powerful and has the potential to tell us much about microbial activity and abundance in complex samples, it has challenges that can undermine the integrity, consistency, and interpretation of results (Behzad et al., 2015). The first significant challenge is complexity and diversity, as microbial communities often comprise uneven and highly variable population sizes, with far too many novel or unculturable species present to know their basic metabolic properties (Handelsman, 2004). Such complexity only exacerbates downstream steps, involving genome assembly, binning, taxonomy classification (especially of low-abundance organisms), and instances of taxonomic profiles appearing incomplete. The second challenge involves the reference databases, which, for the most part, are incomplete or even biased. Too many microbes lack their sequence, thus requiring tools like Kraken2 or MetaPhlAn to make inferences based on biobased samples that are often too small. This often results in too many unknown reads, further contributing to incomplete taxonomic profiles. The sequencing technologies available and used for metagenomics can also present challenges. For example, commonly used short-read sequencing platforms (i.e., Illumina) can limit read length, complicating assembly and gene prediction of repetitive or complex genomic regions. Furthermore, metagenomic bioinformatic workflows are resource-intensive, which can repel smaller labs from performing their bioinformatics due to the need for availability of high-performance computing (HPC) infrastructure. Contamination by possible host DNA, particularly when analyzing clinical or environmental samples, carries similar implications of obfuscating accurate microbial signals. Thus, careful removal steps are required to scrutinize whether the microbial communities are omitted or masked rather than being an actual absence. Because metagenomic bioinformatics fails to employ standardized pipelines or even standardized parameters, results can be disproportionately affected as these methods can lead to more or less recurrent reporting modes of discrepancies across instances of studies, considerably degrading sample reproducibility, and with it, data integrity. Functional annotation has challenges, such as all the hypothetical proteins, horizontal gene transfers, and lack of context, which result in unknown or limited potential functions.

The results of metagenomics are also influenced by some sampling characteristics that influence microbial diversity, such as the source and sample type (soil, gut, water), which may also have compounds that inhibit the extraction of DNA. The steps taken for the DNA extraction process may also carry bias toward differing microorganisms. Sequencing technology also plays a role as short-read platforms take high-quality reads but little resolution. In contrast, longer-read platforms, such as Nanopore/PacBio, make it easier to recreate genomes but have worse quality and higher error rates than short-read platforms. Sequencing depth is important for sensitivity since, depending on how much is sequenced, one may miss or underestimate rare taxa, while sequencing more will increase cost and analysis time. Once results are available, bioinformatics tools and settings implemented play a significant role in the results, as even minor differences such as k-mer size or quality thresholds can change the end quantitative differences.

3.5.1 High Complexity and Diversity

In metagenomics, high complexity and diversity can refer to the sheer number and variety of organisms, including bacteria, archaea, viruses, fungi, and other microbes, found in a single environmental or biological sample, i.e., soil, ocean water, or the human gut (Simon & Daniel, 2011).

High complexity means that many species and strains reside together, each with its own set of genome sizes, GC content, and gene contents. These communities may contain abundant and rare organisms (Scholz et al., 2012). The presence of abundant organisms enhances, but complicates, the tasks of genome assembly, binning, and annotation. High diversity indicates both richness (total number of distinct species) and evenness (how evenly distributed those species are) of the entire microbial community (Lozupone et al.,

2007; Quince et al., 2017). Highly diverse samples may contain many unique microbes that have no or minimal overlap with other samples taken from different environments. This generates another challenge because it is often tough to classify and functionally analyze, since many of the most common microbes may not be well represented or even be found in reference databases, making identification and classification challenging. Thus, a metagenomic sample's high complexity and diversity make downstream sample analysis challenging. One of the key issues is the assembly problem. Short sequencing reads may span several different organisms that may overlap in sequence, similarity, or have many potential places to be assembled, leading to fragmented or incorrect contigs. Binning problems frequently occur as it becomes harder to correctly bin sequences into separate genomes, especially in the case of rare species or species that are closely related. Further, complex data sets can become taxing on analysis pipelines, demanding substantial memory, storage, and processing power on both sides of the sequencing equation, raising both time and monetary costs. The presence of unknown, poorly characterized species is also a further complication, with many microbes having no close relatives in reference databases, leaving much of the data unclassified or improperly annotated. This may also lead to inaccuracies in quantifying the presence of specific species, particularly when sequencing depth is not enough to detect any low-abundance organisms. It renders the data prone to aberrations in microbial abundance estimates. These problems are particularly evident when dealing with high-complexity environments, such as soil, which can contain thousands of species per gram; the human gut microbiome, which has hundreds of species that are highly dynamic; and marine ecosystems, which are also hugely variable depending on the depth, salinity, and temperature, which may drastically alter the microbial diversity.

3.5.2 Computational Demand

Computational demand in metagenomics refers to the substantial processing power, memory (RAM), data storage, and running time required to analyze large, complex sequences. Many metagenomic studies generate millions of DNA sequences from complex mixtures of microbes. As a result, metagenomic bioinformatic tasks such as quality control, assembly, classifying taxonomy, binning, and assigning function are commonly resource-intensive tasks (Quince et al., 2017; Scholz et al., 2012).

High-throughput sequencing produces gigabytes to terabytes of sequencing data. In addition, metagenomics studies that use multiple samples or deep sequencing produce even greater volumes of sequencing data. Some basic API tools, such as Galaxy or RStudio, or programming applications, such as using MEGAHIT or metaSPAdes for genome assembly, or MetaBAT or MaxBin for binning, and the potential for multi-threading/parallel processing, can utilize as many CPUs and as much RAM as available on the system (Nurk et al., 2017).

Even when analysis tools do not use CPU and RAM to the maximum, big sequenced data often requires considerable disk space to store intermediate and final files (i.e., raw reads, contigs, and genomes) for a variety of analysis tasks (Almeida & De Martinis, 2019; Lloyd-Price et al., 2019). For many research labs, computational demand can be beyond reach if researchers cannot access HPC clusters, cloud platforms, or optimized metagenomic pipelines. When the computational demands are substantial, analyses can be limited in scale and depth, take too long to run, crash, or be incomplete.

3.5.3 Incomplete Reference Databases

In metagenomics, incomplete reference databases are a profound challenge because they constrain the ability to accurately classify and analyze sequencing data. The issue often comes from unclassified reads; if many sequences cannot match known genomes in the databases, the resulting microbial profile may be limited, or there may be unidentified species (Segerman, 2020). Also, this gives way to misclassification because if there are reads from a novel organism, these may mistakenly be assigned to the known, closest relations, thus inaccurately assigning taxonomic identifications (Chorlton, 2024). The potential for misclassification exists for functional annotation as well – many genes were assigned as "hypothetical proteins" or not annotated at all without reference matches in another database (like KEGG or eggNOG). This can result in limited insights about the community-level metabolic functions. Additionally, current databases are biased toward well-studied microbes and of clinical or industrial importance, potentially providing a skewed view of microbial ecosystems by underrepresenting many sites of lesser-known but ecologically important species (Marcelino et al., 2020). Finally, incomplete databases limit the potential for discovery by obstructing the identification of novel organisms, genes, or metabolic pathways from rare, extreme, or poorly explored habitats, particularly deep-sea vents or human-associated habitats. Therefore, expanding and updating these reference databases is critical to addressing this shortcoming and fully realizing the potential of metagenomic research.

Incomplete reference databases in metagenomics severely disrupt the accuracy and depth of microbial analysis when taxonomies go unclassified or misidentified (Chorlton, 2024). An example of how generalizations affect microbial work can be seen in taxonomic profiling and functional annotation, which are both limited due to the lack of sequence databases and the potential mistakes in taxonomic assignments that result in inaccurate analysis and interpretations of new biology (e.g., mixtures of homologues). Expanding and updating reference genome collections, especially via genome-resolved metagenomics or long-read sequencing initiatives, are fundamental in addressing these challenges.

3.5.4 False Positives/Negatives in Classification

False positives and negatives in metagenomics classification arise due to limitations in tools, databases, and data distribution and quality. They can substantially influence the microbial profiling data and downstream analysis outcomes (Federhen, 2015).

A false positive occurs when a sequence is assigned to a microbe or function not truly present, and a false negative occurs when a sequence assigned to a present organism or function is identified or goes unclassified, such as mislabeling a read as *Escherichia coli* when it belongs to something like Citrobacter (Sunagawa et al., 2013). False positives can easily arise from incomplete and biased reference databases and low-quality or too-short sequencing reads. Some sequences may belong to closely related species that tools may have difficulty distinguishing. False negatives can also arise from the same limitations noted above, but also from improperly assigned tool parameters or contaminated samples. The consequences of false negatives and positives may include false microbial community profiles, incorrect biological interpretations, failing to detect an important pathogen or pathogen-associated resistance gene, and compromised reproducibility of studies. Minimizing these issues will require more complete reference databases, higher quality, longer sequencing reads, and well-tested and validated bioinformatics pipelines.

3.6 APPLICATIONS

Metagenomics, which refers to studying genetic materials obtained directly from environmental samples, has various applications in many fields.

3.6.1 Gut Microbiome Analysis

The gut microbiome consists of an incredibly diverse and complicated community of trillions of microorganisms, including bacteria, viruses, fungi, and archaea, in the human gastrointestinal tract (Turnbaugh & Gordon, 2009; Weinstock, 2012). They are essential to digestion, nutrient uptake, immune moderation, and overall health. Metagenomics allows researchers to look at the microbial community without growing microorganisms through traditional methods. Using techniques like shotgun metagenomic sequencing, researchers can extract and sequence microbial DNA directly from stool or gut samples and look directly at the microbiota; diversity and functional characteristics are measurable. Metagenomics has changed the level of understanding concerning how the gut microbiome affects human health and disease states. It can differentiate between microbial communities' healthy and dysbiotic (unbalanced) states. Dysbiosis of the gut microbiome has been linked to obesity and type 2 diabetes, inflammatory bowel disease (IBD), colorectal cancer, and neurological disorders like autism, Parkinson's disease, and Alzheimer's disease (Wu et al., 2011). These discoveries show that metagenomics contributes to personalized medicine concerning dietary and therapeutic interventions, such as prebiotics, probiotics, and fecal microbiota transplantation. Metagenomics also aids in understanding how gut microbes shape and modulate the immune system, which may advance new approaches to immunotherapy or address autoimmune disease.

3.6.2 Disease Biomarkers Discovery

Disease biomarkers can be seen as measurable signals, specified genes, proteins, or other microbial species, all of which signal the presence or risk of a particular disease (Segata et al., 2011). Metagenomic sequencing can assist in discovering these biomarkers and uncovering microbial signatures associated with specific disease states. Because metagenomic analysis allows for culture-independent identification of microbial genes or species, it can potentially identify microbial biomarkers that capture early warning signs or are helpful for diagnostic or therapeutic targets. There are ample examples of metagenomics discovering disease biomarkers.

Fusobacterium nucleatum has been found in colorectal cancer tumor tissues and is potentially one microbially derived marker. The influence of gut bacteria on the abundance and functionality of type 2 diabetes was identified in the studies. Also, in individuals with

IBD who have been diagnosed with Crohn's disease and ulcerative colitis (Pérez-Brocal et al., 2015), it is not uncommon to find a depletion of beneficial, anti-inflammatory microbes such as Fecal bacterium *prausnitzii*, where the absence is likely influenced by antibiotic treatment related to the disease.

Metagenomics may be used to create non-invasive diagnostic methods, like stool-based tests that can identify microbial biomarkers with high specificity (Gilbert et al., 2018). Additionally, it facilitates the development of microbiome-based therapies that employ advantageous microorganisms, enzymes, or microbial metabolites to treat or prevent disease and creates opportunities for early disease detection, possibly before clinical symptoms manifest (Lloyd-Price et al., 2016; Rinninella et al., 2019)

3.6.3 Environmental Studies: Soil, Ocean, Extreme Environments

Metagenomics has emerged as a powerful approach in environmental science, allowing researchers to examine, characterize, identify, and understand microbial communities' diversity, functionality, and interactions across various ecosystems, from soils and oceans to extreme environments, entirely without the need for cultivars.

Metagenomics is increasingly used in soil ecosystems to understand complex microbial networks driving important functions such as nutrient cycling, organic matter breakdown, nitrogen fixation, and carbon sequestration. It can be used to evaluate soil health, to understand changes in the microbial population induced by pollution or agricultural applications, and to find beneficial microbes that can be cultivated as biofertilizers or biocontrols to enhance sustainable agricultural practices (Daniel, 2005; Torsvik & Øvreås, 2002)

In marine and oceanographic environments, metagenomics provides powerful insights into the oceans' vast, largely unexplored microbial communities in water and sediment. These microbial communities are responsible for and play an important role in primary global biogeochemical cycles (carbon, sulfur) (Grossart et al., 2020) and facilitate primary production by conducting photosynthesis, such as cyanobacteria. Metagenomics in marine and ocean basin environments has also revealed new and novel genes, novel enzymes, and new metabolic pathways that may be applied in biotechnology (Kennedy et al., 2008). Furthermore, metagenomic approaches can be used to monitor oceanographic changes, which can be used in terms of LC impacts, such as ocean acidification, pollution, and climate change affecting microbial ecosystems.

Metagenomics is essential to understanding extreme environments, which include deep-sea hydrothermal vents, hot springs, arid deserts, polar ice caps, and acidic or alkaline lakes where traditional cultivation techniques cannot be done. Metagenomic studies have demonstrated extremophiles (microorganisms adapted to extreme temperature, pressure, salinity, or pH) (Cowan et al., 2015). Extremophiles are rich in unique genes and enzymes (thermostable polymerases, salt-tolerant proteins) that may be utilized in many different industrial processes (biotechnology) (Kour et al., 2019). The additional study of extremophiles can offer further clues as to the limits of life on Earth and facilitate the search for life in other environments outside of Earth; for those reasons, metagenomics is a significant technology in astrobiology.

3.6.4 Industrial: Bioremediation, Fermentation, Agriculture

3.6.4.1 Bioremediation

Metagenomics has transformed the practice of bioremediation by providing researchers with access to the whole microbial community and functional genes involved in degrading environmental contaminants (Chakraborty et al., 2012). Many traditional methods rely on culturable microbes, which account for only a percentage of the whole microbial world. Metagenomics allows access to the genetic diversity of microbial communities from sites with pollution, such as oil-contaminated soils, industrial discharge, and mining sites with high levels of heavy metals.

Metagenomic studies have found microbes capable of degrading toxic compounds, such as benzene, toluene, xylene (BTX), polychlorinated biphenyls (PCBs), and polycyclic aromatic hydrocarbons (PAHs) (Ufarté et al., 2015). Metagenomic studies have identified the actual genes involved, the degradation genes of enzymes such as monooxygenases, dioxygenases, and dehalogenases, which are crucial in the degradation of toxic organics (Silva et al., 2013). Additionally, metagenomics allows for measuring bioremediation success by tracking evolving microbial community structure and expression of functional genes over time and facilitating the design of customized microbial consortia or bioaugmentation efforts to promote pollution removal.

3.6.4.2 Fermentation Technology

Metagenomics can be a valuable tool for refining industrial fermentation processes fundamental in making food, drink, medicine, biofuels, and other products with value-adding potential. The technology enables either a whole genome analysis or assessment at a functional level for the microbial consortia in traditional and modern fermentation systems, thus supporting traditional applied contextual knowledge of how changes in niche and microbial diversity in fermentation changed from microbial community composition, functional gene pathways, and metabolite production.

For example, in dairy fermentation, metagenomics has identified starter cultures of strains applicable to optimize flavor, inhibit spoilage, or tolerate phage domination (Chen et al., 2017). In alcoholic beverage fermentation (e.g., wine, beer, and kombucha), evidence of the interactions between yeasts (e.g., *Saccharomyces cerevisiae*) and lactic acid bacteria (Tamang et al., 2016) has formed greater credibility of the product with value-adding quality and stabilization, and metagenomics was used to optimize the culture communities. Metagenomics not only documents product quality but also identifies undesired microbes, documented toxins, and undesirable antibiotic resistance genes that inform product safety, which is important as safety compliance is a royalty payment for the vast majority of invested businesses.

The role of metagenomics will only grow as parts of wavelengths for enzyme discovery (or other valuable functional gene) (Li et al., 2009), as metagenomics unearth an abundance of uncultured microorganisms that are only found in extreme conditions in the surrounding environment (Kennedy et al., 2008), and thus can be responsible for the industrially significant application of enzymes: thermostable amylases, proteases, and cellulases that can be used as industrial biocatalysts in various sectors such as textiles, detergents, and pharmaceuticals (Gupta et al., 2002).

3.6.4.3 Agriculture and Agri-Biotechnology

In agriculture, metagenomics serves as a powerful way to allow sustainable and precision farming through the investigation of soil, rhizosphere (the root zone), and phyllosphere (the leaf surface) microbiomes to find beneficial microorganisms that aid crop productivity, disease resistance, and stress tolerance (Kumar et al., 2023).

Metagenomics also allows the discovery and characterization of microbes that are involved with:

- nitrogen fixers (e.g., Rhizobium, Azotobacter)
- phosphate solubilizers
- siderophore producers (to uptake iron)
- biocontrol of pathogens such as Fusarium, Pythium, and Rhizoctonia

Incorporating these insights into biofertilizers and biopesticides reduces reliance on chemical inputs. It uses the metagenomic data to select crop varieties more compatible with the microbiome and improve soil-plant-microbe interactions.

Metagenomics can also be used to track the effect of agricultural practices (e.g., pesticides, crop rotation, and organic farming) on microbial diversity and ecosystem health. This supports the early detection of soil-borne pathogens, allowing farmers to take preemptive steps and helping research into carbon and nitrogen cycling for climate-resilient farming systems.

3.7 RECENT ADVANCES AND FUTURE PERSPECTIVES

Metagenomics is advancing with the advent of technology and computational developments. It has taken huge strides forward, partly because of several wonderful technologies and computational advances. Recently, one of the most significant advances has been the inclusion of artificial intelligence (AI) and deep learning (DL) to improve accuracy and speed up previously time-consuming processes, such as the classification and functional prediction of microbes (e.g., Deep Microbes). The advancement of MAGs has allowed for the first time to reconstruct complete microbial genomes directly from environmental samples, revealing previously unknown microbes with ecological roles. When explored with multi-omics approaches (i.e., metabolomics, transcriptomics, and proteomics), the capacity to represent a microbial community can be done holistically from genetic potential to actual metabolism and gene expression. More recently and just as important have been the attempts to use cloud-based platforms like Galaxy or Google Cloud Life Sciences, allowing scientists to analyze large datasets with social experience, sometimes tens of large datasets, more easily and collaboratively with their associated data, minimizing the need to manipulate their code to include workflows and release their code. Shortly, metagenomics will drive microbial research, leading to innovative developments in precision medicine, sustainable agriculture, environmental conservation, and industrial biotechnology that will uncover new genes, enzymes, and microbial interactions/roles to address global challenges.

3.7.1 Deep Learning and AI in Metagenomics (e.g., DeepMicrobes)

AI, intense learning, is transforming the field of metagenomics through accelerated speed, improved accuracy, and greater scalability to study microbes. Current bioinformatics tools typically underperform in analyzing large, complex, and diverse metagenomic datasets. DL models, such as DeepMicrobes, traditionally employ neural networks to learn patterns from raw DNA sequences and to classify the microbes accordingly (Liang et al., 2020). Unlike traditional bioinformatics tools that rely heavily on a reference database of microbial organisms and meticulously hand-crafted features, DeepMicrobes develops the advanced ability to capture complex, nuanced sequence biases even when the organism is unknown or poorly annotated, allowing for quicker taxonomic assignments, strain-resolved microbial taxonomy, and increased ability to identify novel and rare taxa (Liang et al., 2020). Additionally, AI tools are being developed for functional annotation, gene prediction, and even antibiotic resistance gene detection, allowing for their essential use in processing large, complex datasets generated from environmental, clinical, and agricultural studies (Arango-Argoty et al., 2018). As DL models become more sophisticated, their ability to generate curious knowledge should position them well for future real-time, integrated, high-throughput metagenomic analysis and move toward potential applications for precision medicine, bio-surveillance, and ecological forecasting.

3.7.2 Metagenome-assembled Genomes (MAGs)

MAGs are microbial genomes reconstructed from metagenomic sequencing data without the need to culture the organisms in the lab. When a mixed environmental sample (soil, ocean water, gut content) is sequenced, millions of DNA fragments are produced from different organisms. Advanced bioinformatics processes these fragments into multiple genome bins, which are then refined into MAGs (Albertsen et al., 2013; Parks et al., 2017).

MAGs have entirely changed microbial ecology because they allow studying nebulous, unknown microbes that cannot be cultured. Researchers can study these microbes' genetic potential, metabolic potential, and ecological roles. MAGs can help researchers identify new species, metabolic pathways, and gene clusters for medicine, agriculture, or other biotechnological applications (Parks et al., 2017). By using MAGs, researchers can understand the structure and function of microbial communities, track how they respond to changes in an environment, and study their evolutionary history. The technology of sequencing and sequencing assembly tools is continually improving. Thus, MAGs are expected to be produced more accurately as we explore microbes' responses to environmental changes (Parks et al., 2017).

3.7.3 Integration with Metabolomics and Transcriptomics (Multi-Omics)

A multi-omics approach is when metagenomics is integrated with other omics, such as metabolomics (the study of metabolites) and transcriptomics (the study of gene expression). Metagenomics provides information on the genetic potential of microbial communities (it identifies what microbes are present and what they are capable of doing), but it does not provide information on what genes are actually expressed or what metabolic functions are taking place. Multi-omics is a means to overcome this challenge (Quince et al., 2017).

By adding transcriptomics to metagenomics, we can discover which microbial genes are being actively transcribed in a specific environment or condition. Similarly, adding metabolomics integrates characterizing the small molecules, or metabolic products, that are being produced or being consumed, giving a functional layer of supplementation about microbial activity. Together, these layers provide an expanded understanding of microbial community structure, function, and dynamics.

Multi-omics approaches are increasingly utilized in human health (for example, connecting gut microbes to disease), as well as in agriculture (e.g., plant-microbe interactions), and in the environmental sciences (e.g., watching soils or marine ecosystems). As data integration tools and approaches become more mature and accessible, multi-omics will be an important innovation to precision medicine, ecosystem monitoring, and microbiome-based applications.

3.7.4 Cloud-based Solutions (e.g., Galaxy, Google Cloud Life Sciences)

Cloud-based platforms are cost-effective solutions to the big problem of managing the enormous amounts of sequencing data that are generated with studies of microbial communities in metagenomics (Afgan et al., 2018). As an example, an open-source, web-based platform like Galaxy, and a commercial option like Google Cloud Life Sciences, provide users with a scalable, accessible physical infrastructure to not only find an option of space to store and process their analysis on large metagenomic datasets, but also to analyze metagenomic datasets without the need for expensive computing resources on local machines (Afgan et al., 2018; Taylor et al., 2019). Cloud platforms are able to offer a variety of bioinformatics pipelines that encompass quality control, taxonomic classification, functional annotation, and assembly of metagenomes, all from one location and through web-based systems (Taylor et al., 2019). Cloud-based solutions also provide a variety of opportunities to share and collaborate on documents and workflows among scientists from around the world, and have no access restrictions (Taylor et al., 2019). These platforms also provide a range of automated resource scaling, increasing access to allow scientists time to focus on generating analyses rather than waiting for sequences to be ready, all without sacrificing performance or incurring large costs. As metagenomics expands in all fields of study, cloud-based computing will be a key aspect to manage big data, gain access to real-time analysis of microbe communities, and ultimately contribute to new discoveries made in microbiome research, in health, agriculture, and the environmental sciences.

3.8 FEATURE SELECTION IN METAGENOMIC DATA

Feature selection in metagenomics is a key part of processing and analyzing the high-dimensional sequencing data captured from diverse microbial systems. Metagenomic data may contain anywhere from thousands to millions of features, including microbial taxa (species, genera), gene families, metabolic pathways, and k-mers, many of which are sparsely present, noisy, redundant, or uninformative with respect to the biological question at hand (Knights et al., 2011; Gloor et al., 2017). High dimensionality can be problematic for statistical evaluation, increase computational cost, and can also result in overfitting if machine learning (ML) is used. When feature selection is used, it identifies and retains only the features that are informative and relevant to the biological problem and maximizes model performance, interpretability, and generalization (Pasolli et al., 2016). In metagenomics, feature selection techniques can be divided into three primary groups: filter, wrapper, and embedded methods (Pasolli et al., 2016; Saeys et al., 2007; Wirbel et al., 2021). Filter methods (e.g., mutual information, chi-square tests, correlation) rank features without reference to any model (Saeys et al., 2007), while wrapper methods (e.g., recursive feature elimination, RFE) evaluate subsets of features in conjunction with a predictive model (Pasolli et al., 2016) (e.g., random forests, SVMs). Wrapper methods are often more calibrated to data but require greater computing time and resources (Wirbel et al., 2021). Embedded methods, e.g., LASSO (Least Absolute Shrinkage and Selection Operator) and decision tree methods, select features in conjunction with the modelling process (Wirbel et al., 2021).

In applied sciences, feature selection will allow researchers to discover biologically relevant patterns, such as microbial biomarkers that associate with diseases (e.g., diabetes, cancer), environmental factors (e.g., pollution, climate change), or host characteristics (e.g., gut health, immunity) (Douglas, 2020). For example, in an application seeking the most predictive model of disease status, feature selection would help researchers identify the few taxa or gene pathways most distinctive from the baseline condition. In addition, removing irrelevant noise to focus on core signals increases the reproducibility of outcomes and reduces noise, which allows for more robust studies in metagenomics. Also, if combined with dimensionality reduction methods (e.g., PCA and t-SNE), feature selection has the potential to assist with visualizing non-linear relationships in complex microbial communities (Douglas, 2020). In conclusion, feature selection is essential in converting untidy and large metagenomic data into actionable information in microbiome research, diagnostics, and environmental monitoring.

3.9 CONCLUSION

3.9.1 Summary of the Current Bioinformatics Landscape

Today's bioinformatics environment continues to evolve with the advent of new technologies and disciplines, critically impacting biological research. The tremendous growth of next-generation sequencing (NGS) technologies continues enhancing bioinformatics as an important tool for organizing, analyzing, visualizing, and interpreting biological data in various research areas including – genomics, metagenomics, transcriptomics, proteomics, and metabolomics. In metagenomics, bioinformatics is specifically helpful for the analysis of complex mixed-species datasets through taxonomic classification, functional gene annotation, and reconstruction of MAGs (Albertsen et al., 2013; Parks et al., 2017).

One important change to the bioinformatics landscape is the increasing reliance on ML and DL algorithms, such as DeepMicrobes and MetaPhlAn, to accurately classify microorganisms, predict host-microbe interactions, and detect patterns of antibiotic resistance (Liang et al., 2020). These AI-based computational models are more robust than others, as they respond sensitively to the complex interactions learned from the dataset instead of recovering complex patterns learned directly from a rule-set and context or simply an analytical description.

In addition to the trend toward integrating multi-omics approaches, bioinformatics platforms are developing algorithms that combine genomic, transcriptomic, proteomic, and metabolomic data as part of a systems-level overview of biological processes. Companies such as Galaxy, Google Cloud Life Sciences, Amazon Web Services (AWS), DNAnexus, and other service providers offer scalable computing infrastructure, data-sharing capabilities, and user-friendly interfaces that can provide a means for researchers with limited local capacity to run complex workflows and collaborate on unique data issues with other researchers across the globe.

Moreover, many open-source tools and databases have been created and maintained and are now very prominent in bioinformatics-based research. A few examples include:

- QIIME2, MetaPhlAn, and HUMAnN, for microbiome profiling.
- MEGA, BLAST, and Clustal Omega, for phylogenomics and sequence comparison.
- MG-RAST, IMG/M, and EBI Metagenomics, for submission, analysis, and storage of metagenomic data.
- Ensembl, KEGG, NCBI, and UniProt, for information on genes and proteins.

Bioinformatics also focuses on reproducibility, data standardization, and FAIR (Findable, Accessible, Interoperable, and Reusable) principles to improve research data reliability and usability. Emerging realities of practice include real-time genome surveillance, eDNA tracking, synthetic biology approaches, and portable sequencing capabilities (e.g., Oxford Nanopore MinION) that bring together wet-lab and computational procedures in a field setting (Quick et al., 2017).

3.9.2 Future Direction of Metagenomic Research

The future of metagenomics is assured of a paradigm shift from merely identifying microbes to achieving a deep functional, ecological, and clinical understanding at unprecedented resolution (Quince et al., 2017). Advances in NGS technologies, such as Oxford Nanopore and PacBio HiFi, will enable the accurate assembly of genomes and the generation of MAGs at the strain level. The integration of metagenomics with multi-omics approaches, including transcriptomics, proteomics, and metabolomics, will facilitate a systems-level view of microbial activity, interactions, and real-time functionality across diverse environments. ML and AI will further enhance our ability to predict microbial functions, uncover novel biosynthetic gene clusters, and link microbes to phenotypes or environmental impacts (Liang et al., 2020). The adoption of portable real-time sequencing devices will expand metagenomics into field diagnostics, ecological monitoring, and real-time pandemic surveillance, even in resource-limited settings. Additionally, the convergence of metagenomics with synthetic biology and genome editing tools such as CRISPR/Cas system may enable the design of tailored microbial consortia for applications in agriculture, bioremediation, and industrial biotechnology. Integrated with cloud computing and shared data platforms, these advances will make metagenomics more scalable, collaborative, and translational, driving progress in human health, environmental sustainability, agriculture, and biotechnological innovation.

ACKNOWLEDGMENT

The authors are thankful to Dr. Channa Chikkaputtaiah for their invaluable guidance, corrections, and support throughout the writing of this chapter, and also thankful to CSIR-NEIST, Jorhat

REFERENCES

Afgan, E., Baker, D., Batut, B., van den Beek, M., Bouvier, D., Čech, M., Chilton, J., Clements, D., Coraor, N., Grüning, B. A., Guerler, A., Hillman-Jackson, J., Hiltemann, S., Jalili, V., Rasche, H., Soranzo, N., Goecks, J., Taylor, J., Nekrutenko, A., & Blankenberg, D. (2018). The galaxy platform for accessible, reproducible and collaborative biomedical analyses: 2018 update. Nucleic Acids Research, 46(W1), W537–W544. https://doi.org/10.1093/nar/gky379

Albertsen, M., Hugenholtz, P., & Skarshewski, A. et al. (2013). Genome sequences of rare, uncultured bacteria obtained by differential coverage binning of multiple metagenomes. Nature Biotechnology, 31, 533–538. https://doi.org/10.1038/nbt.2579

Almeida, O. G. G. & De Martinis, E. C. P. (2019). Bioinformatics tools to assess metagenomic data for applied microbiology. Applied Microbiology and Biotechnology, 103, 69–82. https://doi.org/10.1007/s00253-018-9464-9

Alneberg, J., Bjarnason, B., & de Bruijn, I. et al. (2014). Binning metagenomic contigs by coverage and composition. Nature Methods, 11, 1144–1146. https://doi.org/10.1038/nmeth.3103

Arango-Argoty, G., Garner, E., & Pruden, A. et al. (2018). DeepARG: A deep learning approach for predicting antibiotic resistance genes from metagenomic data. Microbiome, 6, 23. https://doi.org/10.1186/s40168-018-0401-z

Behzad, H., Gojobori, T., & Mineta, K. (2015). Challenges and opportunities of airborne metagenomics. Genome Biology and Evolution, 7(5), 1216–1226. https://doi.org/10.1093/gbe/evv064

Blanco-Míguez, A., Carloni, S., & Cardenas, C. et al. (2024). Microbial composition associated with biliary stents in patients undergoing pancreatic resection for cancer. Npj Biofilms Microbiomes, 10, 35. https://doi.org/10.1038/s41522-024-00506-8

Boers, S. A., Jansen, R., & Hays, J. P. (2019). Understanding and overcoming the pitfalls and biases of next-generation sequencing (NGS) methods for use in the routine clinical microbiological diagnostic laboratory. European Journal of Clinical Microbiology & Infectious Diseases, 38, 1059–1070. https://doi.org/10.1007/s10096-019-03520-3

Bolger, A. M., Lohse, M., & Usadel, B. (2014). Trimmomatic: A flexible trimmer for illumina sequence data. Bioinformatics, 30(15), 2114–2120. https://doi.org/10.1093/bioinformatics/btu170

Breitwieser, F. P., Baker, D. N., & Salzberg, S. L. (2018). KrakenUniq: Confident and fast metagenomics classification using unique k-mer counts. Genome Biology, 19, 198. https://doi.org/10.1186/s13059-018-1568-0

Cantalapiedra, C. P., Hernández-Plaza, A., Letunic, I., Bork, P., & Huerta-Cepas, J. (2021). eggNOG-mapper v2: Functional annotation, orthology assignments, and domain prediction at the metagenomic scale. Molecular Biology and Evolution, 38(12), 5825–5829. https://doi.org/10.1093/molbev/msab293

Caporaso, J. G., Lauber, C. L., Walters, W. A., Berg-Lyons, D., Huntley, J., Fierer, N., Owens, S. M., Betley, J., Fraser, L., Bauer, M., Gormley, N., Gilbert, J. A., Smith, G., & Knight, R. (2012). Ultra-high-throughput microbial community analysis on the illumina HiSeq and MiSeq platforms. The ISME Journal, 6(8), 1621–1624. https://doi.org/10.1038/ismej.2012.8

Chakraborty, R., Wu, C. H., & Hazen, T. C. (2012). Systems biology approach to bioremediation. Current Opinion in Biotechnology, 23, 483–490. https://doi.org/10.1016/j.copbio.2012.01.015

Chen, G., Chen, C., & Lei, Z. (2017). Meta-omics insights in the microbial community profiling and functional characterization of fermented foods. Trends in Food Science & Technology, 65, 23–31. https://doi.org/10.1016/j.tifs.2017.05.002

Chiu, C. Y., & Miller, S. A. (2019). Clinical metagenomics. Nature Reviews Genetics, 20, 341–355. https://doi.org/10.1038/s41576-019-0113-7

Chorlton, S. D. (2024). Ten common issues with reference sequence databases and how to mitigate them. Frontiers in Bioinformatics, 4, 1278228. https://doi.org/10.3389/fbinf.2024.1278228

Cowan, D. A., Ramond, J. B., Makhalanyane, T. P., & De Maayer, P. (2015). Metagenomics of extreme environments. Current Opinion in Microbiology, 25, 97–102. https://doi.org/10.1016/j.mib.2015.05.005

Daniel, R. (2005). The metagenomics of soil. Nature Reviews Microbiology, 3, 470–478. https://doi.org/10.1038/nrmicro1160

de Abreu, V. A. C., Perdigão, J., & Almeida, S. (2021). Metagenomic approaches to analyze antimicrobial resistance: An overview. In Frontiers in Genetics (Vol. 11). Frontiers Media S.A. https://doi.org/10.3389/fgene.2020.575592

Di Tommaso, P., Chatzou, M., & Floden, E. et al. (2017). Nextflow enables reproducible computational workflows. Nature Biotechnology, 35, 316–319. https://doi.org/10.1038/nbt.3820

Douglas, A. E. (2020). The microbial exometabolome: Ecological resource and architect of microbial communities. Philosophical Transactions of the Royal Society B, 375(1798), 20190250. https://doi.org/10.1098/rstb.2019.0250

Duan, C., Zang, Z., Xu, Y., He, H., Liu, Z., Li, S., & Li, S. Z. (2024). FGBERT: Function-driven pre-trained gene language model for metagenomics. arXiv preprint arXiv:2402.16901.

Federhen, S. (2015). Type material in the NCBI taxonomy database. Nucleic Acids Research, 43, 28 January D1, D1086–D1098. https://doi.org/10.1093/nar/gku1127

Franzosa, E. A., McIver, L. J., & Rahnavard, G. et al. (2018). Species-level functional profiling of metagenomes and metatranscriptomes. Nature Methods, 15, 962–968. https://doi.org/10.1038/s41592-018-0176-y

Gilbert, J. A., Blaser, M. J., Caporaso, J. G., Jansson, J. K., Lynch, S. V., & Knight, R. (2018). Current understanding of the human microbiome. Nature Medicine, 24(4), 392–400. https://doi.org/10.1038/nm.4517

Gloor, G. B., Macklaim, J. M., Pawlowsky-Glahn, V., & Egozcue, J. J. (2017). Microbiome datasets are compositional: And this is not optional. Frontiers in Microbiology, 8, 2224. https://doi.org/10.3389/fmicb.2017.02224

Greenwald, W. W., Klitgord, N., & Seguritan, V. et al. (2017). Utilization of defined microbial communities enables effective evaluation of meta-genomic assemblies. BMC Genomics, 18, 296. https://doi.org/10.1186/s12864-017-3679-5

Grossart, H. P., Massana, R., McMahon, K. D., & Walsh, D. A. (2020). Linking metagenomics to aquatic microbial ecology and biogeochemical cycles. Limnology and Oceanography, 65, S2–S20. https://doi.org/10.1002/lno.11382

Gupta, R., Beg, Q. K., & Lorenz, P. (2002). Bacterial alkaline proteases: Molecular approaches and industrial applications. Applied Microbiology and Biotechnology, 59(1), 15–32. https://doi.org/10.1007/s00253-002-0975-y

Handelsman, J. (2004). Metagenomics: Application of genomics to uncultured microorganisms. Microbiology and Molecular Biology Reviews, 68. https://doi.org/10.1128/mmbr.68.4.669-685.2004

Kanehisa, M., Furumichi, M., Sato, Y., Kawashima, M., & Ishiguro-Watanabe, M. (2023). KEGG for taxonomy-based analysis of pathways and genomes, Nucleic Acids Research, 51(D1), D587–D592, https://doi.org/10.1093/nar/gkac963

Kang, D. D., Froula, J., Egan, R., & Wang, Z. (2015). MetaBAT, an efficient tool for accurately reconstructing single genomes from complex microbial communities. PeerJ, 3, e1165. https://doi.org/10.7717/peerj.1165

Kennedy, J., Marchesi, J. R., & Dobson, A. D. (2008). Marine metagenomics: Strategies for the discovery of novel enzymes with biotechnological applications from marine environments. Microbial Cell Factories, 7(1), 27. https://doi.org/10.1186/1475-2859-7-27

Knights, D., Kuczynski, J., & Charlson, E. et al. (2011). Bayesian community-wide culture-independent microbial source tracking. Nature Methods, 8, 761–763. https://doi.org/10.1038/nmeth.1650

Köster, J., & Rahmann, S. (2012). Snakemake—A scalable bioinformatics workflow engine. Bioinformatics, 28, 2520–2522. https://doi.org/10.1093/bioinformatics/bts480

Kour, D., Rana, K. L., Kaur, T., Singh, B., Chauhan, V. S., Kumar, A., & Gupta, V. K. (2019). Extremophiles for hydrolytic enzyme production: Biodiversity and potential biotechnological applications. Bioprocessing for Biomolecules Production, 321–372. https://doi.org/10.1002/9781119434436.ch16

Kumar, S., Diksha, & Sindhu, S. S. et al. (2023). Harnessing phyllosphere microbiome for improving soil fertility, crop production, and environmental sustainability. Journal of Soil Science and Plant Nutrition, 4719–4764. https://doi.org/10.1007/s42729-023-01397-y

Langmead, B., & Salzberg, S. (2012). Fast gapped-read alignment with Bowtie 2. Nature Methods, 9, 357–359. https://doi.org/10.1038/nmeth.1923

Li, D., Liu, C.-M., Luo, R., Sadakane, K., & Lam, T.-W. (2015). MEGAHIT: An ultra-fast single-node solution for large and complex metagenomics assembly via succinct de Bruijn graph. Bioinformatics, 31(10), 1674–1676. https://doi.org/10.1093/bioinformatics/btv033

Li, H., & Durbin, R. (2009). Fast and accurate short read alignment with Burrows–Wheeler transform. Bioinformatics, 25(14), 1754–1760. https://doi.org/10.1093/bioinformatics/btp324

Li, L. L., McCorkle, S. R., & Monchy, S. et al. (2009). Bioprospecting metagenomes: Glycosyl hydrolases for converting biomass. Biotechnology for Biofuels, 2, 10. https://doi.org/10.1186/1754-6834-2-10

Liang, Q., Bible, P. W., Liu, Y., Zou, B., & Wei, L. (2020). DeepMicrobes: Taxonomic classification for metagenomics with deep learning. NAR Genomics and Bioinformatics, 2(1), lqaa009. https://doi.org/10.1093/nargab/lqaa009

Lloyd-Price, J., Abu-Ali, G., & Huttenhower, C. (2016). The healthy human microbiome. Genome Medicine, 8(1), 51. https://doi.org/10.1186/s13073-016-0307-y

Lloyd-Price, J., Arze, C., & Ananthakrishnan, A. N. et al. (2019). Multi-omics of the gut microbial ecosystem in inflammatory bowel diseases. Nature 569, 655–662. https://doi.org/10.1038/s41586-019-1237-9

Lozupone, C. A., Hamady, M., Kelley, S. T., & Knight, R. (2007). Quantitative and qualitative β diversity measures lead to different insights into factors that structure microbial communities. Applied and Environmental Microbiology, 73. https://doi.org/10.1128/AEM.01996-06

Marcelino, V.R., Holmes, E. C., & Sorrell, T. C. (2020). The use of taxon-specific reference databases compromises metagenomic classification. BMC Genomics, 21, 184. https://doi.org/10.1186/s12864-020-6592-2

Menzel, P., Ng, K., & Krogh, A. (2016). Fast and sensitive taxonomic classification for metagenomics with kaiju. Nature Communications, 7, 11257. https://doi.org/10.1038/ncomms11257

Navgire, G. S., Goel, N., & Sawhney, G. et al. (2022). Analysis and interpretation of metagenomics data: An approach. Biological Procedures Online, 24, 18. https://doi.org/10.1186/s12575-022-00179-7

Nguyen, L. H., Okin, D., & Drew, D. A. et al. (2023). Metagenomic assessment of gut microbial communities and risk of severe COVID-19. Genome Medicine, 15, 49. https://doi.org/10.1186/s13073-023-01202-6

Nurk, S., Meleshko, D., Korobeynikov, A., & Pevzner, P. A. (2017). metaSPAdes: A new versatile metagenomic assembler. Genome Research, 27(5), 824–834. http://www.genome.org/cgi/doi/10.1101/gr.213959.116.

Olivucci, G., Iovino, E., Innella, G., Turchetti, D., Pippucci, T., & Magini, P. (2024). Long read sequencing on its way to the routine diagnostics of genetic diseases. Frontiers in Genetics, 15, 1374860. https://doi.org/10.3389/fgene.2024.1374860

Parks, D. H., Rinke, C., & Chuvochina, M. et al. (2017). Recovery of nearly 8,000 metagenome-assembled genomes substantially expands the tree of life. Nature Microbiology, 2, 1533–1542. https://doi.org/10.1038/s41564-017-0012-7

Pasolli, E., Truong, D. T., Malik, F., Waldron, L., & Segata, N. (2016). Machine learning meta-analysis of large metagenomic datasets: Tools and biological insights. PLoS Computational Biology, 12(7), e1004977. https://doi.org/10.1371/journal.pcbi.1004977

Pérez-Brocal, V., García-López, R., Nos, P., Beltrán, B., Moret, I., & Moya, A. (2015). Metagenomic analysis of Crohn's disease patients identifies changes in the virome and microbiome related to disease status and therapy, and detects potential interactions and biomarkers. Inflammatory Bowel Diseases, 21(11), 2515–2532. https://doi.org/10.1097/MIB.0000000000000549

Pushpam, P. L., Rajesh, T., & Gunasekaran, P. (2011). Identification and characterization of alkaline serine protease from goat skin surface metagenome. AMB Express, 1(1), 3. https://doi.org/10.1186/2191-0855-1-3

Quick, J., Grubaugh, N. D., Pullan, S. T., Claro, I. M., Smith, A. D., Gangavarapu, K., Oliveira, G., Robles-Sikisaka, R., Rogers, T. F., Beutler, N. A., & Burton, D. R. (2017). Multiplex PCR method for MinION and Illumina sequencing of Zika and other virus genomes directly from clinical samples. Nature Protocols, 12(6), 1261–1276. https://doi.org/10.1038/nprot.2017.066

Quince, C. et al. (2017). Shotgun metagenomics, from sampling to analysis. Nature Biotechnology, 35(9), 833–844. https://doi.org/10.1038/nbt.3935

Quince, C., Walker, A., & Simpson, J. et al. (2017). Shotgun metagenomics, from sampling to analysis. Nature Biotechnology, 35, 833–844. https://doi.org/10.1038/nbt.3935

Rinninella, E., Raoul, P., Cintoni, M., Franceschi, F., Miggiano, G. A. D., Gasbarrini, A., & Mele, M. C. (2019). What is the healthy gut Microbiota composition? A changing ecosystem across age, environment, diet, and diseases. Microorganisms, 7(1), 14. https://doi.org/10.3390/microorganisms7010014

Saeys, Y., Inza, I., & Larrañaga, P. (2007). A review of feature selection techniques in bioinformatics. Bioinformatics, 23(19), 2507–2517. https://doi.org/10.1093/bioinformatics/btm344

Scholz, M. B., Lo, C. C., & Chain, P. S. (2012). Next generation sequencing and bioinformatic bottlenecks: The current state of metagenomic data analysis. Current Opinion in Biotechnology, 23(1), 9–15. https://doi.org/10.1016/j.copbio.2011.11.013

Segata, N., Izard, J., & Waldron, L. et al. (2011). Metagenomic biomarker discovery and explanation. Genome Biology, 12, R60. https://doi.org/10.1186/gb-2011-12-6-r60

Segerman, B. (2020). The most frequently used sequencing technologies and assembly methods are in different time segments of the bacterial surveillance and RefSeq genome databases. Frontiers in Cellular and Infection Microbiology, 10, 527102. https://doi.org/10.3389/fcimb.2020.527102

Silva, C. C., Hayden, H., Sawbridge, T., Mele, P., De Paula, S. O., Silva, L. C., & Oliveira, V. M. (2013). Identification of genes and pathways related to phenol degradation in metagenomic libraries from petroleum refinery wastewater. PloS One, 8(4), e61811. https://doi.org/10.1371/journal.pone.0061811

Simon, C., & Daniel, R. (2011). Metagenomic analyses: Past and future trends. Applied and Environmental Microbiology, 77. https://doi.org/10.1128/AEM.02345-10

Sunagawa, S., Mende, D., & Zeller, G. et al. (2013). Metagenomic species profiling using universal phylogenetic marker genes. Nature Methods, 10, 1196–1199. https://doi.org/10.1038/nmeth.2693

Tamang, J. P. (Ed.). (2016). Ethnic Fermented Foods and Alcoholic Beverages of Asia (Vol. 409). Delhi, India: Springer. https://doi.org/10.1007/978-81-322-2800-4

Taylor, S. J. E. et al. (2019). Enabling cloud-based computational fluid dynamics with a platform-as-a-service solution. IEEE Transactions on Industrial Informatics, 15(1), 85–94. https://doi.org/10.1109/TII.2018.2849558.

Torsvik, V., & Øvreås, L. (2002). Microbial diversity and function in soil: From genes to ecosystems. Current Opinion in Microbiology, 5(3), 240–245. https://doi.org/10.1016/S1369-5274(02)00324-7

Truong, D., Franzosa, E., & Tickle, T. et al. (2015). MetaPhlAn2 for enhanced metagenomic taxonomic profiling. Nature Methods, 12, 902–903. https://doi.org/10.1038/nmeth.3589

Turnbaugh, P. J., & Gordon, J. I. (2009). The core gut microbiome, energy balance, and obesity. The Journal of Physiology, 587(17), 4153–4158. https://doi.org/10.1113/jphysiol.2009.174136

Ufarté, L., Laville, É, Duquesne, S., & Potocki-Veronese, G. (2015). Metagenomics for the discovery of pollutant degrading enzymes. Biotechnology Advances, 33(8), 1845–1854. https://doi.org/10.1016/j.biotechadv.2015.10.009

Uritskiy, G. V., DiRuggiero, J., & Taylor, J. (2018). MetaWRAP—A flexible pipeline for genome-resolved metagenomic data analysis. Microbiome, 6, 158. https://doi.org/10.1186/s40168-018-0541-1

Weinstock, G. (2012). Genomic approaches to studying the human microbiota. Nature, 489, 250–256. https://doi.org/10.1038/nature11553

Wick, R. R., Judd, L. M., Gorrie, C. L., & Holt, K. E. (2017). Unicycler: Resolving bacterial genome assemblies from short and long sequencing reads. PLoS Computational Biology, 13(6), e1005595. https://doi.org/10.1371/journal.pcbi.1005595

Wirbel, J., Zych, K., & Essex, M. et al. (2021). Microbiome meta-analysis and cross-disease comparison enabled by the SIAMCAT machine learning toolbox. Genome Biology, 22, 93. https://doi.org/10.1186/s13059-021-02306-1

Wood, D. E., & Salzberg, S. L. (2014). Kraken: Ultrafast metagenomic sequence classification using exact alignments. Genome Biology, 15, R46. https://doi.org/10.1186/gb-2014-15-3-r46

Wu, G. D., Chen, J., Hoffmann, C., Bittinger, K., Chen, Y. Y., Keilbaugh, S. A., & Lewis, J. D. (2011). Linking long-term dietary patterns with gut microbial enterotypes. Science, 334(6052), 105–108. DOI: 10.1126/science. 1208344

Wu, Y.-W., Simmons, B. A., & Singer, S. W. (2016). MaxBin 2.0: An automated binning algorithm to recover genomes from multiple metagenomic datasets. Bioinformatics, 32(4), 605–607, https://doi.org/10.1093/bioinformatics/btv638

Wu, Y. W., Tang, Y. H., Tringe, S. G., Simmons, B. A., & Singer, S. W. (2014). MaxBin: an automated binning method to recover individual genomes from metagenomes using an expectation-maximization algorithm. Microbiome, 2(1), 26. https://doi.org/10.1186/2049-2618-2-26

Xiong, J. (2006). Essential Bioinformatics. Cambridge University Press. New York, USA.

Yen, S., & Johnson, J. S. (2021). Metagenomics: A path to understanding the gut microbiome. Mammalian Genome, 32, 282–296. https://doi.org/10.1007/s00335-021-09889-x

Zhang, L., Chen, F., Zeng, Z., Xu, M., Sun, F., Yang, L., & Xie, Y. (2021). Advances in metagenomics and its application in environmental microorganisms. Frontiers in Microbiology, 12, 766364, https://doi.org/10.3389/fmicb.2021.766364

CHAPTER 4

Feature Extraction and Selection Methods and Bioinformatics on Omics Data to Identify Signatures for Schizophrenia Mental Health Disorder

Pinju Saikia and Karan Mech

4.1 INTRODUCTION

4.1.1 Overview of Mental Health Disorders

Mental health disorders, such as depression [1], stress [2], schizophrenia, and bipolar disorder, are complex conditions that affect mood, cognition, and behavior which arises from a complex interplay of genetic, molecular, and environmental factors. These disorders are not caused by a single factor but rather arise from a multifactorial interplay of genetic, molecular, and environmental influences. Understanding these factors is crucial for improving diagnosis, treatment, and prevention strategies. The advent of high-throughput omics technologies—including genomics, transcriptomics, proteomics, and metabolomics—has provided new opportunities to explore the underlying molecular mechanisms of these disorders. However, the vast volume and complexity of omics data present significant challenges in identifying robust and interpretable molecular signatures. This chapter introduces computational and bioinformatics tools for feature extraction and selection to facilitate the discovery of molecular biomarkers for schizophrenia, which is a severe psychiatric disorder with hallucinations, delusions, and cognitive dysfunction. It is also a complex disorder with genetic, transcriptomic, epigenomic,

DOI: 10.1201/9781003591658-4

proteomic, and metabolomic influences. Omics technologies help in identifying molecular signatures associated with the disorder: Genomic studies have shown that schizophrenia is primarily a polygenic disorder, which means that multiple genes contribute to an individual's risk.

Schizophrenia (SCZ) is a disorder of the brain in which various individuals have various symptoms, including hallucination—perceiving something (e.g., voices, visual images) that isn't actually present, delusion—a strongly held false belief that persists despite clear evidence to the contrary, flat-talk—a reduced expression of emotions, often seen as a monotone voice or a lack of facial expressions, disorganized thinking—difficulty organizing thoughts logically, which may manifest as incoherent speech, jumping between topics, or difficulty maintaining a conversation., etc. In the long run, it can lead to serious consequences and reduce life expectancy by over a decade. Hence, early and correct diagnosis of SCZ is common, and modalities such as structural magnetic resonance imaging (sMRI), functional MRI (fMRI), diffusion tensor imaging (DTI), and electroencephalogram (EEG) help to witness the patients' brain abnormalities. Moreover, for accurate diagnosis of SCZ, researchers have used machine learning (ML) algorithms for the past decade to distinguish the brain patterns of healthy and SCZ brains using MRI and fMRI images [3]. Around 20 million people worldwide are affected by this disease. Its symptoms are categorized into positive and negative [2]. Positive symptoms include hallucinations, delusions, disorganized thinking, etc., and negative symptoms are the ones that impact a person's workability, lack of motivation, inability to concentrate, reduced range of emotions, etc. Early diagnosis of the SCZ helps slow down the disease's progression by treating it at the initial stage [3].

4.1.2 Role of Genetic, Molecular, and Environmental Factors in Mental Health Disorders

Mental health disorders arise from complex interactions between genetic predispositions, molecular changes, and environmental exposures. These are influenced by a combination of genetic, molecular, and environmental factors. Understanding these influences is crucial for identifying biomarkers and developing targeted treatments.

1. Genetic Factor: Genetics plays a significant role in mental health, with many disorders showing hereditary patterns.
    - Genome-Wide Association Studies (GWAS): Identify genetic variations linked to disorders like schizophrenia and depression.
    - Candidate Gene Studies: Focus on specific genes like COMT, BDNF, and 5-HTTLPR, which are associated with cognitive function and neurotransmitter regulation.
    - Polygenic Risk Scores (PRS): Predict an individual's genetic susceptibility based on multiple genetic variants.

2. Molecular Factors: Molecular-level changes influence brain function and contribute to mental disorders:
   - Epigenetics: DNA methylation and histone modifications regulate gene expression without altering DNA sequences, affecting vulnerability to disorders.
   - Neurotransmitter Imbalance: Altered levels of serotonin, dopamine, and glutamate impact mood and cognition.
   - Proteomics and Metabolomics: Changes in protein expression and metabolic pathways provide insights into disease mechanisms.
3. Environmental Factors: Environmental influences interact with genetic and molecular factors to shape mental health outcomes:
   - Stress and Trauma: Early life stress, abuse, and social adversity increase disorder risk.
   - Diet and Lifestyle: Nutritional deficiencies, lack of physical activity, and substance abuse can contribute to mental health issues.
   - Exposure to Toxins: Pollutants and drugs can trigger epigenetic changes leading to mental disorders.

Integrating omics data from genomics, transcriptomics, and epigenomics can help identify biomarkers and develop personalized medicine approaches for mental health treatment.

4.1.3 Importance of Omics Data

Omics technologies provide a comprehensive and high-resolution approach for studying biological systems. Omics data integration enables a multi-layered approach to mental health research, helping to: Identify biomarkers for early diagnosis, understand disease mechanisms at the molecular level, and develop targeted therapies based on individual genetic and molecular profiles. In mental health research, genomics, transcriptomics, proteomics, and metabolomics help uncover the molecular mechanisms underlying disorders such as schizophrenia, depression, and bipolar disorder.

1. Genomics: Understanding genetic risk genomics focuses on DNA sequences and genetic variations that contribute to mental health disorders.
   - Genome-Wide Association Studies (GWAS): Identify genetic variants linked to psychiatric conditions.
   - Single Nucleotide Polymorphisms (SNPs): Specific SNPs in genes like BDNF, COMT, and DISC1 are associated with disorders.
   - Copy Number Variations (CNVs): Large structural changes in DNA influence neurodevelopment and cognition.
   - Relevance: Helps in identifying genetic risk factors and developing personalized medicine strategies.

2. Transcriptomics: Measuring gene expression transcriptomics studies RNA transcripts to understand gene activity.
   - RNA-Sequencing (RNA-Seq): Identifies differentially expressed genes in psychiatric conditions.
   - Long Non-Coding RNAs (lncRNAs): Regulate gene expression and are implicated in neuropsychiatric disorders.
   - Alternative Splicing Events: Influence neural function and may be dysregulated in schizophrenia and autism.
   - Relevance: Helps identify disease-specific gene expression patterns for diagnostics and treatment.
3. Proteomics: Identifying dysfunctional proteins proteomics examines protein expression, modifications, and interactions.
   - Mass Spectrometry-Based Proteomics: Detects altered protein levels in patients vs. controls.
   - Neurotransmitter-Related Proteins: Dysregulation of dopamine, serotonin, and glutamate pathways is linked to mood disorders.
   - Biomarker Discovery: Identifies proteins in cerebrospinal fluid or blood that serve as biomarkers.
   - Relevance: Provides insights into cellular pathways involved in mental disorders.
4. Metabolomics: Understanding biochemical changes metabolomics investigates small molecules (metabolites) that reflect cellular activity.
   - Neurotransmitter Metabolites: Imbalances in dopamine, serotonin, and GABA are linked to mental health disorders.
   - Lipidomics and Gut-Brain Axis: Metabolic disturbances and gut microbiome interactions affect brain function.
   - Energy Metabolism: Altered glucose metabolism is observed in schizophrenia and depression.
   - Relevance: Reveals metabolic biomarkers for disease progression and drug response.

4.1.4 Challenges in Handling High-Dimensional Omics Data

Since Omics data is high-dimensional, it contains thousands to millions of features (e.g., genes, proteins, metabolites). The handling of such complex data presents several challenges, such as data complexity, noise, and computational burden.

Data Complexity: The number of features (e.g., genes) far exceeds the number of samples, leading to overfitting in ML models. Redundant or irrelevant features can obscure meaningful patterns. Feature selection and dimensionality reduction (e.g., PCA, UMAP, NMF) help retain essential information.

- Noisy and incomplete data experimental errors, sequencing biases, and missing values can affect data quality. It leads to incorrect conclusions or reduced model accuracy. Data preprocessing techniques such as normalization, imputation, and batch effect correction help mitigate noise.
- Data integration complexity combining multiple omics layers (e.g., genomics + proteomics) requires handling different data types and scales. Integrating heterogeneous data without information loss is challenging. Computational tools like multi-omics factor analysis (MOFA) and similarity network fusion (SNF) help in multi-omics integration.
- Computational and storage demands of large-scale omics datasets require high-performance computing (HPC) resources. Processing gigabytes to terabytes of data is time-consuming and costly. Cloud computing, parallel computing, and efficient algorithms help manage computational loads.
- Interpretability of results: ML models applied to omics data may lack biological interpretability. Difficult to translate computational findings into biological insights. Using explainable AI (XAI) and biological pathway analysis (e.g., Kyoto Encyclopedia of Genes and Genomes [KEGG], gene ontology [GO] analysis) improves interpretability.
- Reproducibility and Standardization: Different datasets may produce inconsistent results due to variations in sample collection, sequencing platforms, or preprocessing methods. Lack of reproducibility makes it hard to validate findings across studies. Establishing standardized protocols, data sharing practices, and benchmarking datasets ensures reproducibility.

Effective management of high-dimensional omics data relies on advanced computational strategies, high-quality data preprocessing, and biological verification. These challenges will be addressed to uncover good biomarkers and novel therapeutic targets for studying mental illness.

4.2 HIGH-DIMENSIONAL DATA AND FEATURE EXTRACTION

4.2.1 Importance of Feature Extraction in Omics Data Analysis

Omics data (e.g., genomics, transcriptomics, proteomics) is typically high-dimensional, because they contain thousands to millions of features (e.g., genes, proteins, or metabolites). Directly analyzing such data can lead to challenges like:

- Curse of dimensionality: As the number of features increases, computational complexity and noise also increase, making analysis difficult.
- Overfitting: ML models may learn noise rather than meaningful patterns if too many features are used.
- Interpretability issues: High-dimensional data can obscure biologically meaningful patterns.

Feature extraction reduces dimensionality while preserving key information, improving model performance, reducing computational costs, and aiding interpretation. Common techniques include principal component analysis (PCA), uniform manifold approximation and projection (UMAP), and non-negative matrix factorization (NMF), each with distinct strengths and weaknesses.

4.2.2 Dimensionality Reduction Techniques

4.2.2.1 Principal Component Analysis (PCA)

PCA is a linear dimensionality reduction method that originated from the notion of a linear subspace in 1901 by Karl Pearson. It is a method of converting high-dimensional data that is represented by the vectors transformed to the new coordinate system, uncorrelated principal components (PCs). These PCs contain most of the variance in the data; the remaining part can be safely removed. In simple words, PCA is the technique which is used to make a large but complex dataset simple by changing its way and hence its features will become less but at the same time, it will keep the main. Maybe important information as it is.

The steps are as follows:

1. Data Preprocessing (Standardization): If X is a collection of m tests and n classes, then one would have to normalize the data to have a zero mean and a unit variance such that:

$$X_{\text{scaled}} = \frac{X - \mu}{\sigma}$$

where μ is the average of all the unrepeatable data:

$$\mu = \{\mu_1, \mu_2, \ldots\ldots, \mu m\}$$

σ is a typical deviation of the uncorrelated features:

$$\sigma = \{\sigma_1, \sigma_2, \ldots\ldots, \sigma m\}$$

$$X_{\text{scaled}} = \frac{X - \mu}{\sigma}$$

where X_{scaled} is the scaled version of the original data matrix X.
And by making the features have a mean equal to zero, standardization guarantees that all features have specified weight in PCA.
They are transformed this way, respectively:

$$\text{Mean} = 0, \quad \text{Standard Deviation} = 1$$

2. Compute the Covariance Matrix: The covariance matrix stores some features' relation. It calculates the data of 2–8 items to reveal how they all move together. The covariance calculator allows us to measure the amount of a proportional relationship

in two or more datasets, their mutual inclination with respect to each other being stated. The covariance can be calculated using the following formula. The covariance for two features x_1 and x_2, the covariance is:

$$cov(x_1, x_2) = \frac{\sum_{i=1}^{n}(x_{1i} - \overline{x_1})(x_{2i} - \overline{x_2})}{n-1}$$

where $\overline{x_1}$ and $\overline{x_2}$ are the means of x_1 and x_2, respectively, and n is the number of data points.

x_{1i} and x_{2i} are values of feature 1 and feature 2 for sample i.

The value of covariance can be:

- Positive: when x_1 increases, x_2 also increases.
- Negative: when x_1 increases, x_2 decreases.
- Zero: when no direct relationship between x_1 and x_2.

3. Find the Magic Directions (Principal Components): PCA identifies new axes (like rotating a camera) where the data spreads out the most:

- 1st Principal Component (PC1): The direction of maximum variance (most spread).
- 2nd Principal Component (PC2): The next best direction, perpendicular to PC1, and so on.

These directions are calculated using Eigenvalues and Eigenvectors where: eigenvectors (math tools that find these axes), and their importance is ranked by eigenvalues (how much variance each captures). For a square matrix A, an eigenvector X (a non-zero vector) and its corresponding eigenvalue λ (a scalar) satisfy:

$$AX = \lambda X$$

This means:

- When A acts on X, it only stretches or shrinks X by the scalar λ.
- The direction of X remains unchanged (hence, eigenvectors define "stable directions" of A). It can also be written as:

$$AX - \lambda X = 0$$

$$(A - \lambda I)X = 0$$

where I is the identity matrix of the same shape as matrix A. The above conditions will be true only if $(A - \lambda I)$ is non-invertible (i.e., a singular matrix). That means,

$$A - \lambda I| = 0$$

This determinant equation is called the characteristic equation. Solving it gives the eigenvalues λ, and the corresponding eigenvectors can be found using the equation:

$$AX = \lambda X$$

It connects with PCA by the following:

- In PCA, the covariance matrix C (from Step 2) acts as matrix A.
- Eigenvectors of C are the PCs.
- Eigenvalues represent the variance captured by each PC.

4. Pick the Top Directions and Transform Data: Keep only the top 2–3 directions (or enough to capture ~95% of the variance).

   Project the data onto these directions to get a simplified, lower-dimensional version. Let's see with help of visual image where x-axis (Radius) and y-axis (Area) represent two original features in the dataset as depicted in Figure 4.1 [4].

   Principal Components (PCs):

- PC_1 (First Principal Component): The direction along which the data has the maximum variance. It captures the most significant information.
- PC_2 (Second Principal Component): The direction orthogonal (perpendicular) to PC_1. It captures the remaining variance but is less significant.

  Now, the red dashed lines indicate the spread (variance) of data along different directions. The variance along PC_1 is greater than PC_2, which means that PC_1 carries more useful information about the dataset.
- The data points (blue dots) are projected onto PC_1, effectively reducing the dataset from two dimensions (radius, area) to one dimension (PC_1).
- This transformation simplifies the dataset while retaining most of the original variability.

  Why PCA chooses the direction with the biggest variance (PC_1) is illustrated graphically in the image. We preserve important information while reducing redundancy by eliminating PC_2. Data compression, visualization, and enhanced model performance are all aided by the transformation.

4.2.2.2 Uniform Manifold Approximation and Projection (UMAP)

UMAP is a dimension reduction method that facilitates visualization and analysis of high-dimensional data by capturing both local and global data structures in a lower-dimensional space. It has been compared to t-SNE (t-distributed Stochastic Neighbor Embedding) because of its similar usage in data visualization but with a number of advantages, such as improved global data structure preservation and faster computation.

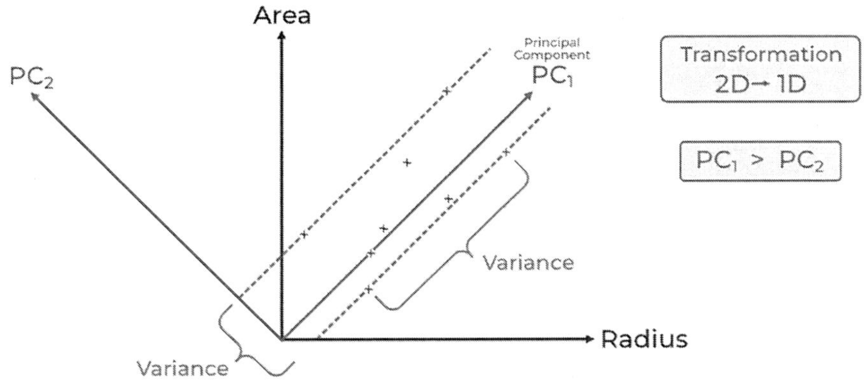

FIGURE 4.1 Transforming a 2D dataset into a 1D representation while maximizing variance preservation using principal component analysis (PCA).

The UMAP algorithm can be broken down into two main phases:

1. Constructing the Fuzzy Topological Representation:
   - Nearest Neighbor Search: The UMAP goes about finding nearest neighbors of every point. This will typically be done using approximate nearest neighbor algorithms in an effort to speed it up. Simplicial Set: The fuzzy simplicial set is built out of the closest neighbors. The set preserves the local connectivity of the points.
   - Fuzzy Membership Strengths: Membership strengths are located on the edges of the simplicial set, depicting the likelihood that two points are connected.
2. Optimizing the Low-Dimensional Embedding:
   - Entropy Optimization: Optimization of low-dimensional embedding to reduce the cross-entropy of the fuzzy simplicial set in the high-dimensional space with respect to the low-dimensional space. Stochastic Gradient Descent: UMAP optimizes using stochastic gradient descent. It does so by iteratively moving the points in the low-dimensional space to reach a better fit to the fuzzy topological structure.

4.2.2.3 Non-Negative Matrix Factorization (NMF)

NMF is a technique of factorizing a large dataset into small informative fragments in such a way that all the values are non-negative. It helps in the extraction of informative features from data and data analysis and data processing simplification. It is a dimensionality reduction technique that factorizes a non-negative data matrix into two low-dimensional non-negative matrices. It is particularly useful in the discovery of hidden patterns in high-dimensional datasets.

For a matrix A of dimensions $m \times n$ where each element is non-negative $(A \geq 0)$, NMF factorizes it into two matrices as in Figure 4.2 [5].

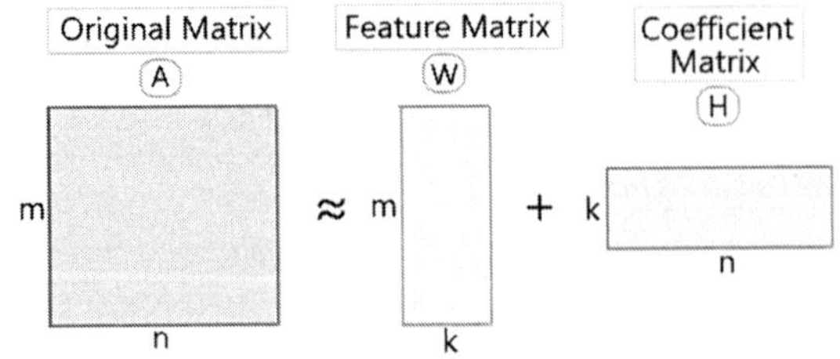

FIGURE 4.2 NMF intuition.

W and H with dimensions $m \times k$ and $k \times n$ respectively, where both matrices contain only non-negative elements:

$$A_{m \times n} \approx W_{m \times k} H_{k \times n}$$

where:

- $A \rightarrow$ Original input matrix (linear combination of W and H.
- $W \rightarrow$ Basis elements of feature matrix.
- $H \rightarrow$ Coefficient matrix (the weights for W).
- $k \rightarrow$ rank (dimension of the compressed description where $k \leq \min(m,n)$.

NMF helps identify hidden patterns in data by assuming that each data point can be represented as a combination of fundamental features found in W.

TABLE 4.1 Comparison of Feature Extraction Techniques

Technique	Advantages	Limitations
Principal component analysis (PCA)	• Linear transformation—finds axes that maximize variance. • Fast and computationally efficient—works well for large datasets. • Interpretable—Eigenvalues show variance explained by each component.	• Assumes linear relationships—cannot capture complex nonlinear structures. • Sensitive to scaling—needs data standardization. • May lose local structure—focuses only on maximizing variance.
Uniform manifold approximation and projection (UMAP)	• Captures nonlinear patterns—preserves both local and global structures. • Better visualization—often produces clearer clusters than PCA. • Faster than t-SNE—scales well with large datasets.	• Hyperparameter sensitive—different settings can lead to different embeddings. • Not deterministic—May give slightly different results on repeated runs. • Difficult to interpret, Embeddings do not directly correspond to variance.
Non-negative matrix factorization (NMF)	• Non-negativity constraint—produces meaningful, interpretable components. • Useful for parts-based learning. • Common in gene expression and text analysis. • Can be used for clustering coefficient matrix can help classify data.	• Requires non-negative data, cannot handle negative values. • May converge to local minima—depends on initialization. • Slower than PCA—iterative optimization process.

4.2.2.4 Comparative Advantages and Limitations of Each Technique

The section below summarizes the key advantages and limitations of each technique for clear comparison (Table 4.1).

4.3 FEATURE SELECTION METHODS

4.3.1 Statistical Filtering Approach

Statistical filtering approaches are particularly useful for reducing the dimensionality of large datasets in schizophrenia research. For example, genome-wide association studies (GWAS) and neuroimaging studies often generate thousands of features (e.g., SNPs, brain regions), making dimensionality reduction essential.

4.3.1.1 t-test

The t-test is used to identify whether the mean value of a characteristic differs significantly between two groups (e.g., schizophrenia patients vs healthy controls).

It can be represented as $t = \frac{\left\{\overline{\{X\}_1} - \overline{\{X\}_2}\right\}}{\left\{\sqrt{\left\{\frac{\{s_1^2\}}{\{n_1\}} + \frac{\{s_2^2\}}{\{n_2\}}\right\}}\right\}}$

where

- $\overline{x1}, \overline{x2}$, is the mean values of the feature in two groups.
- s_1^2, s_2^2 are variance of the feature in each group.
- n_1, n_2 are sample sizes in each group.
- It Works by:
  - The null hypothesis states that the means of the two groups are equal.
  - High t-value and low p-value indicate that the feature is significantly different between groups.
  - Features with significant differences are retained for further analysis. Example in Schizophrenia:
  - Identifying differences in gray matter volume between patients with schizophrenia and healthy controls.
  - Finding differences in gene expression levels between affected and unaffected individuals.

4.3.1.2 ANOVA

ANOVA extends the t-test to compare means across three or more groups. In schizophrenia research, this can be useful to compare characteristics between multiple diagnostic subgroups or treatment responses.

It is represented as

$$F = \frac{\{Between - group\ variance\}}{\{Within - group\ variance\}}$$

It works by:

- The null hypothesis states that the means of all groups are equal.
- A high F-statistic and low p-value indicate significant differences between groups.
- Features with significant differences are retained.

Example in Schizophrenia:

- Comparing brain connectivity patterns across schizophrenia subtypes (e.g., paranoid, disorganized, catatonic).
- Comparing cognitive test scores between medicated, unmedicated, and control groups.

4.3.1.3 Chi-square

The Chi-square test measures the association between two categorical variables. It evaluates whether the presence of one feature is dependent on the value of the target variable.

It is represented as

$$\chi^2 = \frac{\sum\{(O_i - E_i)^2\}}{\{E_i\}}$$

where O_i = observed frequency and E_i = expected frequency.

It works by:

- The null hypothesis states that the two variables are independent.
- A high Chi-Square value and low p-value indicate dependency between the feature and the target variable.

Example in Schizophrenia:

- Testing the association between genetic mutations (e.g., SNPs) and schizophrenia diagnosis.
- Analyzing the relationship between smoking status and schizophrenia onset.

4.3.1.4 Mutual Information

Mutual information measures how much information one variable provides about another. It captures both linear and nonlinear relationships.

$$I(X;Y) = \sum_{x \in X} \sum_{y \in Y} P(x,y) \log \frac{\{P(x,y)\}}{\{P(x)P(y)\}}$$

How it works:

- High mutual information indicates that knowing the value of one variable reduces uncertainty about the other.
- Features with high mutual information are retained.

Example in Schizophrenia:

- Identifying brain regions whose activity is predictive of schizophrenia symptoms.
- Finding genetic markers that explain significant variance in schizophrenia risk.

4.3.2 Wrapper-based Methods

Wrapper methods use a predictive model to evaluate subsets of features, selecting those that optimize model performance. These methods are computationally expensive but are more effective at capturing feature interactions.

4.3.2.1 Recursive Feature Elimination (RFE)

RFE is a backward elimination method that recursively removes the least important feature based on model performance. How it works:

- Train a model (e.g., SVM, Logistic Regression) using all features.
- Rank features based on their contribution to model accuracy.
- Remove the least important feature.
- Repeat until the optimal feature set is found.

Example in Schizophrenia:

- Using RFE with an SVM to identify the most predictive combination of brain activity patterns for diagnosing schizophrenia.
- Applying RFE with logistic regression to find genetic variants linked to schizophrenia.

4.3.3 Embedded Methods

Embedded methods select features during model training by incorporating feature importance into the learning process. They are often more efficient than wrapper methods.

4.3.3.1 Least Absolute Shrinkage and Selection Operator (LASSO)

LASSO adds L1 regularization to the loss function, shrinking some coefficients to zero and effectively removing those features.

$$L = \sum (y_i - \hat{y}_i)^2 + \lambda \sum |w_i|$$

Example in Schizophrenia:

- Selecting the most important genetic markers from high-dimensional SNP data.
- Identifying brain regions whose activity patterns are most predictive of schizophrenia.

4.3.3.2 Random Forest Feature Selection

Random Forest ranks features based on their contribution to reducing impurity (e.g., Gini index).

How it works:

- Train a Random Forest model.
- Compute feature importance based on how much each feature improves node splits.
- Retain the top-ranked features.

Example in Schizophrenia: Identifying the most predictive SNPs from GWAS data, Selecting brain regions most associated with schizophrenia symptoms.

4.3.4 Importance of Selecting Relevant Biomarkers for Mental Health Disorders

Biomarkers are measurable biological, psychological, or behavioral indicators that provide insight into the underlying mechanisms of mental health disorders.

Selecting relevant biomarkers is essential for improving diagnosis, treatment, and understanding of complex psychiatric conditions like schizophrenia, depression, bipolar disorder, and anxiety.

Effective feature selection methods help identify the most meaningful biomarkers from high-dimensional datasets, including genetic, neuroimaging, and clinical data.

4.3.4.1 Understanding the Biological Basis of Mental Health Disorders

Mental health disorders arise from complex interactions between genetic, environmental, and neurobiological factors. Identifying relevant biomarkers helps uncover the biological underpinnings of these disorders:

- Genetic Biomarkers: Genome-wide association studies (GWAS) have linked variations in genes like DISC1 and COMT to schizophrenia risk.
- Neuroimaging Biomarkers: Structural and functional MRI studies reveal reduced gray matter in the prefrontal cortex and altered functional connectivity in the default mode network (DMN) in schizophrenia patients.
- Neurochemical Biomarkers: Elevated dopamine activity in the striatum is a hallmark of psychosis in schizophrenia.

Understanding these biological pathways allows researchers to classify subtypes of disorders and identify targets for treatment.

4.3.4.2 Improving Diagnosis and Early Detection

Mental health diagnoses are often based on subjective clinical assessments, leading to misdiagnosis and delayed treatment. Biomarkers provide objective measures that improve diagnostic accuracy and consistency:

- Abnormal functional connectivity patterns in the DMN distinguish schizophrenia from other psychotic disorders.
- Low brain-derived neurotrophic factor (BDNF) levels are linked to depression severity.
- Genetic risk scores combining multiple SNPs improve early prediction of schizophrenia onset. Accurate early diagnosis enables timely intervention, reducing long-term symptom severity.

4.3.4.3 Personalizing Treatment and Predicting Response

Not all patients respond equally to the same treatment. Biomarkers can predict treatment outcomes, enabling personalized medicine:

- Genetic variations in the CYP2D6 enzyme influence metabolism of antipsychotic drugs.
- Dopamine receptor density, measured through PET imaging, predicts the effectiveness of antipsychotic medications.
- EEG patterns (e.g., P300 amplitude) predict response to transcranial magnetic stimulation (TMS) in depression.
- Tailoring treatments to an individual's biomarker profile improves efficacy and reduces side effects.

4.4 APPLICATION OF HIGH-THROUGHPUT OMICS TECHNOLOGIES IN UNDERSTANDING SCHIZOPHRENIA

High-throughput omics technologies have transformed our understanding of complex psychiatric disorders like schizophrenia by enabling the large-scale analysis of biological molecules. These technologies provide insights into the genetic, transcriptomic, epigenetic, proteomic, and metabolomic alterations associated with schizophrenia, helping to identify biomarkers, disease mechanisms, and potential therapeutic targets. Here's a structured overview of how different omics approaches are applied to schizophrenia research.

4.4.1 Genomics: Identifying Genetic Risk Factors

Genomics focuses on DNA-level variations that contribute to schizophrenia risk.

4.4.1.1 Genome-Wide Association Studies (GWAS)

- Large-scale GWAS have identified over 100 loci associated with schizophrenia.
- Variants in genes related to synaptic function (e.g., CACNA1C, GRIN2A) and immune response (e.g., MHC region) have been linked to schizophrenia.
- Polygenic risk scores (PRS) help estimate genetic vulnerability based on cumulative risk from multiple loci.

4.4.1.2 *Whole-Genome and Whole-Exome Sequencing*

- Rare variants and de novo mutations have been identified in genes involved in synaptic plasticity and neurodevelopment (e.g., NRXN1, SETD1A).
- Structural variations like copy number variants (CNVs) have been linked to schizophrenia risk (e.g., 22q11.2 deletion).

4.4.2 Transcriptomics

Transcriptomics explores the expression of genes in different tissues and cell types.

4.4.2.1 *RNA Sequencing (RNA-seq)*

- Altered expression of genes involved in synaptic transmission, glutamate signaling, and inflammatory pathways has been reported in schizophrenia.
- Differential gene expression in the prefrontal cortex and hippocampus highlights disrupted neural circuitry.
- Single-cell RNA sequencing (scRNA-seq) reveals cell-type-specific changes in gene expression, showing altered activity in GABAergic interneurons and glutamatergic neurons.

4.4.2.2 *Alternative Splicing and Non-Coding RNA*

- Dysregulation of alternative splicing patterns and increased levels of specific microRNAs (miRNAs) have been implicated in schizophrenia.
- For example, miR-137 is a key regulator of synaptic plasticity and a GWAS-identified risk locus.

4.4.3 Epigenomics

Epigenomics examines modifications that regulate gene expression without altering the DNA sequence.

4.4.3.1 *DNA Methylation*

- Hypermethylation and hypomethylation patterns have been found in schizophrenia-associated genes (e.g., Brain-Derived Neurotrophic Factor (BDNF), Catechol-O-Methyltransferase (COMT)).
- Differences in methylation profiles have been identified in postmortem brain tissue, particularly in genes regulating neurodevelopment and stress response.

4.4.3.2 *Histone Modifications*

- Changes in histone acetylation and methylation have been linked to altered gene expression in schizophrenia.
- Targeting histone-modifying enzymes is being explored as a therapeutic strategy.

4.4.4 Proteomics

Proteomics focuses on identifying and quantifying protein expression levels and post-translational modifications.

4.4.4.1 Mass Spectrometry-Based Proteomics

- Abnormal levels of proteins involved in synaptic function, oxidative stress, and inflammation have been identified in cerebrospinal fluid and brain tissue.
- Reduced levels of parvalbumin and altered levels of calcium-binding proteins suggest impaired inhibitory signaling.

4.4.4.2 Protein-Protein Interaction Networks

- Mapping protein interaction networks reveals disrupted signaling pathways, such as the dopaminergic, glutamatergic, and GABAergic systems.
- Dysregulated immune-related proteins suggest a link between schizophrenia and neuroinflammation.

4.4.5 Metabolomics

Metabolomics identifies metabolic changes associated with schizophrenia, reflecting underlying biochemical disturbances.

4.4.5.1 Nuclear Magnetic Resonance (NMR) and Mass Spectrometry

- Altered levels of metabolites related to energy metabolism (e.g., glucose, lactate) and lipid metabolism have been observed in schizophrenia.
- Reduced levels of glutathione and increased markers of oxidative stress reflect mitochondrial dysfunction.
- Dysregulation of the kynurenine pathway suggests an imbalance in tryptophan metabolism linked to neuroinflammation.

4.4.5.2 Multi-Omics Integration

Combining omics data (genomics, transcriptomics, proteomics, metabolomics, and epigenomics) provides a more comprehensive view of schizophrenia.

- eQTL analysis links genetic variants to gene expression changes.
- Proteogenomics integrates protein and genomic data to identify functional variants.
- ML approaches are used to identify multi-omic signatures predictive of schizophrenia onset, progression, and treatment response.

4.5 MONITORING DISEASE PROGRESSION AND RELAPSE IN SCHIZOPHRENIA

Schizophrenia is a severe and chronic mental health disorder characterized by disturbances in thought, perception, emotion, and behavior. While antipsychotic medications and psychosocial interventions can help manage symptoms, schizophrenia remains a highly relapsing and progressive condition. Monitoring disease progression and relapse is critical for improving long-term outcomes, personalizing treatment, and preventing functional decline. This chapter explores the biological and clinical markers used to track disease progression and predict relapse in schizophrenia. It also discusses the role of advanced computational methods, including feature selection and ML, in identifying the most relevant biomarkers for effective monitoring.

4.5.1 Understanding Disease Progression and Relapse in Schizophrenia

4.5.1.1 What Is Disease Progression in Schizophrenia?

Schizophrenia typically emerges in late adolescence or early adulthood and progresses through several phases:

1. Prodromal Phase: Subtle cognitive and behavioral changes before the onset of full psychotic symptoms.
2. Acute Phase: Characterized by positive symptoms such as hallucinations, delusions, and disorganized thinking.
3. Residual/Chronic Phase: Persistent negative symptoms (e.g., anhedonia, social withdrawal) and cognitive deficits. Progression in schizophrenia is associated with:
    - Progressive gray matter loss in the prefrontal cortex and hippocampus.
    - Worsening functional connectivity between brain regions involved in executive function and emotional regulation.
    - Cognitive decline, particularly in working memory, attention, and executive function.

4.5.1.2 What Is Relapse in Schizophrenia?

Relapse refers to the return or worsening of psychotic symptoms after a period of remission or symptom control. Relapse is common in schizophrenia, with relapse rates as high as 80% within five years of diagnosis if untreated. Clinical features of relapse:

- Return of hallucinations and delusions.
- Increased social withdrawal and emotional blunting.
- Agitation, hostility, and disorganized behavior.

Triggers for relapse:

- Nonadherence to antipsychotic medications.
- High levels of life stress and substance abuse.
- Inadequate social support and high expressed emotion (EE) in family dynamics.

4.5.2 Biological Markers of Disease Progression and Relapse

4.5.2.1 Structural and Functional Brain Changes

Progressive structural and functional brain changes are well-documented in schizophrenia. These changes serve as critical biomarkers for tracking disease progression:

- Gray Matter Loss:
  - Progressive reduction in gray matter volume in the prefrontal cortex, temporal lobes, and hippocampus is associated with cognitive decline and negative symptom severity.
  - Longitudinal MRI studies show that gray matter loss is most pronounced in the early years after diagnosis.
- White Matter Integrity:
  - Reduced fractional anisotropy (FA) in DTI reflects impaired white matter integrity in the corpus callosum and cingulum.
  - White matter disruptions are linked to impaired cognitive processing speed and working memory.
- Functional Connectivity:
  - Hypoconnectivity within the default mode network (DMN) and hyperconnectivity in the salience network are hallmarks of schizophrenia.
  - Changes in functional connectivity are correlated with symptom severity and cognitive dysfunction.

4.5.2.2 Neurochemical Markers

Neurotransmitter dysfunction plays a central role in schizophrenia progression and relapse:

- Dopamine Dysfunction:
  - Elevated dopamine synthesis capacity in the striatum is linked to positive symptoms (e.g., hallucinations and delusions).
  - Positron Emission Tomography (PET) imaging studies show increased presynaptic dopamine function during acute psychotic episodes.

- Glutamate Hypofunction:
    - Reduced glutamate activity in the prefrontal cortex and hippocampus contributes to negative symptoms and cognitive deficits.
    - N-Methyl-D-Aspartate (NMDA) receptor dysfunction is implicated in working memory and executive function impairments.
- GABAergic (Gamma-Aminobutyric Acid) Deficits:
    - Reduced cortical GABA levels are associated with impaired cognitive control and sensory processing.

4.5.2.3 Inflammatory and Immune Markers

Chronic inflammation and immune dysregulation are increasingly recognized as contributors to schizophrenia progression:

1. Cytokine Levels:
    - Elevated levels of pro-inflammatory cytokines (e.g., IL-6, TNF–α) are linked to symptom severity and relapse risk.
    - Increased peripheral inflammation is correlated with cognitive decline and emotional dysregulation.
2. Microglial Activation:
    - PET imaging studies show increased microglial activation in the brains of individuals with schizophrenia.
    - Microglial overactivation may contribute to synaptic pruning and gray matter loss.

4.5.2.4 Electrophysiological Markers

Abnormalities in brain electrical activity provide real-time insights into disease progression and relapse risk:

- EEG Biomarkers:
    - Reduced P300 amplitude reflects impaired cognitive processing and attention.
    - Increased theta power and reduced gamma synchrony are linked to impaired working memory.
- Event-Related Potentials (ERPs):
    - Deficits in mismatch negativity (MMN) and auditory steady-state response (ASSR) reflect impaired sensory processing.
    - ERP abnormalities predict poor functional outcomes and increased relapse risk.

4.5.3 Behavioral and Cognitive Markers

In addition to biological markers, changes in behavior and cognition often precede clinical relapse:

- Sleep Disturbances:
  - Disrupted sleep architecture, including reduced slow-wave sleep, is linked to increased psychotic symptoms.
  - Actigraphy studies show reduced sleep efficiency before relapse episodes.
- Social Withdrawal:
  - Decreased social engagement and increased isolation are early indicators of relapse risk.
  - Digital monitoring of social activity through smartphones can help detect these changes.
- Impaired Cognitive Function:
  - Decline in working memory and executive function often precedes relapse.
  - Computer-based cognitive testing can identify early cognitive deterioration.

4.5.4 Examples of Multi-Omics Approaches in Mental Health Research

Multi-omics means combining different types of biological data (DNA, RNA, proteins, and metabolites) to understand mental health disorders better. Here are some examples of how this approach is used in research:

- Genes + RNA: Finding Risk Factors for Schizophrenia: Schizophrenia is a complex mental health disorder, and scientists want to understand what causes it. One way to do this is by studying DNA (genomics) and gene activity (transcriptomics). Scientists found that certain genes, like Disrupted in Schizophrenia 1 (DISC1) and Neurexin 1 (NRXN1), behave differently in schizophrenia patients. These genes are important for brain development and communication between brain cells. If they don't work properly, they might contribute to schizophrenia.

  How it works:

  - DNA variations: Some people may have small changes (mutations) in their genes that increase the risk of schizophrenia.
  - Gene activity: Even if the DNA is the same, some genes might be more or less active in people with schizophrenia compared to those without it.

  By identifying these risk genes, scientists can:

  - Understand what causes schizophrenia.
  - Develop better treatments by targeting these genes.
  - Find ways to diagnose schizophrenia earlier.

- Epigenetics + RNA: How Stress Affects Depression: Epigenetics is like a switch that turns genes on or off without changing the actual DNA. Things like stress, diet, and environment can control these switches. Scientists found that stress can change the activity of a gene called SLC6A4 (Solute Carrier Family 6 Member 4), which helps control serotonin (a brain chemical linked to mood). People with depression often show these changes. This helps us understand how life experiences—like stress—can impact mental health at a molecular level. It also opens the door for new treatments based on reversing these changes
- Proteins + Metabolites: Biomarkers for Bipolar Disorder: Proteomics (protein study) and metabolomics (chemical analysis) help find blood markers for bipolar disorder. People with bipolar disorder often have lower levels of BDNF (Brain-Derived Neurotrophic Factor)—a protein that supports brain cell growth and health. Since BDNF helps regulate mood and emotions, its decrease may contribute to mood swings in bipolar disorder. If doctors can measure these proteins and chemicals in the blood, they could detect bipolar disorder earlier and monitor how well treatments are working.
- Brain Scans + Genes: Understanding Autism: Scientists use MRI brain scans and genetic data to study autism. Research shows that people with autism have weaker connections between brain regions, and this is linked to genes like SHANK3, which play a role in brain development. Understanding these connections helps explain how genetic differences affect brain function in autism.

4.6 METHODOLOGY FOR FEATURE EXTRACTION AND SELECTION IN SCHIZOPHRENIA MENTAL HEALTH DISORDER

Schizophrenia is a chronic psychiatric illness with delusions, hallucinations, speech disorganization, mood, and abnormal behavior. The development of early schizophrenia depends on the nature of the onset of the disorder; its symptoms are complex and diverse and cannot be distinguished individually from other neurological disorders. Therefore, its Early diagnosis is quite difficult [6].

To identify meaningful features from multi-omics and clinical data in schizophrenia research, various methods can be used for feature extraction (obtaining relevant information) and feature selection as we already discussed in Section 2.

By combining statistical, ML, and domain-specific feature extraction and selection methods, researchers can identify key molecular signatures for schizophrenia, leading to better diagnostics and therapeutic targets.

ML algorithms have figured prominently in the recent years crucial role in the diagnosis of diseases like tumors, cancer, neurodegenerative diseases, and psychiatric diseases. Exploration of brain function and precise predictions become possible when ML methods are applied to neuroimaging data. Having originated in artificial intelligence, ML creates algorithms that enable computers to "learn" new items by learning about existing ones [3].

In recent years, ML approaches have been successfully applied for analysis of neuroimaging data, to help in the context of disease diagnosis.

Detecting schizophrenia using ML typically involves multi-modal features from different biological and clinical sources. Some key feature types include:

1. Neuroimaging Features (Brain Scans):
   - Structural MRI (sMRI)
     - brain volume
     - Cortical thickness
     - gray/white matter density
   - Functional MRI (fMRI):
     - brain activity,
     - connectivity patterns,
     - resting-state networks
   - Diffusion Tensor Imaging (DTI):
     - white matter integrity, FA
2. Genetic and Transcriptomic Features:
   - Single Nucleotide Polymorphisms (SNPs): Genetic risk variants (e.g., in DISC1, NRXN1, SHANK3)
   - Gene Expression (RNA-seq, Microarrays): Altered expression levels in neurotransmitter-related genes.
   - Epigenetics (DNA Methylation, miRNA): Environmental effects on gene regulation.
3. Electroencephalogram (EEG) Features:
   - Power Spectral Density (PSD): Differences in alpha, beta, and gamma wave activity.
   - Event-Related Potentials (ERPs): Delayed or reduced P300 response.
   - Functional Connectivity: Synchronization between different brain regions.
4. Clinical and Behavioral Features:
   - Cognitive Tests: Working memory, executive function, and attention scores.
   - Symptom Severity (PANSS Score): Positive and Negative Syndrome Scale for schizophrenia.
   - Speech and Language Patterns: Disorganized speech, coherence, vocabulary use.

5. Proteomic and Metabolomic Biomarkers:
   - Neurotrophic Factors (e.g., BDNF): Brain-derived neurotrophic factor levels.
   - Inflammatory Markers: Cytokines (IL-6, TNF-α) linked to neuroinflammation.
   - Neurotransmitter Metabolites: Dopamine, serotonin, and glutamate levels.

ML Methods for Schizophrenia Detection: ML techniques are widely used to analyze neuroimaging, genetic, and clinical data for schizophrenia detection. DL is subset of ML. ML methods require manual feature extraction, whereas the DL technique automatically derives feature since it is made up of many concealed layers. DL algorithms are data-hungry and require a huge amount of data for classification problems [3]. Here are some key methods:

- Support Vector Machine (SVM) algorithm can be utilized to differentiate between healthy patients and schizophrenia patients. Utilized with MRI, EEG, and genetic features for high-accuracy classification. It selects the hyperplane that maximally separates the dataset from the decision boundary. The higher the margin better will be the result obtained. The geometric representation of SVM is shown in Figure 4.3. Thus, multidimensional datasets fit with maximal margin into different classes. If the data is linearly separable, a linear-type hyperplane is used. If the dataset is linearly

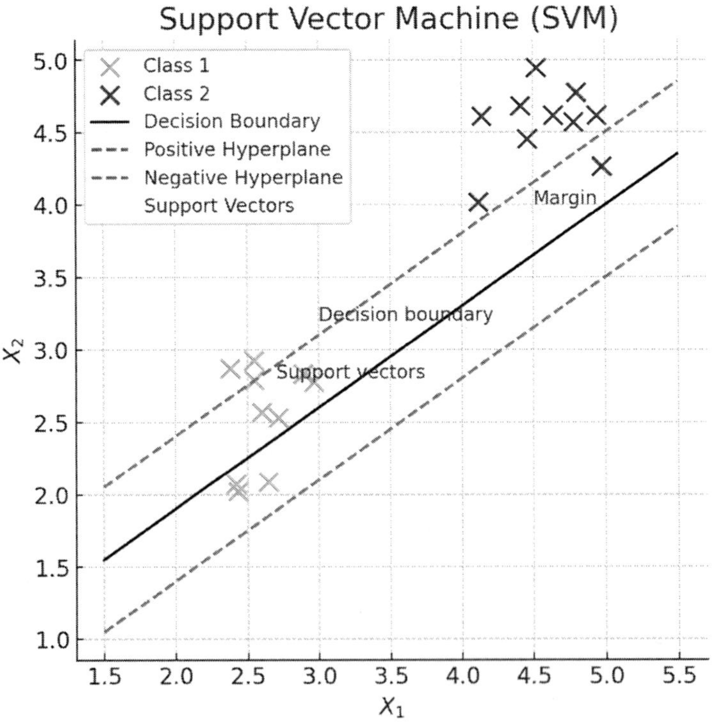

FIGURE 4.3 The geometric representation of SVM.

inseparable or non-linear different kinds of kernels were offered to classify the dataset perfectly. The classifier's performance largely relies on precisely choosing the kernel type and its parameters. The decision function of the high-dimensional hyperplane can be represented as f (x) = wTX + b, where w is the orthogonal vector to the hyperplane, b is bias or the constant, and X are the variables.

- Random Forest (RF): Ensemble ML algorithm used in gene expression and metabolic biomarker analysis. Provides feature importance scores to identify important biomarkers for schizophrenia. Random forest (RF) utilizes an ensemble of decision trees (DT) to improve the accuracy of DT. Bagging or bootstrap aggregation is used to train the RF produced algorithm. Bagging is a meta-algorithm used to improve the ACC of ML techniques by putting them together. It determines the outcome based on DT predictions as shown in Figure 4.4 [7].
- Deep Learning (CNN, RNN): Used for analyzing brain imaging and speech data. Convolutional neural networks (CNN) can be used for analyzing MRI, fMRI, and PET brain imaging data. Recurrent neural networks (RNN) can be applied to speech patterns and time-series EEG data for schizophrenia detection. DL algorithms are data-hungry and demand a huge quantity of data for classification problems. The fundamental components of CNN are Input, convolutional, max pooling, full-connected, and classification layers, as shown in Figure 4.5 [8]. Every pixel of the

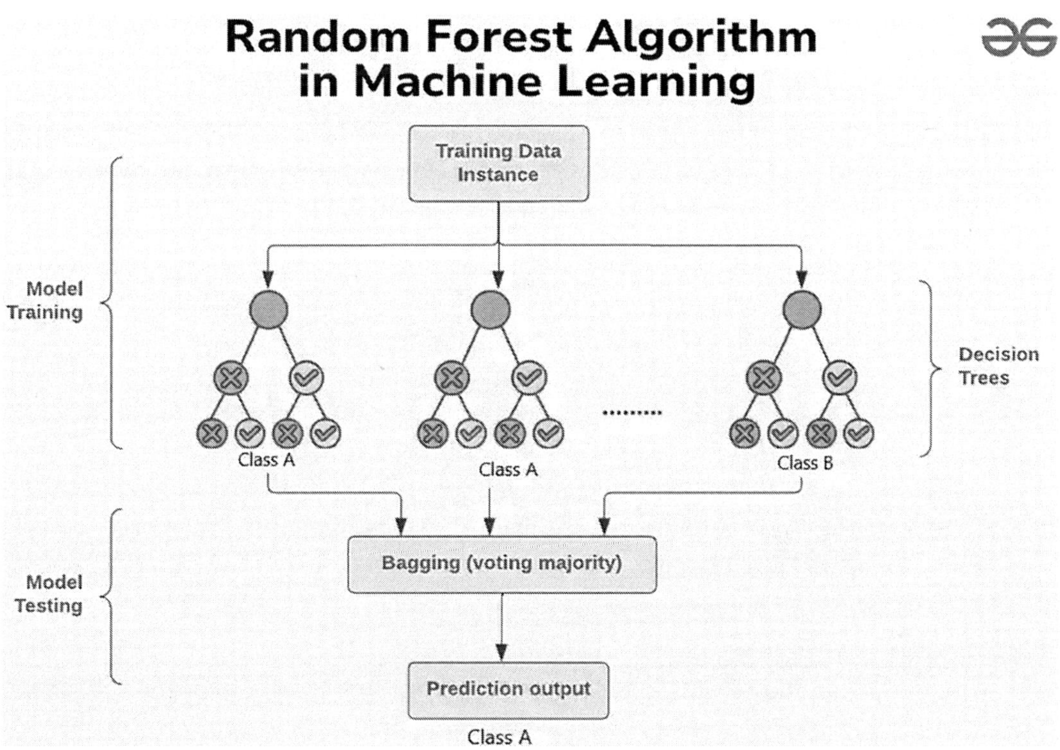

FIGURE 4.4 The geometric representation of RF.

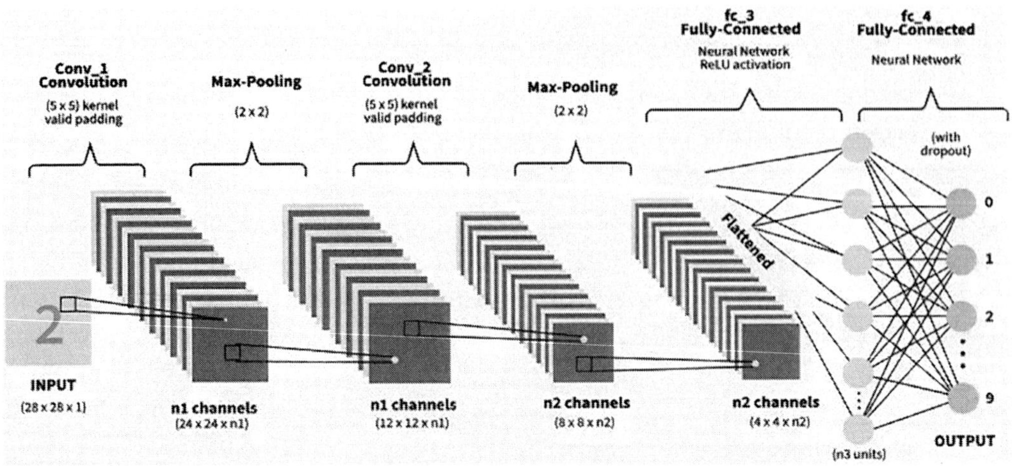

FIGURE 4.5 The geometric representation of CNN.

image is multiplied by the filter or convolution kernel. Shifting and multiplication create the final complicated feature map. ReLU layers discard negative pixels. The convolution layer extracts high-level input image attributes like color, boundaries, etc. The pooling layer down-samples some data to retain it. Max pooling gathers the maximum pixel value of filters, whereas average pooling is averaging the feature map. After feature mapping, having a fully connected layer converts the feature matrix to a vector, then a softmax or sigmoid activation function maps the input. There are different types of CNN models are there such as LeNet, AlexNet, ResNet, etc.

- PCA and UMAP: It is used for dimensionality or Feature reduction in neuroimaging and genetic datasets. Helps improve classification performance by removing redundant information as we already discussed in Section 2.2.1.
- SHAP (SHapley Additive Explanations): A feature attribution method used to interpret ML models in schizophrenia research. It identifies the most important genetic, metabolic, and imaging features influencing predictions.

These methods contribute to early detection, prognosis, and personalized treatment by analyzing complex biological data. In [9], they primarily focus on neuroimaging data, likely derived from: Structural MRI (sMRI): Brain volume, Cortical thickness, Gray matter density, and functional MRI (fMRI): Functional connectivity between brain regions, resting-state activity differences. Pattern Recognition Features: Spatial and temporal patterns in brain activity.

4.7 SEARCH STRATEGY

According to [3], a total of 657 papers are searched on PubMed and Google Scholar with the keywords SCZ (Schizophrenia) and neuroimaging. In the initial screening, papers with small datasets that were not according to the current model- ity, feature extraction, and ML or DL methods are rejected. After initial screening, twenty-nine papers on DL and

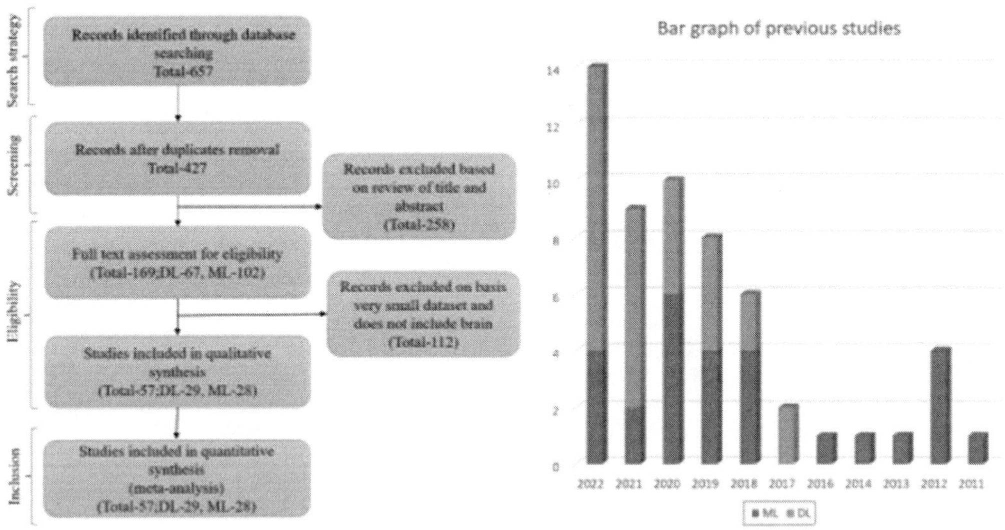

FIGURE 4.6 Flowchart of search strategy and statistics of papers on schizophrenia.

twenty-eight ML papers were chosen. The flow chart of the search strategy and year-wise statistics of paper bar graph is presented in Figure 4.6.

4.8 RESULT AND DISCUSSION

ML and deep learning (DL) have demonstrated significant potential in diagnosing schizophrenia using both neuroimaging data and social media text analysis. Verma et al. [3] reviewed various ML and DL techniques for schizophrenia diagnosis through neuroimaging, as summarized in Tables 4.2 and 4.3. Structural MRI (sMRI) revealed abnormalities in regions such as the hippocampus and prefrontal cortex, while fMRI detected disruptions in

TABLE 4.2 Machine Learning Methods

Input Type	References	Dataset	Modalities	Feature Extracted	Methods	Performance (%)
2D scans	Park et al. [10]	Clinical/SCZ-86, HC66	sMRI	HP LR	AB, XGBoost, SVM	ACC-80.4
	Febles [11]	Clinical/SCZ-54, HC-54	EEG	–	Multi-kernel SVM	ACC-83
	Rosa [12]	Clinical/SCZ-20, HC-20	Postmortem brain	HP and DLPFC	RF	AUC-95
3D scans	Chen et al. [13]	COBRE/SCZ-34, HC-34	sMRI	GM and WM	Linear SVM	ACC-85
	Lei et al. [14]	COBRE/SCZ295, HC- 452	sMRI,fMRI	GM and WM	SVM	ACC-90.83
	Yang et al. [15]	OpenfMRI/SCZ44 HC-44	fMRI	Whole brain	SVM	ACC-99.46
	Tanveer et al. [16]	COBRE/SCZ72	HC-74 sMRI	GM and WM	SVM and RF	ACC-80.71

TABLE 4.3 Deep Learning Methods

Input Type	Reference	Dataset	Modalities	Methods	Performance (%)
2D scans	Li et al. [17]	Clinical/SCZ-89, HC-83	sMRI	CNN	ACC-99.72
	Sharma et al. [18]	Clinical/SCZ-45, HC-39	EEG	CNN and LSTM	ACC-96
	Wu et al. [19]	Clinical/SCZ-14, HC-14	EEG	RAE	ACC-81.81
3D scans	Zhang et al. [20]	Clinical/SCZ-437, HC-450	sMRI	3D CNN	ACC-92
	Masoudi et al. [21]	COBRE/SCZ- 64, HC-81	sMRI, fMRI, and DTI	3D CNN	ACC-99.35

brain activity using blood oxygenation level-dependent (BOLD) signals. Diffusion tensor imaging (DTI) assessed white matter integrity, and electroencephalography (EEG) identified irregular brain wave patterns. Preprocessing methods such as image realignment, normalization, and registration were essential for improving data quality. Support vector machines (SVM) achieved up to 99.46% accuracy, while random forest (RF) was effective for feature selection and classification. Artificial neural networks (ANN) simulated brain function but were prone to overfitting. Deep learning methods, including CNNs (up to 99.72% accuracy), long short-term memory (LSTM) networks, 3D CNNs, and hybrid CNN-LSTM models, enabled automatic feature extraction and improved classification accuracy.

Public datasets such as SchizConnect (1,298 scans), COBRE (184 scans), BrainGluSchi (175 scans), and NUSDAST (451 scans) were used for model training and validation. Limitations included the focus on neuroimaging data alone, the lack of experimental validation, and challenges with model interpretability. Future directions involve integrating clinical and genetic data, improving feature selection, developing lightweight models for real-time diagnosis, and applying federated learning for multicenter studies.

Bae et al. [22] explored ML-based detection of schizophrenia using social media text analysis. Natural language processing (NLP) techniques such as linguistic inquiry and word count (LIWC) and latent Dirichlet allocation (LDA) were employed to extract linguistic and psychological features and identify discussion themes. Supervised ML algorithms, including SVM, logistic regression (LR), Naïve Bayes (NB), and RF, were trained to classify schizophrenia-related posts, while unsupervised clustering methods such as DBSCAN and t-SNE identified semantic patterns. Shapley Additive Explanations (SHAP) determined the most influential linguistic features. The dataset included 13,156 posts from the r/schizophrenia subreddit and 247,569 control posts from non-mental health subreddits, balanced for classification through down sampling and trained using 10-fold cross-validation. Limitations included the lack of clinical validation, Reddit user bias, and the focus on classical ML methods without exploring deep learning.

Future work should integrate audio, video, and behavioral data, address privacy and ethical issues, and investigate transformer-based models like BERT and GPT for improved classification.

Zhang et al. [20] applied 3D CNNs to classify schizophrenia patients using structural MRI scans. Techniques included squeeze-and-excitation (SE) blocks for enhanced feature selection, batch normalization (BN) for training stability, and gradient class activation mapping (Grad-CAM) for visualizing key brain regions involved in classification.

The proposed SE-VGG-11BN model achieved an AUC of 0.987, outperforming the benchmark model (AUC = 0.938), with an accuracy of 92.1%, sensitivity of 94.9%, and specificity of 94.6%. Regional analysis identified abnormalities in the hippocampus, thalamus, amygdala, and ventricles, with ventricular enlargement correlating with schizophrenia.

The model showed strong generalization when tested on BrainGluSchi data (AUC = 0.913) and outperformed ResNet and DenseNet models. Challenges included the high computational complexity of 3D CNNs, small dataset sizes, and the need for external validation. Future research should focus on multi-class classification, larger datasets, XAI, and real-world clinical implementation.

4.9 CONCLUSION

Integrating bioinformatics and ML techniques with multi-omics data allows researchers to uncover important biological patterns in schizophrenia. These approaches enable the identification of molecular signatures, which are specific biomarkers at the genetic, transcriptomic, proteomic, and metabolic levels that can help in understanding the disorder. However, challenges remain in standardizing data across studies, improving model interpretability, and validating findings through large-scale clinical trials.

By studying genomic data, researchers can identify genetic mutations and schizophrenia-related variations, e.g., COMT (Catechol-O-Methyltransferase) and BDNF (brain-derived neurotrophic factor) gene variation. Similarly, epigenetic modifications, like DNA methylation changes, provide insight into how environmental factors contribute to the disease. Transcriptomic data (gene expression levels) reflect the activation or inhibition of different genes and their functions in schizophrenic patients. For example, genes that regulate neurotransmitter systems such as dopamine and glutamate would typically show abnormal expression levels that affect brain function and behavior.

Proteomic and metabolomic analyses further refine the understanding of schizophrenia by identifying changes in proteins and metabolites related to brain function, energy metabolism, and inflammation. Low levels of BDNF and altered levels of key metabolites have been validated in schizophrenia patients, pointing toward potential therapeutic and diagnostic targets.

Neuroimaging findings (MRI, fMRI, PET scans) together with ML algorithms identify structural and functional impairments in schizophrenic patients' brains. Reduced connectivity of the prefrontal cortex, hippocampus, and temporal lobes is a common finding in accordance with the idea that schizophrenia is linked to disrupted neural circuits.

Despite these developments, challenges remain in translating these outcomes into the clinic. Across-study comparability requires standardization of the data collection process. Furthermore, most ML models, especially deep learning, are not explainable, making it difficult for clinicians to trust and apply these methods in real-world settings. Finally, validation of biomarkers identified in large-scale clinical trials must ensue prior to their application for routine diagnosis and treatment planning.

Looking ahead, future research must focus on improving computational approaches, integrating more than one layer of omics in an improved manner, and utilizing AI-assisted feature selection to improve schizophrenia diagnosis. Personalized medicine approaches where treatment is tailored based on the genetic and biomarker profile of the patient can become a reality with advancements.

REFERENCES

1. B. Taylor, M. Hobensack, S. N. de Rivera, Y. Zhao, R. M. Creber, K. Cato *et al*, "Identifying depression through machine learning analysis of omics data: Scoping review," *JMIR Nursing*, vol. 7, no. 1, p. e54810, 2024.
2. X.-L. Liu and L.-S. Chang, "Deciphering the genetic links between psychological stress, autophagy, and dermatological health: Insights from bioinformatics, single-cell analysis, and machine learning in psoriasis and anxiety disorders," *International Journal of Molecular Sciences*, vol. 25, no. 10, p. 5387, 2024.
3. S. Verma, T. Goel, M. Tanveer, W. Ding, R. Sharma, and R. Murugan, "Machine learning techniques for the schizophrenia diagnosis: A comprehensive review and future research directions," *Journal of Ambient Intelligence and Humanized Computing*, vol. 14, no. 5, pp. 4795–4807, 2023.
4. https://medium.com/pythoneers/the-math-behind-principal-component-analysis-pca-17744f047271
5. https://medium.com/@vlad.andreichuk/non-negative-matrix-factorization-nmf-for-the-grouping-of-articles-titles-a73e654b6244
6. H. Göker, "1d-convolutional neural network approach and feature extraction methods for automatic detection of schizophrenia," *Signal, Image and Video Processing*, vol. 17, pp. 2627–2636, 2023.
7. https://www.geeksforgeeks.org/machine-learning/random-forest-algorithm-in-machine-learning/
8. https://www.geeksforgeeks.org/deep-learning/convolutional-neural-network-cnn-in-machine-learning/
9. E. Veronese, U. Castellani, D. Peruzzo, M. Bellani, and P. Brambilla, "Machine learning approaches: From theory to application in schizophrenia," *Computational and Mathematical Methods in Medicine*, vol. 2013, no. 1, p. 867924, 2013. [Online]. https://onlinelibrary.wiley.com/doi/abs/10.1155/2013/867924
10. Y. W. Park, D. Choi, J. Lee, S. S. Ahn, S.-K. Lee, S.-H. Lee, and M. Bang, "Differentiating patients with schizophrenia from healthy controls by hippocampal subfields using radiomics," *Schizophrenia Research*, vol. 223, pp. 337–344, 2020.
11. E. Santos Febles, M. Ontivero Ortega, M. Valdés Sosa, and H. Sahli, "Machine learning techniques for the diagnosis of schizophrenia based on event-related potentials," *Frontiers in Neuroinformatics*, vol. 16, p. 893788, 2022.
12. A. De Rosa, A. Fontana, T. Nuzzo, M. Garofalo, A. Di Maio, D. Punzo, M. Copetti, A. Bertolino, F. Errico, A. Rampino *et al*, "Machine learning algorithm unveils glutamatergic alterations in the post-mortem schizophrenia brain," *Schizophrenia*, vol. 8, no. 1, p. 8, 2022.
13. Z. Chen, T. Yan, E. Wang, H. Jiang, Y. Tang, X. Yu, J. Zhang, and C. Liu, "Detecting abnormal brain regions in schizophrenia using structural MRI via machine learning," *Computational Intelligence and Neuroscience*, vol. 2020, no. 1, p. 6405930, 2020.
14. D. Lei, W. H. Pinaya, J. Young, T. Van Amelsvoort, M. Marcelis, G. Donohoe, D. O. Mothersill, A. Corvin, S. Vieira, X. Huang *et al*, "Integrating machine learning and multimodal neuroimaging to detect schizophrenia at the level of the individual," *Human Brain Mapping*, vol. 41, no. 5, pp. 1119–1135, 2020.

15. H. Yang, X. Di, Q. Gong, J. Sweeney, and B. Biswal, "Investigating inhibition deficit in schizophrenia using task-modulated brain networks," *Brain Structure and Function*, vol. 225, pp. 1601–1613, 2020.
16. M. Tanveer, J. Jangir, M. Ganaie, I. Beheshti, M. Tabish, and N. Chhabra, "Diagnosis of schizophrenia: A comprehensive evaluation," *IEEE Journal of Biomedical and Health Informatics*, vol. 27, no. 3, pp. 1185–1192, 2022.
17. Z. Li, W. Li, Y. Wei, G. Gui, R. Zhang, H. Liu, Y. Chen, and Y. Jiang, "Deep learning based automatic diagnosis of first-episode psychosis, bipolar disorder and healthy controls," *Computerized Medical Imaging and Graphics*, vol. 89, p. 101882, 2021.
18. G. Sharma and A. M. Joshi, "Szhnn: A novel and scalable deep convolution hybrid neural network framework for schizophrenia detection using multichannel EEG," *IEEE Transactions on Instrumentation and Measurement*, vol. 71, pp. 1–9, 2022.
19. Y. Wu, M. Xia, X. Wang, and Y. Zhang, "Schizophrenia detection based on EEG using recurrent auto-encoder framework," in *International Conference on Neural Information Processing.* Springer, 2022, pp. 62–73.
20. J. Zhang, V. M. Rao, Y. Tian, Y. Yang, N. Acosta, Z. Wan, P.-Y. Lee, C. Zhang, L. S. Kegeles, S. A. Small *et al*, "Detecting schizophrenia with 3d structural brain MRI using deep learning," *Scientific Reports*, vol. 13, no. 1, p. 14433, 2023.
21. B. Masoudi, S. Daneshvar, and S. N. Razavi, "Multi-modal neuroimaging feature fusion via 3d convolutional neural network architecture for schizophrenia diagnosis," *Intelligent Data Analysis*, vol. 25, no. 3, pp. 527–540, 2021.
22. Y. J. Bae, M. Shim, and W. H. Lee, "Schizophrenia detection using machine learning approach from social media content," *Sensors*, vol. 21, no. 17, p. 5924, 2021.

CHAPTER 5

Feature Selection through Brownian Motion Search

A Case Study for Breast Cancer Prediction

Abrar Yaqoob, R. Vijaya Lakshmi, Navneet Kumar Verma, GVV Jagannadha Rao, Rabia Musheer Aziz, Tejaswini Pradhan and Pranjali Malviya

5.1 INTRODUCTION

Breast cancer is a widespread and serious health condition marked by the uncontrolled multiplication of cells within the breast tissue. It is among the most frequently diagnosed cancers in women across the globe and carries major consequences for both personal health and healthcare systems. This disease can appear in different forms, from tumors that remain limited to the breast ducts or lobules to invasive types that extend into nearby tissues and other parts of the body. Detecting the disease at an early stage is vital, as it greatly enhances treatment effectiveness and increases survival chances. Common tools for diagnosing breast cancer include mammograms, breast ultrasounds, MRI scans, and tissue biopsies. Treatment strategies may involve surgical removal, radiation therapy, chemotherapy, hormone-based treatment, targeted drugs, or a combination of these methods, depending on the cancer stage, specific subtype, and patient-specific characteristics. Recent advancements in scientific research and medical technologies have significantly improved the ways in which breast cancer is identified, diagnosed, and treated, giving patients better prospects and improved quality of life. Nevertheless, ongoing initiatives in awareness, screening programs, early diagnosis, and scientific investigation remain crucial for continued progress in fighting breast cancer and minimizing its impact on individuals and society. Gene expression datasets hold extensive and valuable information essential for interpreting biological mechanisms. However, their high dimensionality and complexity make it difficult to pinpoint the most significant genes due to the presence of redundant and irrelevant features. As a result, developing robust feature selection (FS) methods

DOI: 10.1201/9781003591658-5

has become increasingly important to effectively filter this vast biological data and uncover meaningful patterns. In recent years, numerous FS approaches have been investigated by researchers to improve the accuracy and reliability of gene selection [1]. These techniques aim to identify subsets of genes that are most relevant to specific biological phenomena, such as disease classification or treatment response prediction. Among these approaches, the integration of optimization algorithms with machine learning models has shown promise in addressing the challenges of gene selection. The Brownian Motion Search Algorithm (BMSA) is one such optimization technique inspired by the random motion of particles observed in Brownian motion. By simulating the stochastic behavior of particles, BMSA effectively explores solution spaces and identifies optimal solutions. Combining BMSA with machine learning models, such as support vector machine (SVM), presents a novel approach to gene selection and classification tasks [2].

In this study, we introduce an innovative framework that integrates BMSA with SVM to perform gene selection for breast cancer classification. Using the breast cancer dataset, the proposed method seeks to effectively pinpoint the most informative gene subsets that enhance the accurate identification of different breast cancer subtypes. By integrating BMSA for FS and SVM for classification, we aim to achieve high classification accuracy (CA) while providing valuable insights into breast cancer-associated gene biomarkers. In the subsequent sections, we provide a detailed overview of the methodology, experimental setup, results, and discussion, elucidating the effectiveness and implications of our proposed approach for gene selection and breast cancer classification. Figure 5.1 shows various techniques to visualize breast cancer images.

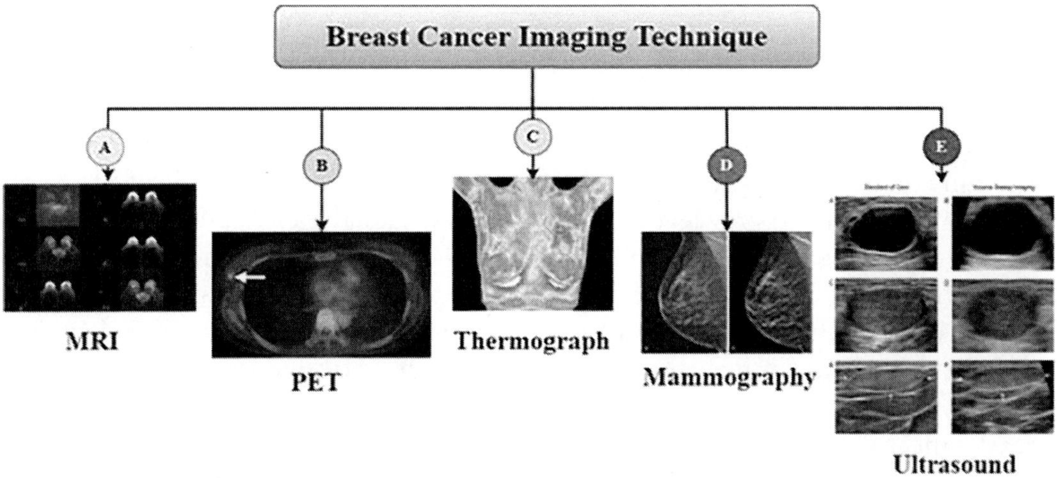

FIGURE 5.1 Different techniques to visualize breast cancer images.

5.1.1 Objective of the Paper

By utilizing diagnostic information alongside machine learning methods, the proposed approach seeks to overcome key challenges in cancer classification and support enhanced patient care and clinical decision-making. Through the creation of a precise and dependable classification system, our goal is to equip healthcare practitioners with effective tools for early tumor detection and timely intervention, ultimately improving treatment effectiveness and boosting patient survival rates.

5.2 RELATED WORK

Tehnan I. A. Mohamed et al. propose a CNN model for breast cancer detection using gene expression data from TCGA. The dataset includes 1208 samples with 19,948 genes, comprising 113 normal and 1095 cancerous samples. Preprocessing involves AAIC for outlier removal and normalization to mitigate biases. Gene reduction is achieved through filtration with a threshold of 0.25. The data is then converted into grayscale images to meet the model's requirements [3]. Amel Ali Alhussan et al. present a novel framework that integrates metaheuristic optimization with deep learning and FS to classify breast cancer from ultrasound images effectively. The methodology comprises five stages: data augmentation to enhance CNN model learning, transfer learning using GoogleNet for feature extraction, FS using a hybrid optimization algorithm combining dipper throated and particle swarm techniques, and classification of selected features using CNN optimized by the proposed algorithm [4]. Ammar Abdulrahman Ahmed et al. introduce an innovative approach for classifying breast thermography images into normal, benign, or malignant categories. The method incorporates bio-inspired algorithms, namely, ant colony optimization (ACO) and particle swarm optimization (PSO), for FS. The proposed method consists of four phases: image preprocessing, feature extraction, FS, and classification. Evaluation is conducted using a standard thermography dataset, revealing promising performance of the proposed method [5]. David González-Patiño et al. investigate the application of meta-heuristics for mammographic image segmentation. Unlike traditional usage for optimization problems, this study focuses on segmenting mammograms using the Dunn index as the optimization function and grey levels to represent individuals. The update of grey levels aims to maximize the Dunn's index, resulting in improved segmentation quality. The findings demonstrate lower error rates using these meta-heuristics compared to the widely used Otsu method for segmentation [6]. Lingxi Peng et al. blend cutting-edge life science research with artificial intelligence, introducing a semi-supervised learning algorithm to minimize reliance on labeled data. Utilizing two renowned breast cancer datasets from the UCI machine learning repository, extensive experiments confirm the effectiveness and efficiency of the proposed algorithm. These findings underscore the algorithm's potential as a promising automated diagnosis method for breast cancer [7]. Ali Al Bataineh et al. introduce a novel application of CSA for training MLP architectures in solving diverse real-world problems, including breast cancer diagnosis, active sonar target classification, wheat classification, and flower classification. CSA optimizes the weights and biases of MLPs to enhance CA significantly. Comparative analysis against other popular training methods, such as GA, ACO, PSO, HHO, MFO, FPA, and BP, is conducted using benchmark datasets including Iris Flower, Sonar, Wheat Seeds, Breast Cancer Wisconsin, and Haberman's Survival. Results demonstrate the superior performance of CSA in improving MLP performance across various disciplines, establishing it as a competitive approach for solving real-world applications [8]. Selvakumar Thirumalaisamy et al. propose an advanced breast cancer detection method, combining a synthesized CNN, an optimized optimization algorithm (EACO), and transfer learning. Experimental analysis on MIAS and CBIS-DDSM datasets demonstrates superior accuracy, sensitivity, and specificity compared to conventional methods, highlighting the effectiveness of the proposed EACO–ResNet101 model [9]. Abrar Yaqoob et al. propose a novel gene selection approach named the Sine Cosine and Cuckoo Search Algorithm (SCACSA), designed to enhance SVM classifiers. Using a breast cancer dataset, the hybrid SCACSA algorithm's performance is rigorously evaluated and compared to other FS methods. Initially, the minimum redundancy maximum relevance (mRMR) filtering strategy is employed to improve the quality of the feature set. Subsequently, the SCACSA method optimizes the gene selection process. Finally, the dataset is classified using the SVM classifier based on the selected genes. Given the critical role of gene selection in understanding complex biological datasets, SCACSA emerges as a valuable tool for cancer dataset classification, aiding medical practitioners in making informed decisions about cancer diagnosis [10].

5.3 PROPOSED METHODOLOGY

In this section, we will explore the use of the BMSA for FS. Following that, we will delve into the role of SVM within this hybrid algorithm for classification.

5.3.1 Stage 1: Feature Selection via Brownian Motion Search Algorithm

BMSA is a stochastic optimization method modeled after the unpredictable movement of particles in Brownian motion. In this algorithm, a set of candidate solutions called particles is defined by their respective positions within the search space. Let x_i represents the position of the i-th particle in the d-dimensional search space, where I =1, 2…, N and d represents the dimensionality of the problem. Each particle's position is updated iteratively according to equation 5.1 [11].

$$x_i^{(t+1)} = x_i^{(t)} + \alpha.\Delta x_i \quad (5.1)$$

where $x_i^{(t)}$ is the position of the i-th particle at iteration t, α is a step size parameter, and Δx_i represents a random perturbation vector that follows a Gaussian distribution with mean 0 and variance σ^2. This random perturbation allows particles to explore the search space in a stochastic manner, akin to the random movement of particles in Brownian motion. The fitness of each particle is evaluated using an objective function $f(x_i)$which quantifies the quality of the solution associated with the particle's position. The objective function guides the search process by providing feedback on the performance of candidate solutions. At each iteration, particles update their positions based on the evaluation of their fitness where $x_i^{(t+1)}$ represents the updated position of the i-th particle at iteration t+1. BMSA dynamically balances exploration and exploitation by adjusting the step size α and the variance σ^2 of the random perturbation vector. This allows the algorithm to effectively explore the search space while exploiting promising regions identified through the fitness evaluations. The iterative optimization process continues until a termination criterion is met, such as reaching a maximum number of iterations or achieving convergence. In summary, the BMSA provides a powerful framework for stochastic optimization, leveraging the principles of Brownian motion to efficiently explore solution spaces and find optimal solutions in diverse optimization problems [12, 13].

In the feature selection (FS) phase using BMSA, the objective is to determine the most relevant subset of features from a high-dimensional dataset. BMSA, derived from the random and stochastic behavior seen in Brownian motion, provides an effective mechanism for thoroughly exploring the search space. Unlike conventional FS techniques, BMSA can simultaneously balance local exploitation with global exploration, making it highly suitable for complex optimization problems. At the beginning of the FS procedure, BMSA generates an initial population of candidate feature subsets within the specified search space. Each candidate solution represents a subset of features from the original dataset. The algorithm then evaluates the fitness of each candidate subset based on a predefined objective function, which typically measures the relevance or discriminative power of the features for the given classification task. During the iterative optimization process, BMSA simulates Brownian motion by introducing random perturbations to the candidate solutions. These perturbations enable the algorithm to explore the solution space in a stochastic manner, gradually refining the candidate feature subsets toward more optimal solutions. Throughout the optimization process, BMSA dynamically adjusts the exploration-exploitation balance, allowing it to escape local optima and discover globally optimal feature subsets. As BMSA iterates, it continuously updates the best feature subset found so far based on the fitness evaluation. This iterative refinement process continues until a termination criterion is met, such as a maximum number of iterations or convergence of the algorithm [14].

Overall, the feature selection phase using BMSA provides a strong and effective method for determining the most meaningful features for a classification problem. By utilizing the stochastic behavior inherent in Brownian motion, BMSA efficiently navigates the search space, enabling the identification of feature subsets that greatly improve the performance of machine learning classifiers. The pseudocode for BMSA is presented in Algorithm 1.

Algorithm 1: Pseudocode of Brownian Motion Search Algorithm

```
Initialize population of gene subsets randomly within search space
best_subset = None
best_accuracy = -Infinity
for iteration = 1 to max_iterations:
  for each gene subset:
    Evaluate fitness using SVM classification accuracy on training dataset
    if accuracy > best_accuracy:
      Update best_subset
      Update best_accuracy
  for each gene subset:
    Simulate Brownian motion to perturb gene subset composition
    Clip gene subset to maintain within search space bounds
return best subset
```

5.3.2 Stage 2: Classification by Using Support Vector Machine

SVM is a robust algorithm for classification and regression, designed to find the optimal hyperplane that separates data points of different classes in high-dimensional space. By maximizing the margin between the decision boundary and the nearest data points (support vectors), SVM improves model generalization. Using kernel functions, SVM transforms data into a higher-dimensional space to create non-linear decision boundaries. It employs soft margin classification to handle non-linearly separable data, balancing margin maximization and classification error through the regularization parameter (C). SVM is efficient with high-dimensional data and suitable for small to medium-sized datasets. It can manage multi-class problems using one-vs-one or one-vs-all strategies and has applications in text classification, image recognition, bioinformatics, and financial forecasting [15]. In summary, SVM constructs optimal decision boundaries, offers robustness against overfitting, and is effective across various domains, making it a popular choice for diverse machine learning tasks [16].

5.3.3 Parameter Settings for the Hybrid BMSA-SVM Approach

For the efficient application of the hybrid BMSA–SVM framework in gene selection and classification using breast cancer datasets, certain parameter configurations are recommended. In BMSA, the initial population typically consists of 50 feature subsets, with the algorithm running for a maximum of 100 iterations. The perturbation rate—which determines how often feature subsets are modified to explore the search space—generally lies between 0.1 and 0.3. The fitness evaluation is usually based on the classification accuracy (CA) obtained from the SVM. The algorithm is considered to have converged when no notable improvement is detected over the previous 20 iterations.

In the SVM component, the Radial Basis Function (RBF) kernel is commonly chosen, although linear, polynomial, and sigmoid kernels are also options. The regularization parameter (C), which balances the trade-off between low training error and model complexity, is often set between 1 and 10. The kernel coefficient (γ), defining the influence range of a single training example, typically falls between 0.1 and 1 for the RBF kernel. To handle class imbalance, the class weight parameter can be set to "balanced", automatically adjusting weights inversely proportional to class frequencies. Cross-validation, using 10 folds, is employed to fine-tune the SVM parameters. Performance metrics for evaluating the hybrid algorithm include the confusion matrix, which visualizes the classification performance, and the precision-recall (PR) curve, depicting the trade-off between precision and recall at various thresholds. The ROC curve and the area under the curve (AUC) provide insights into the classifier's diagnostic ability and overall performance. By carefully setting these parameters, the proposed hybrid BMSA-SVM approach can effectively select relevant gene subsets and achieve high CA in breast cancer datasets, offering valuable insights into associated gene biomarkers.

5.4 EXPERIMENTAL SETUP

This study examined breast cancer microarray datasets to assess the performance of the proposed gene selection method. The datasets, related to Stage II lung cancer, were sourced from the publicly available repository (https://csse.szu.edu.cn/staff/zhuzx/Datasets.html), and their key details are summarized in Table 5.1. To ensure reliable evaluation, we used Leave-One-Out Cross-Validation (LOOCV), where each sample is tested once while the remaining samples are used for training. This provides a thorough and unbiased assessment of model performance. The effectiveness of the method was evaluated based on the classification accuracy (CA) of the classifier and the number of selected genes. The CA is calculated using the following formula:

$$Classification = \frac{CC}{N} \times 100$$

where CC denotes correctly classified samples and N represents the total number of samples within a specific class. The Experiments were conducted on a desktop with a 64-bit Windows 10 OS, an Intel Core i7-3770 CPU at 3.40 GHz, and 8 GB of RAM, using Python 3.12.0. The fitness function of the proposed approach is based on the CA of the SVM classifier. If the current fitness score surpasses the previous one, it replaces the previous result;

TABLE 5.1 Detailed Overview of Used Dataset

Dataset	Classes	Genes	No. of Samples	Short Description
Breast cancer	2	24481	97	Breast cancer is a type of cancer that originates in the cells of the breast. It can occur in both men and women, but it is far more common in women. The disease typically begins in the milk-producing glands (lobules) or the ducts that carry milk to the nipple.

otherwise, the previous solution is retained. The solution with the highest fitness score is selected as the optimal gene subset for prediction.

5.5 RESULTS

Table 5.2 summarizes the classification accuracy (CA) of the proposed BMSA-SVM approach on the breast cancer dataset using different numbers of selected genes. For each subset size, the table reports the best, mean, and worst accuracies obtained. With 12 selected genes, the method achieved a best CA of 93.23%, a mean of 85.18%, and a worst accuracy of 79.34%. Increasing the subset to 14 genes further improved performance to a best accuracy of 95.45%, a mean of 88.89%, and a worst accuracy of 83.5%. The highest performance was observed with 16 genes, where the best accuracy reached 99.14%, the mean was 92.05%, and the worst was 87.81%. However, selecting 18 genes resulted in a slight decline, with best, mean, and worst accuracies of 97.09%, 89.81%, and 84.7%, respectively. Choosing 20 genes produced a best accuracy of 96.15%, a mean of 88.73%, and a worst of 82.45%. With 24 genes, the best accuracy was 94.91%, while the mean and worst accuracies dropped to 86.07% and 79.33%. Finally, selecting 28 genes further reduced performance, with a best accuracy of 93.37%, a mean of 82.19%, and a worst of 74.49%. Overall, the results indicate that the optimal accuracy is achieved with 16 genes, and increasing the subset beyond this size does not enhance performance and may even reduce it. This demonstrates that the BMSA-SVM approach effectively identifies a compact and highly informative set of genes for accurate breast cancer classification.

We assess the algorithm's performance using diverse metrics, including the confusion matrix, which provides a detailed breakdown of true positives (TP), true negatives (TN), false positives (FP), and false negatives (FN), allowing us to evaluate the classifier's accuracy and error distribution. The PR curve is used to measure the balance between precision and recall across different thresholds, giving insights into the trade-offs between FP and FN. Additionally, the ROC curve, along with the AUC, evaluates the diagnostic ability of the classifier, illustrating its performance across various threshold settings and helping to identify the optimal balance between sensitivity and specificity. These metrics collectively

TABLE 5.2 Classification Accuracy of Proposed Method for Breast Cancer Dataset

	Proposed Method		
No. of Genes	**Best**	**Mean**	**Worst**
12	93.23	85.18	79.34
14	95.45	88.89	83.5
16	**99.14**	**92.05**	**87.81**
18	97.09	89.81	84.7
20	96.15	88.73	82.45
24	94.91	86.07	79.33
28	93.37	82.19	74.49

Note: The best values are highlighted in bold.

offer a comprehensive evaluation of the proposed method's effectiveness in classifying breast cancer data.

5.5.1 Descriptions for Each Metric Based on the Provided Confusion Matrix

1. Accuracy: The model correctly classified 97 samples, resulting in an accuracy of 99.01%, indicating that the vast majority of predictions were correct.
2. Precision: The precision of the model is 99.01%, meaning that out of all the samples predicted as positive (97 total), 42 were actually positive, reflecting high reliability in positive predictions.
3. Sensitivity (Recall): The sensitivity (or recall) of the model is 100%, indicating that the model successfully identified all 42 actual positive cases without missing any.
4. F1-Score: The F1-Score is 99.72%, showing a strong balance between precision (99.01%) and recall (100%), making it a robust measure of the model's performance in handling positive predictions accurately.

5.5.2 Confusion Matrix

Figure 5.2 shows the confusion matrix, an essential tool for evaluating classification performance. This matrix displays the model's predicted labels against the actual labels in a structured grid consisting of four categories: true positives (TP), true negatives (TN), false positives (FP), and false negatives (FN). TP represent correctly predicted positive cases, TN refer to correctly identified negative cases, FP indicate negative samples incorrectly classified as positive, and FN denote positive samples that the model failed to identify.

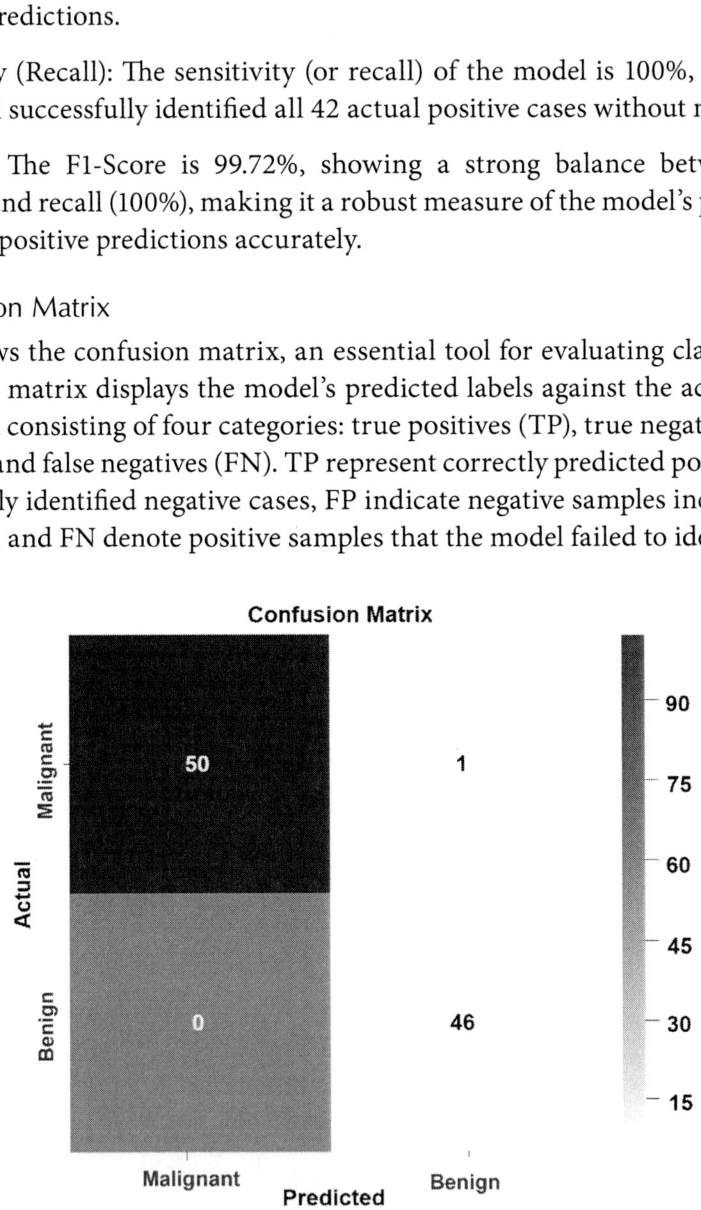

FIGURE 5.2 Confusion matrix based on the breast cancer dataset.

The confusion matrix provides a clear summary of the model's strengths and weaknesses, aiding in understanding its predictive abilities. By examining TP, TN, FP, and FN, performance metrics like accuracy, precision, recall, and F1-score can be calculated. These metrics offer a comprehensive evaluation of the model's classification effectiveness. The confusion matrix is essential for refining models and making informed deployment decisions. Calculations for these performance metrics are based on specific equations derived from the confusion matrix values

$$ACC = \frac{TP + TN}{TP + TN + FP + FN} \ (y)$$

$$P = \frac{TP}{TP + FP} \ (j)$$

$$Sn = \frac{TP}{TP + FN} \ (k)$$

$$F - score = 2 \times \frac{P \times Sn}{P + Sn}$$

5.5.3 Precision Recall Curve

To enhance its findings, the study utilizes visual aids like Figure 5.3, which presents the PR graph. This graph evaluates the proposed approach's performance by illustrating the balance between precision (accuracy of positive predictions) and recall (proportion

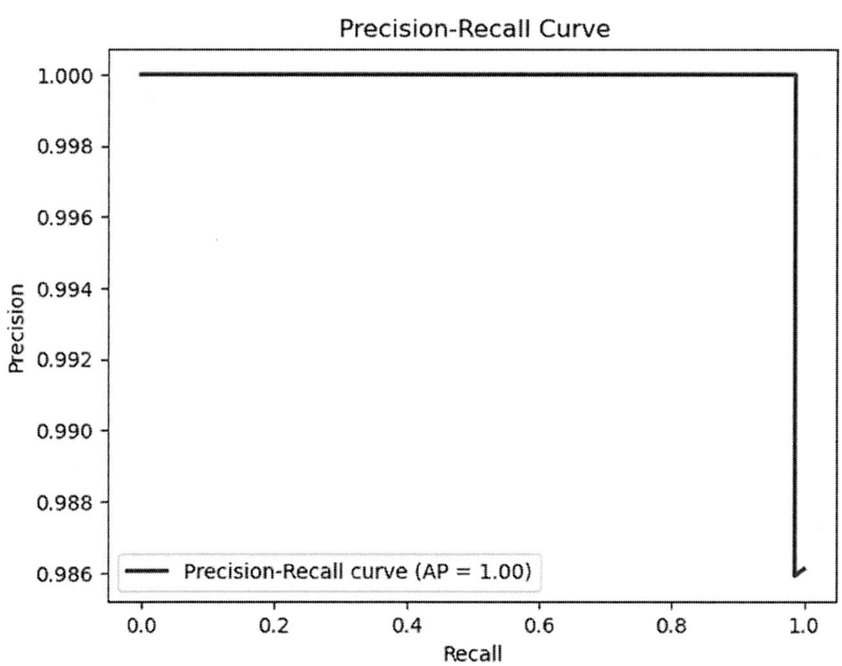

FIGURE 5.3 Precision recall curve on the breast cancer dataset.

of actual positives correctly predicted). The PR graph provides deeper insights into the approach's efficacy, contributing to a comprehensive understanding of its predictive capabilities.

5.5.4 Area Under the Curve

A larger AUC value reflects stronger model performance in differentiating between positive and negative samples. In the PR curve, a higher AUC further emphasizes the robustness and effectiveness of the proposed approach, especially in correctly detecting positive instances within imbalanced datasets. This comprehensive evaluation demonstrates the method's reliability and practical usefulness, particularly in applications where accurate identification of positive cases is essential.

Figure 5.4 depicts the receiver operating characteristic (ROC) curve, illustrating the balance between TP rate (sensitivity) and FP rate (1-specificity). A higher AUC-ROC value indicates the model's superior ability to distinguish between classes. The ROC curve from the proposed method demonstrates strong discriminatory capability. High AUC values for both the PR and ROC curves confirm the method's effectiveness and reliability in accurately classifying cancer types. These visual representations provide compelling evidence of the method's proficiency and potential for real-world clinical applications.

Table 5.3 compares the CA and standard deviation (STD) of decision tree, K-nearest neighbors (KNN), and random forest (RF) classifiers using the proposed FS algorithm for breast cancer classification. The RF classifier consistently achieves the highest

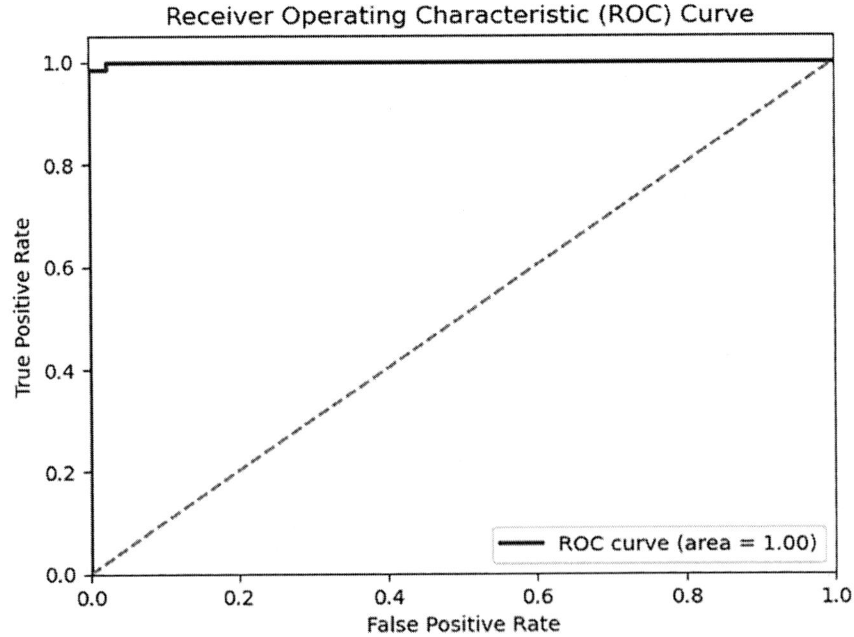

FIGURE 5.4 ROC curve of the breast cancer dataset.

TABLE 5.3 The Comparison between the Different Classifiers with Proposed Algorithm for Breast Cancer Classification

	Decision Tree	KNN	RF
Selected Genes	Mean CA (±STD)	Mean CA (±STD)	Mean CA (±STD)
12	97.22 ± 0.04	96.62 ± 2.0	99.03 ± 0.76
14	91.01± 1.10	98.23 ± 3.10	96.43 ± 3.15
16	82.30 ± 3.08	99.31 ± 1.01	99.01 ±1.01
18	90.10 ± 1.24	97.25 ± 2.06	96.54 ±1.06
20	83.12 ± 2.70	90.13 ± 1.34	99.03 ±0.11
24	97.53 ± 3.10	93.22 ± 1.98	98.40 ± 0.98

accuracy across different gene sets with low variability. For example, with 12 genes, RF achieves 99.03% CA, compared to 97.22% for Decision Tree and 96.62% for KNN. As the number of selected genes varies, RF maintains superior performance, highlighting its robustness. In contrast, Decision Tree and KNN show more variability in their performance. These results emphasize the effectiveness of the proposed FS algorithm and the importance of selecting the appropriate classifier for optimal breast cancer classification. Figure 5.5 shows more clearly the comparison of different classifiers used on the same dataset.

Based on the data presented in Table 5.3, RF consistently outperforms both the Decision Tree and KNN classifiers across all listed measurements. The accuracy values for RF are higher than those for the other two classifiers in most cases, which is indicative of RF's robustness and better performance.

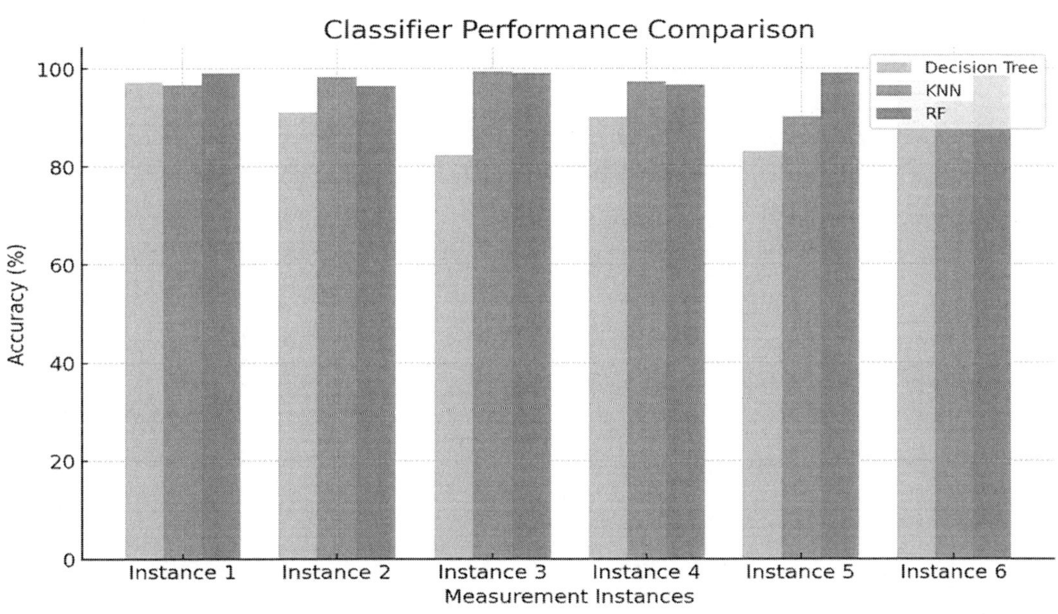

FIGURE 5.5 Histogram comparison of different classifiers.

To illustrate this, we can create a histogram that visually compares the performance of the three classifiers. In this histogram:

- The x-axis will represent the different measurement instances.
- The y-axis will represent the accuracy percentages.
- Different colors or patterns will distinguish the bars representing decision tree, KNN, and RF.

Each group of bars at each measurement instance will show how RF (with the highest bars) performs better than decision tree and KNN, emphasizing the superior accuracy of the RF classifier in this context. Figure 5.5 shows the comparison of different classifiers.

The histogram clearly shows that the RF classifier consistently outperforms both the decision tree and KNN classifiers across all six measurement instances. Each group of bars represents the performance of the three classifiers for a specific instance, with RF (red bars) achieving higher accuracy percentages in most cases. This visual representation highlights the superior performance of RF, demonstrating its robustness and effectiveness compared to decision tree and KNN. The consistently higher bars for RF across the instances reinforce the conclusion that RF is the best-performing classifier among the three in this particular dataset.

Table 5.4 compares the performance of the proposed BMSA-SVM algorithm with several advanced FS and classification methods based on different references. Each method is evaluated on key metrics, including the number of selected genes, CA, training time, and testing time. The proposed BMSA-SVM method stands out with an impressive CA of 99.01% using only 16 genes. It demonstrates efficient performance with a training time of 3.03 seconds and a testing time of 250.9 milliseconds. Reference [17] describes a method combining mRMR with KNN and CNN, achieving an accuracy of 98.19% with 33 genes, a training time of 3.37 seconds, and a testing time of 324.59 milliseconds. Reference [18] presents the SVM-mRMRe approach, which selects 65 genes and achieves an accuracy of 91.17%, with a training time of 3.88 seconds and a testing time of 248.01 milliseconds. The M-FS method in Reference [19] matches the proposed method's accuracy of 99.01% but requires 26 genes and has longer training (4.19 seconds) and testing times

TABLE 5.4 Comparison of Proposed Algorithm with Advanced Methods

References	Algorithms	Gene no.	Accuracy	Training Time	Testing Time
Proposed method	BMSA-SVM	16	99.14	3.03 s	250.9 Ms
[18]	mRMR+KNN+CNN	33	98.19	3.37 s	324.59 Ms
[19]	SVM-mRMRe	65	91.17	3.88 s	248.01 Ms
[20]	M-FS	26	99.01	4.19 s	348.89 Ms
[21]	HER+MLT	34	82.6	3.42 s	222.9 Ms
[22]	FR+SE	14	99.45	2.29 s	352.9 Ms
[23]	PSO+DT	20	85.48	3.72 s	243.2 Ms

(348.89 milliseconds). Reference [20] describes the HER+MLT method, which achieves an accuracy of 82.6% with 34 genes, with training and testing times of 3.42 seconds and 222.9 milliseconds, respectively. The FR+SE method from Reference [21] achieves the highest accuracy of 99.45% with just 14 genes, but it has the shortest training time (2.29 seconds) and the longest testing time (352.9 milliseconds). Lastly, the PSO+DT method in Reference [22] achieves an accuracy of 85.48% using 20 genes, with a training time of 3.72 seconds and a testing time of 243.2 milliseconds. In summary, the proposed BMSA-SVM algorithm excels in balancing high CA, efficient gene selection, and competitive training and testing times. Compared to other advanced methods, it demonstrates significant efficiency and effectiveness in classifying breast cancer data, highlighting its potential as a robust tool for gene selection and disease classification.

5.6 CONCLUSION

In summary, the hybrid framework integrating BMSA with SVM proves highly effective for gene selection and classification in the breast cancer dataset. As evidenced by the accuracy results, the method achieves its highest classification accuracy of 99.14% using 16 genes, along with a mean accuracy of 92.05% and a worst-case accuracy of 87.81%. This performance surpasses that of several other feature selection approaches in terms of both accuracy and computational efficiency. BMSA successfully identifies the most informative genes from high-dimensional data, addressing issues of redundancy and irrelevant features, while SVM provides precise classification of the selected gene subsets. The evaluation using the confusion matrix, PR curve, and ROC curve further confirms the model's robustness and stability. Overall, these results demonstrate that the BMSA-SVM strategy not only delivers superior accuracy but also offers meaningful insights into potential gene biomarkers for breast cancer. This approach represents a valuable advancement in bioinformatics and cancer research, providing an effective tool for identifying key genes within complex biological datasets.

5.7 FUTURE DIRECTIONS

Looking ahead, this work provides a solid platform for further developments in cancer classification and gene expression analysis. Several potential research directions arise from the effectiveness of the proposed method:

Direction 1 – Multi-Omics Integration: Combining information from various omics layers—such as genomics, transcriptomics, proteomics, and epigenomics—may offer a more holistic understanding of cancer biology. Integrating these diverse datasets can improve the accuracy and reliability of classification models.

Direction 2 – Deep Learning Architectures: Applying advanced deep learning techniques, including recurrent neural networks (RNNs) and convolutional neural networks (CNNs), to gene expression data could help uncover complex patterns and interactions, enabling more sophisticated and precise cancer classification.

DIRECTION 3 (Interpretable AI): Developing methods that not only attain a high level of accuracy but also offer insights into the underlying biological processes responsible for classification decisions is crucial. Enhancing the interpretability of AI models could increase their adoption in clinical settings.

DIRECTION 4 (Personalized Medicine): The utilization of gene expression profiles for patient-specific treatment recommendations is a promising avenue. Future research could focus on tailoring classification models to individual patients, considering their unique genetic signatures and characteristics.

DIRECTION 5 (Benchmarking and Generalization): Extending the validation and benchmarking of hybrid approaches across diverse datasets and cancer types can provide a broader understanding of their performance and generalizability.

DATA AVAILABILITY STATEMENT

The datasets used and/or analyzed during the current study are available from the corresponding author on reasonable request.

CONFLICT OF INTEREST DISCLOSURE

The author confirms that they have no conflicts of interest to disclose.

FUNDING

No funding has been received.

AUTHORS CONTRIBUTION

In this study, Abrar Yaqoob made significant contributions by preparing materials, meticulously collecting data, and conducting data analysis. Navneet Kumar Verma provided valuable expertise through comprehensive literature reviews, critical insights, and substantial contributions to the manuscript's intellectual content and expertly designing experiments to address key research questions, and supervising the research process to maintain integrity and coherence. This collaborative effort, with each team member leveraging their specific expertise, led to a comprehensive and impactful research project.

REFERENCES

1. R. Poli, J. Kennedy, and T. Blackwell, "Particle swarm optimization: An overview," pp. 33–57, 2007, doi: 10.1007/s11721-007-0002-0.
2. R. A. Khurma, I. Aljarah, A. Sharieh, M. A. Elaziz, R. Damaševičius, and T. Krilavičius, "A review of the modification strategies of the nature inspired algorithms for feature selection problem," *Mathematics*, vol. 10, no. 3, 2022. doi: 10.3390/math10030464.

3. T. I. A. Mohamed, A. E. Ezugwu, J. V. Fonou-Dombeu, A. M. Ikotun, and M. Mohammed, "A bio-inspired convolution neural network architecture for automatic breast cancer detection and classification using RNA-Seq gene expression data," *Sci. Rep.*, vol. 13, no. 1, pp. 1–19, 2023. doi: 10.1038/s41598-023-41731-z.
4. A. A. Alhussan, M. M. Eid, S. K. Towfek, and D. S. Khafaga, "Breast cancer classification depends on the dynamic dipper throated optimization algorithm," *Biomimetics*, vol. 8, no. 2, pp. 1–20, 2023. doi: 10.3390/biomimetics8020163.
5. A. A. Ahmed, M. A. S. Ali, and M. Selim, "Bio-inspired based techniques for thermogram breast cancer classification," *Int. J. Intell. Eng. Syst.*, vol. 12, no. 2, pp. 114–124, 2019. doi: 10.22266/IJIES2019.0430.12.
6. D. González-Patiño, Y. Villuendas-Rey, A. J. Argüelles-Cruz, and F. Karray, "A novel bio-inspired method for early diagnosis of breast cancer through mammographic image analysis," *Appl. Sci.*, vol. 9, no. 21, 2019. doi: 10.3390/app9214492.
7. L. Peng, W. Chen, W. Zhou, F. Li, J. Yang, and J. Zhang, "An immune-inspired semi-supervised algorithm for breast cancer diagnosis," *Comput. Methods Programs Biomed.*, vol. 134, no. 61472092, pp. 259–265, 2016. doi: 10.1016/j.cmpb.2016.07.020.
8. A. Al Bataineh, D. Kaur, and S. M. J. Jalali, "Multi-layer perceptron training optimization using nature inspired computing," *IEEE Access*, vol. 10, pp. 36963–36977, 2022. doi: 10.1109/ACCESS.2022.3164669.
9. S. Thirumalaisamy *et al*, "Breast cancer classification using synthesized deep learning model with metaheuristic optimization algorithm," *Diagnostics*, vol. 13, no. 18, pp. 1–21, 2023. doi: 10.3390/diagnostics13182925.
10. A. Yaqoob, N. Kumar, V. Rabia, and M. Aziz, "Optimizing gene selection and cancer classification with hybrid sine cosine and cuckoo search algorithm," *J. Med. Syst.*, 2024, doi: 10.1007/s10916-023-02031-1.
11. S. M. Almufti, "Historical survey on metaheuristics algorithms," *Int. J. Sci. World*, vol. 7, no. 1, p. 1, 2019. doi: 10.14419/ijsw.v7i1.29497.
12. M. Azizi, S. Talatahari, and A. H. Gandomi, *Fire Hawk Optimizer: A novel metaheuristic algorithm*, vol. 56, no. 1. Springer Netherlands, 2023. doi: 10.1007/s10462-022-10173-w.
13. P. Agrawal, H. F. Abutarboush, T. Ganesh, and A. W. Mohamed, "Metaheuristic algorithms on feature selection: A survey of one decade of research (2009-2019)," *IEEE Access*, vol. 9, pp. 26766–26791, 2021. doi: 10.1109/ACCESS.2021.3056407.
14. Yang, X.-S., editor. Nature-inspired algorithms and applied optimization. Vol. 744. Cham: Springer, 2018.
15. H. Mohsen, E.-S. A. El-Dahshan, E.-S. M. El-Horbaty, and A.-B. M. Salem, "Classification using deep learning neural networks for brain tumors," *Futur. Comput. Inform. J.*, vol. 3, no. 1, pp. 68–71, 2018. doi: 10.1016/j.fcij.2017.12.001.
16. A. K. Shukla, P. Singh, and M. Vardhan, "A new hybrid wrapper TLBO and SA with SVM approach for gene expression data," *Inf. Sci.*, vol. 503, pp. 238–254, 2019. doi: 10.1016/j.ins.2019.06.063.
17. M. Toğaçar, B. Ergen, and Z. Cömert, "Detection of lung cancer on chest CT images using minimum redundancy maximum relevance feature selection method with convolutional neural networks," *Biocybern. Biomed. Eng.*, vol. 40, no. 1, pp. 23–39, 2020. doi: 10.1016/j.bbe.2019.11.004.
18. P. El Kafrawy, H. Fathi, M. Qaraad, A. K. Kelany, and X. Chen, "An efficient SVM-based feature selection model for cancer classification using high-dimensional microarray data," *IEEE Access*, vol. 9, pp. 155353–155369, 2021. doi: 10.1109/ACCESS.2021.3123090.
19. L. Venkataramana, S. G. Jacob, and R. Ramadoss, "A parallel multilevel feature selection algorithm for improved cancer classification," *J. Parallel Distrib. Comput.*, vol. 138, pp. 78–98, 2020. doi: 10.1016/j.jpdc.2019.12.015.

20. K. R. Pradeep, and N. C. Naveen, "Lung cancer survivability prediction based on performance using classification techniques of support vector machines, C4.5 and Naive Bayes Algorithms for healthcare analytics," *Procedia Comput. Sci*, vol. 132, pp. 412–420, 2018. doi: 10.1016/j.procs.2018.05.162.
21. P. Kurniawati, "No title التواصل ال» طفرة على تتغذى جرائم. الإلكتروني الابتزاز," *Univ. Nusant. PGRI Kediri*, vol. 1, pp. 1–7, 2017.
22. K. H. Chen *et al*, "Gene Selection for cancer identification: A decision tree model empowered by particle swarm optimization algorithm," *BMC Bioinform.*, vol. 15, no. 1, pp. 0–9, 2014. doi: 10.1186/1471-2105-15-49.

CHAPTER 6

Feature Extraction and Selection Methods and Bioinformatics Approach on Omics Data to Identify Molecular Signatures for Specific Diseases

Muskan Syed, Anushka Gupta, Priyanka Narad and Abhishek Sengupta

6.1 INTRODUCTION

The revolution of high-throughput omics technologies has revolutionized biomedical research and made large-scale profiling of biological molecules like DNA, RNA, proteins, and metabolites possible [1]. High-throughput omics technologies have helped provide an understanding of the mechanisms of diseases, their progression, and therapeutic response. Nevertheless, the immense and intricate nature of omics datasets poses a strong challenge in deriving meaningful information [2]. Uncovering meaningful molecular signatures from such high-dimensional datasets demands sound computational methods that can separate biologically relevant features from the noise and redundant information [3, 4]. Feature selection and extraction techniques are the focus of such datasets' enrichment through the removal of the most informative variables, the improvement of model performance, and the interpretation [5]. These range from conventional statistical techniques to complex machine learning (ML) and deep learning (DL) methodologies, which enhance dimensionality reduction without losing relevant biological information. Bioinformatics-aided strategies, such as network-based approaches, clustering analysis, and integrative multi-omics studies, further narrow down the choice of significant molecular markers for diseases [6]. This chapter overviews the fundamental concepts of feature extraction and

DOI: 10.1201/9781003591658-6

selection in the analysis of omics data, with a focus on their application in biomarker identification and precision medicine [7]. With the use of bioinformatics strategies, researchers can unravel disease-associated molecular patterns, improve diagnostic accuracy, and design targeted therapeutic strategies [8].

6.1.1 Overview of Omics Data

6.1.1.1 Definition and Types (Genomics, Transcriptomics, Proteomics, Metabolomics)

Omics data is the extensive datasets derived from high-throughput technologies that profile different biological molecules in an organism [9]. These datasets enable a system-wide view of cellular functions and interactions at several molecular levels [10]. The principal areas of omics sciences are [11]:

- **Genomics:** The study of an organism's complete genome, including the structure, function, evolution, and interactions of genes. Genomic analysis relies on whole-genome sequencing (WGS), whole-exome sequencing (WES), and genome-wide association studies (GWAS) to sequence the genetic variation and its effect on biological functions and disease [12].
- **Transcriptomics:** Examination of all the RNA molecules that are transcribed from the genome under different conditions. Transcriptomic profiling, using the technology of RNA sequencing (RNA-Seq) and microarrays, allows one to identify patterns of gene expression, regulatory events, and transcriptional differences among biological states [13].
- **Proteomics:** Global analysis of proteins, structure, modification, and functional properties. Protein expression, interaction, and post-translational modification information using proteomic approaches such as MS and 2D-GE are used in identification of protein biomarkers of disease [14].
- **Metabolomics:** Research on small molecules (metabolites) in a biological system that indicate metabolic functions and physiological status. Methods like nuclear magnetic resonance (NMR) spectroscopy and liquid chromatography-mass spectrometry (LC-MS) can be used for the identification and quantification of metabolites to aid in biomarker discovery and disease diagnosis [15].
- **Interactomics:** The investigation of biomolecular interactions, such as protein-protein interactions (PPIs) and signaling pathways, that are important for deciphering cellular communication, regulatory networks, and disease mechanisms [16].

6.1.1.2 Significance of Omics Data in Elucidating Biological Systems and Disease

Omics technologies have transformed biomedical research through a deep understanding of biological systems and disease mechanisms [17]. Multi-omics data integration gives us deeper insight into intricate molecular interactions, revealing new directions for precision medicine, drug discovery, and biomarker discovery [18].

- **Interpreting the Complexity of Biological Systems:** Omics data paints a complete picture of molecular interactions and networks inside cells, tissues, and organisms [19]. By analyzing different layers of biology, researchers can discover intrinsic regulatory mechanisms and identify molecular signatures of physiological and pathological conditions [20].
- **Biomarker Discovery:** Reliable biomarkers for preclinical diagnosis, prediction, and monitoring of disease therapy to be found using omics-based profiling Multiple omics improve the accuracy and specificity of the models by means of associations among genomic, transcriptomic, proteomic, and metabolomic changes [21].
- **Drug Development and Mechanistic Insights:** Omics data help to identify new pharmacological targets and mechanisms of action by which drugs may be developed. The efficacy of targeted drugs, the ability to predict pharmacological response, and tailoring treatment regimens to patient needs all rely on scientists providing understanding of molecular interactions and metabolic processes [22].
- **Precision and Personalized Medicine:** Personalized medicine seeks to customize treatment plans based on genetic and molecular profiles using patient-specific omics data. Pharmacogenomics, being a subcategory of genomics, facilitates prediction of a patient's response to drugs, hence reducing drug-induced side effects and improving efficacy [23].
- **Environmental and Agricultural Uses:** Omics technologies have an important role to play in environmental monitoring by assessing microbial populations, ecosystem dynamics, and effects of environmental stressors. Genomics and metabolomics are employed in agriculture for improved crop yield, nutritional quality, and resistance against disease and pests [24].

6.1.2 Importance of Omics Data Analysis

Biological systems are complex in nature with several regulatory characteristics affecting gene expression, protein function, and metabolic processes [25]. The human genome, for example, is made up of about 3.2 billion nucleotides that code for 20,000–25,000 protein-coding genes, which through alternative splicing and post-translational modifications produce more than a million protein isoforms [26]. Additionally, microbiome interaction and epigenetic regulation impact the expression of proteins and genes in various contexts [27]. Whereas multi-omics data does inform us regarding the biological systems, its handling isn't trivial [28]. The complexity as well as dimensionality in an omics dataset necessitate solid bioinformatics pipelines, proper statistical models, as well as computing strategies incorporated and integrated in synergy [28]. Additional factors that contribute to this complexity include heterogeneity of the data, variation in experiments, and the demand for large-scale comparisons [29]. The study of omics also calls for substantial inputs in terms of personnel with technical expertise, computation facilities, as well as a financial outlay [30]. Consequently, meticulous study design, strategies for integrating the data, and sophisticated ML methods are mandatory to optimize biological understanding from studies of omics [31]. As omics technologies advance, their application in biomedical research and in the practice of medicine increases, generating new knowledge and new healthcare solutions. Bioinformatics resources utilized to collect, analyze, and interpret massive omics datasets are vital in developing our understanding of human health and disease [32].

6.1.3 Significance of Molecular Signatures

Molecular signatures are central to contemporary biomedical research, providing essential information on disease mechanisms, biomarker identification, and personalized medicine [33]. These signatures—consisting of genes, proteins, metabolites, or other biomolecules—are specific markers of physiological or pathological conditions, allowing researchers to distinguish between health and disease at the molecular level. By combining multi-omics data, researchers can reveal subtle molecular patterns that precede disease onset, progression, and treatment response [34]. For example, transcriptomic and proteomic signatures can identify dysregulated pathways in cancer, whereas metabolomic profiles can offer insights into metabolic disorders like diabetes [35]. The identification of molecular signatures is also key to the discovery of targeted therapies, in that it is possible to create drugs that interact precisely with molecules implicated in the disease, making treatments more effective and less likely to cause unwanted effects [36]. The signatures are also incredibly useful for early disease diagnosis, prognosis, and monitoring treatment outcomes, hence constituting essential personalized medicine tools [37]. With the advances in bioinformatics and ML, identification and validation of molecular signatures have been made more effective, enabling their translation from bench to bedside. Molecular signatures are thus revolutionizing the future of diagnostics, therapeutics, and precision medicine, making patient management more effective and personalized [38].

Recent advances in molecular signature research have enhanced knowledge of disease etiology, diagnosis, and treatment approaches, and even changed their direction. Of these, among others [39]:

- **Alzheimer's Disease Detection:** University of Pittsburgh researchers developed a biomarker test capable of identifying early tau tangles, a hallmark of Alzheimer's disease. This test involves analyzing cerebrospinal fluid (CSF) obtained via a spinal tap, potentially enabling earlier and more effective interventions for individuals at risk of Alzheimer's [40].
- **Pediatric Health Risk Assessment:** Researchers at King's College London have developed a blood test that examines lipid profiles in kids to determine those at risk of developing major diseases, such as type 2 diabetes, liver disease, and heart disease. It is an early marker of health risks in assessing a wide range of lipid molecules, allowing it to prompt timely lifestyle interventions to avoid disease development [41].
- **Cancer Risk and Inflammation:** In a study conducted by Dr. Laura Kellman at Stanford University, almost 400 single-nucleotide variants have been found to

potentially trigger and affect the onset of 13 prevalent cancer types, including breast, lung, colorectal, and prostate cancers. This research emphasizes the critical role inflammation plays in cancer and identifies emerging genetic targets for new therapies and prevention [42].

- **Blood Protein Screening to Prevent Illness:** The UK Biobank has launched the largest ever investigation of blood proteins, screening 300,000 blood samples for thousands of proteins. The project will revolutionize the treatment and prevention of disease by being able to better understand the role proteins play in diseases like Parkinson's, Alzheimer's, and schizophrenia, which will open new targets for drugs and tailored approaches to treatment [43].

These cases hold the potential to revolutionize by molecular signatures driving personalized medicine, early diagnosis of disease, and the design of targeted treatment for many fields in medicine [44].

6.1.3.1 Definition and Role in Disease Diagnosis, Prognosis, and Treatment

Molecular signatures are distinct sets of genes, proteins, metabolites, or other biomolecules that define a particular biological state, disease condition, or response to therapy [45]. Molecular signatures, the most important biomarker in precision medicine, are defined by high-throughput omics technologies including genomics, transcriptomics, proteomics, and metabolomics. The identification and characterization of molecular signatures have transformed disease diagnosis, risk estimation, prognosis, and therapeutic choice, leading to more individualized and efficient treatment strategies [46].

6.1.3.1.1 Role in Disease Diagnosis and Prognosis Identification of molecular signatures is an important element in the early detection of diseases and accurate diagnosis [47]. Classical diagnosis techniques like imaging and histopathology can identify the disease only in advanced stages, but molecular biomarkers enable identification of the disease even before it generates clinical symptoms [48]. For example, in the case of cancer, circulating tumor DNA (ctDNA) isolated from liquid biopsies is gaining traction as it is being used to recognize the genetic mutation associated with several forms of cancer and allows non-invasive highly sensitive early diagnosis [49]. In the same way, transcriptomic signatures have been determined for autoimmune conditions like rheumatoid arthritis and lupus, which facilitate improved discrimination among disease subtypes and the selection of appropriate treatments [50]. Moreover, in neurodegenerative diseases such as Alzheimer's disease, molecular signatures from CSF proteins like amyloid-beta and tau have played a critical role in differentiating affected subjects from healthy controls. Identification of these biomarkers has enabled early diagnosis and enhanced patient selection for clinical trials to develop disease-modifying treatments [51].

Molecular signatures are also useful in predicting disease progression and patient outcomes. In cancer, gene expression profiling has also provided prognostic tests like Oncotype DX and MammaPrint, which determine the risk of recurrence in breast cancer patients [52]. These tests measure the expression of various genes that are implicated in tumor progression and enable clinicians to decide if a patient will require chemotherapy or if hormone therapy would be enough [53]. Likewise, in cardiovascular diseases, plasma proteomics-derived molecular signatures can predict myocardial infarction risk and heart failure, enabling early intervention and tailored management approaches [54]. In infectious diseases, transcriptomic profiling has been employed to uncover immune response signatures that can be used to forecast infection severity, such as in SARS-CoV-2 and other viral infections [55].

6.1.3.1.2 Role in Treatment Selection and Monitoring One of the most important uses of molecular signatures is in personalized medicine, in which the treatment is customized based on the molecular signature of an individual [56]. Pharmacogenomics, the study of how genetic diversity affects drug response, has already discovered several molecular markers that dictate treatment. For instance, in non-small cell lung cancer (NSCLC), epidermal growth factor receptor (EGFR) gene mutations predict sensitivity to targeted agents like erlotinib and Osimertinib [57]. Equally, the availability of PD-L1 expression on tumors dictates treatment with immune checkpoint inhibitors such as pembrolizumab, which have changed the face of cancer therapy by leveraging the body's immune response. Molecular signatures are also implicated in monitoring treatment response and drug resistance [58]. In chronic myeloid leukemia (CML), quantitation of BCR-ABL fusion transcript is employed to assess the response of tyrosine kinase inhibitors (TKIs) like imatinib. Resistance-associated mutations such as T790M in EGFR-mutant lung cancer can be detected by liquid biopsy, with resultant therapy adaptation early on [29]. Furthermore, molecular signatures have recently also begun to be tested for enhancing non-oncology treatments of autoimmune diseases, diabetes, and mental health disorders [60]. For example, transcriptomic and proteomic profiling is being applied to determine patient subgroups in rheumatoid arthritis that are likely to benefit from biologic therapies like TNF inhibitors [61].

6.1.3.2 Objectives of Identifying Molecular Signatures

Identification and utilization of molecular signatures are designed to address several major objectives in biomedical discovery and clinical application [62]. Systematic investigation of high-dimensional omics data enables scientists to elicit meaningful information that improves disease understanding, improves diagnostic accuracy, and personalized therapeutic treatment [63].

6.1.3.2.1 Early Disease Detection and Risk Assessment One of the key objectives of molecular signature discovery is to enable early disease diagnosis, which will greatly enhance the prognosis of the patient by enabling interventions at the appropriate time [64]. For instance, in ovarian cancer, where the signs only manifest in late stages, proteomic and metabolomic patterns have been found to detect the disease at an early, curable stage [65]. Similarly, epigenetic signatures such as DNA methylation patterns have also been identified in early cancers, providing new avenues for non-invasive screening methods [66].

Apart from cancer, molecular profiles are also employed to determine disease risk factors in what appear to be healthy individuals. GWAS-derived polygenic risk scores (PRS) can assess a patient's risk for complex conditions like cardiovascular disease, type 2 diabetes, and neurodegenerative disease [67]. With the application of these risk genes in the clinic, doctors are now able to advise patients regarding lifestyle modifications and preventive interventions to address individual-specific needs [68].

6.1.3.2.2 Disease Mechanism Understanding Molecular signatures give a broader understanding of disease pathogenesis, and how biological pathways are involved in disease development [69]. Such an understanding is fundamental to the identification of new therapeutic targets. In neurodegenerative diseases, such as Parkinson's and Alzheimer's, metabolomic signatures have disclosed dysregulation in lipid metabolism and mitochondrial activity, potential new targets for drug action. In infectious disease, transcriptomic studies have revealed unique host immune response signatures to bacterial and viral infection [70]. For instance, scientists have discovered gene expression signatures in blood that distinguish between bacterial and viral infection and can help ensure proper use of antibiotics and the limitation of antimicrobial resistance [71].

6.1.3.2.3 Stratifying Patient Populations for Personalized Medicine Molecular signatures enable patients to be stratified into subgroups based on their biological characteristics, and precision medicine strategies become easier. In this case, this manifests itself by dividing breast cancers into HER2-positive, triple-negative, and hormone receptor-positive subtypes each requiring its own set of therapies [72]. Outside of cancer, molecular signatures have also played a key role in neurological and psychiatric illness. Transcriptome and proteome profiling of schizophrenia and major depressive disorder (MDD), for instance, has yielded molecular subtypes that can account for drug-response heterogeneity, opening the door to more targeted psychiatric therapies [73].

6.1.3.2.4 Facilitating Drug Discovery and Repurposing As a blueprint, disease-associated molecular markers are used to expedite drug discovery and re-engineer drugs with new indications [74]. Multi-omics technologies have disclosed new targets for drugs in inflammatory, metabolic, and cancer diseases. For example, through the integration of genomics and proteomics, it became possible to recognize PD-1 and CTLA-4 as checkpoint immune molecules, producing checkpoint inhibitors which have revolutionized cancer treatment. In the same manner, metabolomic research has also discovered dysregulated pathways in non-alcoholic fatty liver disease (NAFLD), and this has guided the design of therapeutic interventions in lipid metabolism [75].

6.1.3.2.5 Tracking Response to Treatment and Drug Resistance Molecular biomarkers allow physicians to track patients' drug response and detect emerging drug resistance [76]. As an example, real-time minimal residual disease (MRD) monitoring based on next-generation sequencing (NGS) has improved ALL relapse prediction. In antibiotic resistance, proteomic and transcriptomic signatures of bacteria are being used to find mechanisms of resistance and to design new approaches for targeting multidrug-resistant pathogens. Use of rapid molecular diagnostics from these signatures can also inform appropriate use of antibiotics and enhance treatment efficacy [76].

6.1.3.2.6 Enhancing Predictive Modelling in Systems Biology Technological advances in artificial intelligence and ML have enabled the incorporation of multi-omics data to allow predictive models of disease progression to be developed [77]. Molecular signatures are being incorporated more and more into computational models to simulate intricate biological systems, predict patient outcomes, and maximize therapeutic interventions [78]. For example, models based on artificial intelligence using genomic and proteomic data are utilized to predict response to immunotherapy in oncology patients, enabling clinicians to select the most appropriate regimens. Such predictive models hold great promise to redefine personalized medicine and enhance clinical decision-making [79].

6.1.3.3 Stages of Molecular Signature Discovery

With the development of high-throughput omics technologies, molecular signature discovery has become increasingly important for disease-specific diagnosis and phenotypic classification [80]. This encompasses several steps, ranging from setting the scientific and clinical context through to testing signatures on independent datasets [81].

6.1.3.3.1 Establishing the Scientific and Clinical Context The initial step in molecular signature discovery is the establishment of its scientific and clinical significance [82]. This involves exact characterization of the target clinical phenotypes such that the molecular signature identified possesses pragmatic value in diagnostics, prognostics, or therapeutic decisions [83]. Researchers need to ensure that they select the right omics data types that suit the research question most optimally and define sensitivity and specificity levels for ascertaining the clinical applicability of the discovered signature [84]. For instance, in the study of cancer, determination of a molecular signature to estimate recurrence risk is achieved by identifying the phenotypic endpoint (e.g., recurrence vs. non-recurrence) and having the discovered biomarkers be meaningful clinically, such as influencing the decision on treatment or stratifying the patients [85]. The identification of molecular signatures from omics data has many promising applications including omics-based tests for disease-specific diagnostics and accurate phenotype classification [86].

6.1.3.3.2 Omics Data Collection Collection of data plays an important role in obtaining strong and reproducible molecular signatures [87]. Having a good amount of sample size is important to have sufficient statistical power, and regulation of experimental and analytical processes ensures that batch effects that introduce bias are minimized [88]. Researchers also need to be careful to avoid confounding factors that may skew results. While working with available datasets, it is crucial to confirm sample sizes and verify the existence of sufficient patient data [89]. Datasets should be normalized to be able to compare across studies, and combining various datasets, though useful, demands caution to ensure heterogeneity management. Major datasets like The Cancer Genome Atlas (TCGA) and Gene

Expression Omnibus (GEO) have played significant roles in deriving stable molecular signatures, especially in cancer research [90].

6.1.3.3.3 Model Building and Feature Selection Once the data have been collected, the next step is to choose interesting features and create prediction models. Omics data consist of many thousands of features, but just a few will be useful for classification or prediction. Feature selection can be supervised, in which features most correlated with the clinical phenotype are selected (e.g., top 20% by statistical significance), or unsupervised, in which features with large variance across samples are selected [91]. For example, in transcriptomics research, DEGs are discovered through statistical analysis like t-tests or ANOVA [92]. Following feature selection, computational models are developed to predict clinical outcomes. The choice of model depends on the research aim—Cox proportional hazards models for survival analysis, logistic regression, support vector machines (SVM), and neural networks for classification, while principal component analysis (PCA) and clustering algorithms aid in reducing dimensions and unsupervised learning [93]. For example, in breast cancer research, ML models have been trained with RNA sequencing data and have identified patients as belonging to high- and low-risk groups based on molecular signatures [94].

6.1.3.3.4 Model Performance Evaluation Model comparison is very important to ensure reliability and accuracy. Cross-validation is commonly employed to evaluate performance, and feature selection needs to be done in every training fold to avoid data leakage [95]. Doing otherwise, like feature selection prior to cross-validation, has the tendency of producing unduly optimistic error rates. An excellent example in the field of Alzheimer's disease studies demonstrates this principle, where ML algorithms that had been trained on proteomic information were tested with independent cohorts to ensure generalizability and reproducibility of predictions [96].

6.1.3.3.5 Testing on Independent Datasets A good molecular signature must be cross-validated in independent sets to validate its reproducibility and clinician usability [97]. Ideally, independent test sets should be of various sources, i.e., multi-center studies, to improve generalizability. Testing in varying populations obviates bias and promotes similarity in predictiveness. For instance, genomic signatures used for colorectal cancer prognostication are cross-validated across geographically diverse location-based datasets, which proves their validity in prediction in varying patient groups [98].

6.1.3.3.6 Increasing Reproducibility and Transparency To maintain scientific integrity, scientists are required to make all the information concerning their research work, such as the criteria for patient enrolment, the experimental and analytical methods, data processing procedures, feature selection procedures, and model-building procedures, public [99]. Source code availability of computational models is more transparent and permits independent validation. FAIR (Findable, Accessible, Interoperable, Reproducible) principles are increasingly being adopted to promote data sharing and improve reproducibility of omics studies [100].

6.1.3.3.7 Reporting Performance Results Objectively To reduce publication bias and enhance transparency, positive and negative results should be published by authors [101]. These encompass publication of correct and incorrect predictions, reporting study limitations, and reporting potential confounding variables. As an example, failed predictions in the identification of biomarkers in Parkinson's disease are reported systematically to aid in the improvement of subsequent models as well as make them more robust. Recognition of these limitations and challenges helps in the creation of more accurate and clinically relevant molecular signatures [102].

6.2 ADVANCED STRATEGIES FOR FEATURE SELECTION IN MOLECULAR SIGNATURE DISCOVERY

Feature selection is an essential step in the identification of stable molecular signatures from high-dimensional omics data [103]. Due to the enormous number of variables (genes, proteins, metabolites) and the generally small sample sizes, the most informative features need to be selected for enhancing model performance, interpretability, and biological significance. Effective feature selection enhances predictive performance, reduces computational complexity, and avoids overfitting [104].

6.2.1 Types of Feature Selection Methods

Feature selection techniques can be categorized as a general three-way category [105]:

- **Filter Methods**—These methods evaluate individual characteristics based on statistical indicators such as t-tests, ANOVA, chi-square tests, and correlation coefficients [106]. They are extensively utilized in transcriptomics for the detection of differentially expressed genes (DEGs) in conditions such as cancer [107]. For example, in breast cancer studies, DEGs detected by fold-change analysis and false discovery rate (FDR) adjustments have been employed to differentiate patients according to molecular subtypes [108].

- **Wrapper Methods**—Wrapper methods iteratively test feature subsets with predictive models. Recursive feature elimination (RFE) and Genetic Algorithms (GA) are used to identify the optimum feature sets [109]. While computationally costly, wrapper methods are more precise in ML model feature selection [110]. In proteomics-assisted Alzheimer's disease biomarker discovery, RFE with SVM has been effective in eliminating potential proteins for early diagnosis [111].

- **Embedded Methods**—Embedded methods incorporate feature selection as a part of learning. Feature importance of LASSO (Least Absolute Shrinkage and Selection Operator) and tree-based algorithms (random forest, XGBoost) is assigned automatically upon training [112]. LASSO regression has been extensively used in metabolomics studies of gestational diabetes mellitus (GDM) to select metabolites that play a significant role in disease pathogenesis [113].

6.2.2 Machine Learning and Network-Based Feature Selection

The integration of ML and DL models has revolutionized feature selection in molecular signature identification [114]. Models such as random forests, gradient boosting machines (GBM), and deep neural networks (DNNs) are capable of ranking biomarkers based on their predictive capability [115]. For example, in multi-omics cancer classification, extreme gradient boosting (XGBoost) has been utilized to rank genes associated with treatment response. Network-based feature selection has gained attention because it can incorporate biological interactions [115]. Graph neural networks (GNNs) and network regularized regression (NRR) both use molecular interaction networks (for instance, PPI databases) to find biologically meaningful features. Neurodegenerative disorders like Parkinson's, for instance, have had network-based approaches identify disease progression-related key gene hubs [116].

6.2.3 Dimensionality Reduction Techniques

In high-dimensional data, standard unsupervised methods like PCA and PLS-DA are used for dimensionality reduction of feature space maintaining informative data. In metabolomics profiling of patients with sepsis, PLS-DA has been applied to find critical metabolic signatures with the ability to distinguish survivors and non-survivors [117].

6.2.4 Multi-Omics Integration

Novel feature selection paradigms integrate data across several layers of omics to improve biomarker stability. Bioinformatics tools such as DIABLO (data integration analysis for biomarker discovery) and MOFA (multi-omics factor analysis) enable simultaneous feature selection across transcriptomic, proteomic, and metabolomic data. With emerging precision medicine, future research will target self-supervised learning and federated learning-based feature selection to allow scalable and privacy-preserving biomarker discovery across multi-center cohorts [118]. Explainable AI (XAI) approaches, including SHAP (Shapley Additive Explanations), will be used increasingly to increase the interpretability of the selected features to ensure clinical utility and relevance [119].

6.3 DATA PREPROCESSING AND NORMALIZATION

Discovery of molecular signatures depends on high-throughput omics approaches like genomics, transcriptomics, proteomics, and metabolomics to determine biomarkers linked with disease states, drug reactions, and physiological conditions [120]. The raw omics data tend to be noisy, batch-effect plagued, and have missing values, which may dramatically affect the integrity of subsequent analysis [121]. Thus, preprocessing and normalization are the most essential steps that help to guarantee data quality, increase the comparability of samples, and make the ML model more accurate [122]. In this section, there is an in-depth description of the methods of preprocessing, quality control (QC) procedures, handling missing data approaches, and transformation techniques with new real-world examples [123].

6.3.1 Preprocessing Techniques

Preprocessing is the first step in data analysis, which cleans raw data into a structured, clean, and technical artifact-free format. Different omics datasets have specific preprocessing approaches depending on their nature, but common goals are to remove unwanted variations, correct for systematic biases, and normalize data for comparison [124].

6.3.1.1 Preprocessing for Genomics and Transcriptomics Data

Preprocessing genomic and transcriptomic data encompasses several stages beginning with quality checking through the FastQC tool that assesses per-base quality scores, GC content, and possible adapter contamination [125]. Following quality checking, low-quality reads and adapter content are trimmed through software such as Trim Galore or Trimmomatic to prevent downstream errors [126]. Once cleaned, the sequencing reads are aligned against a reference genome using STAR (for RNA-Seq) or BWA-MEM (for DNA-Seq) to assign gene expression or variant calls in the right way. Duplicate reads, in general, generated during PCR amplification, are eliminated by Picard Mark Duplicates, and bias correction is performed by software like DESeq2 to adjust for depth variation in sequencing [127]. A Pan-cancer transcriptomics analysis utilized these preprocessing techniques to quantify RNA-Seq from more than 10,000 tumor samples using feature counts and incorporating cross-sample comparability. These methods allowed reliable cancer subtype classification from transcriptomic profiles [128].

6.3.1.2 Proteomics Data Preprocessing

Preprocessing in proteomics begins with peak detection and deconvolution of mass spectrometry (MS) data [129]. Tools like MaxQuant identify peptide signals, eliminate background noise, and attribute fragment peaks to proteins. For reproducibility between runs, MS Convert and OpenMS synchronize retention times, compensating for variations that occur during LC-MS. Batch effects, typically an issue in large-scale proteomics studies, are corrected by ComBat (sva R package), improving the consistency of protein quantifications [130]. A proteomics study of breast cancer drug resistance, published in 2024, used DIA-NN (Data-independent acquisition) pipelines to preprocess and quantify low-abundance proteins, improving the accuracy of drug response predictions [131].

6.3.1.3 Metabolomics Data Preprocessing

Metabolomics data preprocessing features extraction, peak annotation, and normalization for technical variation correction [132]. XCMS (for LC-MS) and MZmine 3 detect and quantify metabolite features and correct instrumental drift. Metabolite annotation against known databases is performed by MetFrag and GNPS, which map detected peaks to reference fragmentation patterns [133]. Normalization methods, e.g., probabilistic quotient normalization (PQN) and log transformation, correct sample dilution effects and variance stabilization. A 2023 multi-cohort analysis of metabolic biomarkers in diabetes used XCMS and MetaboAnalyst to preprocess more than 5,000 LC-MS samples so that metabolic features were kept consistent across various patient cohorts [134].

6.3.2 Data Cleaning and Quality Control

Data cleaning and data QC must be performed to maintain the integrity and reliability of the omics data [135]. Without them, batch effects, systematic errors, and low-quality measurements will introduce extreme biases and lead to false conclusions. Data cleaning is also a key process while preprocessing omics data, which makes the data accurate, consistent, and devoid of technical or biological artifacts [136]. High-throughput biological data contains errors in terms of sequencing artifacts, instrument malfunction, missing data, or improper merging of data, making it a challenge to analyze. Low-quality data can significantly impact downstream analysis, leading to incorrect conclusions and volatile biomarker identification. Data cleaning is the procedure of identifying and correcting such errors with standardized methods such as error detection, batch effect correction, and removal of inconsistent data points [137].

Key Aspects of Data Cleaning:

- **Error Detection and Correction:** The majority of omics dataset errors result from limitations in technology such as sequencing misreads, incorrect peak identification in metabolomics, or incorrect gene annotation. Phred scores (in sequencing data), signal-to-noise ratio (in proteomics), and MS/MS spectral quality (in metabolomics) are QC measures used to eliminate low-confidence data [137].
- **Dealing with Outliers:** The outliers may be of biological (other than rare mutation or outlier phenotype) or of technical (removed by batch effects, contamination, or misannotation) origin. Several techniques like Z-score method, interquartile range (IQR) method, and Mahala Nobis distance can be used in order to mark and remove the technical outliers [138].
- **Standardization of Sample and Metadata Annotations:** Omics data would normally be gathered from multiple sources, and different naming conventions for samples, file format, and organization of metadata can cause inconsistencies. Standardization by adoption of MIAME for transcriptomics or mzML/mzTab for metabolomics allows standard representation of data [139].
- **Batch Effect Correction:** Batch effects result from systematic variation according to various experimental conditions (e.g., sequencing platform, reagent batches, time of processing). ComBat (sva R package), RemoveBatchEffect (limma R package), batch effect correction through DL is some software normalizing the data. Cross-cohort reproducibility of DEGs in a multi-centre RNA-Seq analysis of breast cancer was enhanced using batch correction with ComBat-seq [140].

By following good data cleaning practices, researchers can ensure that their datasets are biologically relevant and suitable for proper downstream analysis, such as feature selection and ML modeling [141].

6.3.2.1 *Genomics and Transcriptomics Quality Control*

Genomics and transcriptomics QC involve outlier detection, filtering of low-quality genes, and batch effect correction. Hierarchical clustering and PCA identify samples that have high variability and are either removed or corrected [142]. Low-expressed genes, regarded as background noise, are removed with HTSeq by discarding genes that have less than 10 reads across all the samples. Batch effects, which occur due to varying sequencing platforms or experimental conditions, are adjusted by surrogate variable analysis (SVA) or PEER (probabilistic estimation of expression residuals). The GTEx (genotype-tissue expression) Project implemented these QCs to eliminate confounders such as tissue heterogeneity, enhancing the reproducibility of gene expression data between different tissue types [143].

6.3.2.2 *Proteomics Quality Control*

Proteomics data need to have stringent QC procedures implemented to minimize noise and enhance peptide quantification [144]. Variable peptides with a coefficient of variation (CV) of > 20% are removed to have consistency in LFQ experiments. To control false discovery rates (FDR), the Benjamini-Hochberg method is utilized through the Perseus software to remove unreliable protein identifications [145, 146].

6.3.2.3 *Metabolomics Quality Control*

Metabolomics data are subjected to rigorous QC to correct for instrumental drift and biological variation. Internal standard normalization with stable isotope-labeled metabolites normalizes for instrument response variation [147]. Pooled QC samples are also injected at regular intervals to track LC-MS performance, allowing for drift correction in MZmine and XCMS pipelines. A metabolomics study of colorectal cancer utilized pooled QC samples to guarantee reproducible feature detection across various analytical runs, enhancing the reproducibility of metabolite-driven cancer biomarkers [148].

6.3.3 Handling Missing Data

Missing values in omics experiments may occur due to low detection limits, technical failure, or sample dropout. Several imputation methods are employed based on the missing data mechanism [149].

6.3.3.1 *Missing Data Mechanisms*

Missing values may be of three kinds:

- **Missing Completely at Random (MCAR):** Occurs because of random errors such as instrument noise and does not reflect any discernible pattern.
- **Missing at Random (MAR):** It is related to other observed variables that need data-driven imputation processes.
- **Missing Not at Random (MNAR):** Systematic missing values as is often found in proteomics, where low-abundance peptides are missing [150].

6.3.3.2 Imputation Techniques

Varying imputation methods are applied based on the richness of missing values. K-nearest neighbors (KNN) fills in missing values using similar examples, whereas probabilistic PCA (PPCA) is optimal for large-scale proteomics data. Random forest imputation is commonly applied for metabolomics, whereas DL strategies such as Deep Impute (RNA-Seq) and autoencoders are becoming viable options for large transcriptomics datasets [151]. An AI-aided metabolomics study of Parkinson's disease used random forest imputation to restore 85% missing metabolite features with high improvements in pathway enrichment accuracy.

6.3.4 Data Transformation and Scaling

Scaling and data transformation are essential preprocessing steps that render omics data more comparable, interpretable, and statistically robust [152]. High-throughput platforms such as RNA sequencing (RNA-Seq), proteomics, and metabolomics generate high-dimensional data with complex distributions, e.g., heavy skewness, heteroscedasticity (variance differences between samples), and batch effects [153]. Use of proper transformation and scaling strategies guarantees that downstream statistical analyses like differential expression analysis, clustering, and ML models yield stable and biologically relevant conclusions [154].

6.3.4.1 Common Transformation Methods

Most omics data have a high dynamic range, in that values range over several orders of magnitude [155]. This is especially true in transcriptomics, where expression levels for the highly expressed housekeeping genes and the lowly expressed regulatory genes may differ by many orders of magnitude [156]. Likewise, for proteomics and metabolomics, MS peak intensities can be greatly skewed by the occurrence of dominant proteins or metabolites. For these issues to be solved, transformation methods are utilized to stabilize variance, lower skewness, and enhance normality assumptions needed for parametric statistical tests [156].

- **Log2 or Log10 Transformation:** Used extensively throughout transcriptomics and proteomics, log transformation minimizes right-skewing of distributions by decreasing large values without loss of relative difference between features. Log2 transformation is commonly applied in RNA-Seq data to enable significant fold changes in gene expression with outlier value resistance [157].
- **Variance Stabilizing Transformation (VST):** VST in DESeq2 R package fixes heteroscedasticity by transforming count data to achieve closer-to-equal variance across expression levels. It is particularly useful for RNA-Seq and scRNA-Seq studies [158].
- **Regularized Log Transformation (rlog):** Another DESeq2 tool, rlog is helpful with small-sample data by shrinking genes with low expression to reduce noise while preserving biologically meaningful variation [159].

- **Box-Cox and Yeo-Johnson Transformations:** These kinds of power transformations are often used in metabolomics to standardize variance and achieve normality in zero and negative value-containing data. Unlike log transformations, Box-Cox requires strictly positive values, whereas Yeo-Johnson supports zero and negative values and is better suited for clinical metabolomics data. For example, in a 2023 Cancer Transcriptomics Study, researchers found that applying VST to RNA-Seq data preprocessing significantly improved cancer subtype clustering accuracy to more effectively detect tumor-specific gene expression patterns [160].

6.3.4.2 Scaling Methods

Scaling puts features of an omics dataset onto a common numerical scale to avoid the influence of large-magnitude variables in analysis, including PCA, clustering, and ML models [161]. Omics datasets differ in terms of distribution and biological variability and therefore need specialized scaling methods accordingly [162]:

- **Z-Score Standardization:** It normalizes all the features by taking away the mean and dividing by the standard deviation, so the mean is 0 and the standard deviation is 1. This is particularly prevalent when working with ML where models are presuming data that is normally distributed [163].
- **Min-Max Scaling:** Used broadly across proteomics and metabolomics, min-max scaling scales values onto a common range (e.g., 0 to 1) at the cost of retaining relationships between points at the loss of extreme variation [164].
- **Quantile Normalization:** This technique, widely used in microarray and RNA-Seq data analysis, normalizes all samples to one distribution, which minimizes batch effects and technical variation between samples. It is very useful in cross-sample comparisons of large-scale transcriptomics experiments [165].
- **Pareto Scaling:** Pareto scaling is metabolomics-specific since all the features are normalized by dividing the features with the square root of the feature standard deviation. Pareto scaling preserves biological variation without reducing the impact of the most concentrated metabolites and is suitable for analysis having complex metabolic profiles [166].
- **Total Ion Current (TIC) and Probabilistic Quotient Normalization (PQN):** Metabolomics-specific scaling methods applied to offset sample dilution variation and instrument drift, enhancing the reproducibility of metabolite quantitation [167].

A multi-omics cancer study demonstrated that quantile normalization of transcriptomics and Pareto scaling of metabolomics significantly enhanced biomarker discovery by making it easier to integrate heterogeneous omics data. By ensuring that feature distributions are uniform across platforms, researchers were able to identify robust molecular signatures associated with cancer development [168].

To offer a more detailed perspective, Table 6.1 below presents a comparative analysis of various normalization methods frequently employed in omics data preprocessing.

6.3.5 Challenges in Feature Extraction from Omics Data

6.3.5.1 High Dimensionality

Omics datasets have a very high feature-to-sample ratio. For example, in genomics, a dataset might contain expression data for tens of thousands of genes but only a few hundred samples. It is difficult to identify which features are associated with the biological problem under high dimensionality, and overfitting is more likely. In addition, analyzing and storing these kinds of large datasets require a great number of computational resources [169].

Solution Strategies:

- **Feature selection programs:** Methods like RFE, Lasso regression, or mutual information-based approaches help identify the most useful features [170].
- **Dimensionality reduction:** Techniques like PCA, t-SNE, and Uniform Manifold Approximation and Projection (UMAP) reduce variables while maintaining variance in the data [171].

6.3.5.2 Missing Data

Omics datasets are likely to include missing or incomplete data due to various reasons like measurement errors, sample loss, or insensitivity in detection for some molecules. Missing data are likely to be biased and lead to erroneous feature extraction [172].

Missing Data Handling:

- **Imputation:** KNN, Bayesian imputation, or regression-based imputation are used to impute missing values [173].
- **Data exclusion:** Tastes or samples with excessive missing data may be excluded from analysis in certain cases [174].

6.3.5.3 Noise and Variability

Omics data is prone to biological and technical noise. Biological variability can conceal true underlying biological signals by individual differences (e.g., genetic diversity) [175]. Technical noise can be caused by limitations of instruments, variability in sample preparation, or batch effects (systematic variations between experimental runs) [176].

Noise Mitigation Techniques:

- **Normalization:** Methods such as Z-score normalization, quantile normalization, or batch effect correction (e.g., ComBat) diminish technical noise and correct batch effects [177].
- **Replicates:** Biological and technical replicates can make patterns observed reproducible and stable [178].

TABLE 6.1 Comparison of Normalization Methods in Omics Data Preprocessing

Normalization Method	Description	Formula/Method	Range	Use Case	Advantages	Disadvantages
Min-Max normalization	Rescales the data to a fixed range, usually [0, 1] or [−1, 1]	$x_{\text{norm}} = \frac{x - \min(x)}{\max(x) - \min(x)}$	[0, 1] (or custom range)	Suitable when you know the data range and need to bound it to a specific scale	Simple to implement Effective for algorithms like neural networks	Sensitive to outliers Assumes a known range of data
Z-Score normalization (standardization)	Converts data to have zero mean and unit variance	$x_{\text{std}} = \frac{x - \mu}{\sigma}$ where μ is the mean and σ is the standard deviation of x.	Unborn but usually [−3, 3] or similar	Used for algorithms assuming Gaussian distributions (e.g., SVM, logistic regression)	Not sensitive to outliers Does not assume a fixed range	Can be misleading with skewed data Assumes normal distribution
Robust scaler	Scales data using the median and interquartile range (IQR)	$x_{\text{robust}} = \frac{x - \text{median}(x)}{\text{IQR}(x)}$ where $\text{IQR}(x) = Q_3(x) - Q_1(x)$ (interquartile range).	Unbounded	Best for data with outliers, as it's less sensitive to them	Handles outliers well More robust than Min-Max or Z-Score	May not work well for data that is normally distributed
Decimal scaling	Scales data by a factor of 10 based on the maximum absolute value of the feature	$x_{\text{decimal}} = \frac{x}{10^j}$ where j is the smallest integer such that $\max(\lvert x \rvert) < 1$	Depends on data	For features with large values, scaling by powers of 10	Simple to understand Avoids changing data distribution	Less commonly used Can result in loss of precision
Unit vector normalization (L2 normalization)	Scales each data point such that its norm (length) is 1	$x_{\text{L2}} = \frac{x}{\lVert x \rVert_2}$ where, $\lVert x \rVert_2 = \sqrt{\sum_{i=1}^{n} x_i^2}$	Varies, but norm = 1	Common in text processing or when comparing vectors (e.g., cosine similarity)	Useful for measuring angles between vectors Works well with sparse data	Not suitable for algorithms that assume distribution

6.3.5.4 High Complexity of Biological Systems

Biological systems are highly complex, and there are intricate interactions between genes, proteins, and metabolites [179]. It is hard to find features that truly represent these interactions without loss of information or over-simplification. For instance, gene expression levels may not be indicative of the total biological activity; protein post-translational modifications, metabolite-metabolite interactions, or regulatory networks could be equally important [180].

Solution Approaches:

- **Pathway-based analysis:** Pathway analysis places features in clusters based on whether they contribute to recognized biological processes or pathways instead of single genes or proteins [181].
- **Multi-omics integration:** Integration of data from various layers of omics (e.g., genomics, transcriptomics, and proteomics) will provide a full understanding of the biological system but requires tight integration of data as well as analysis strategies [182].

6.3.5.5 Lack of Ground Truth

There cannot be "ground truth" for what features are most important in most biological settings, especially when in disease conditions like cancer. A specific expression pattern of genes, for example, will hardly give a good explanation of disease because there are complex interactions [183]. Such lack of gold-standard biologic knowledge compels feature extraction methods to validation challenges [184].

Validation Approaches:

- **Cross-validation:** Without ground truth, cross-validation protocols help to estimate the stability of feature extraction and selection algorithms [185].
- **External datasets:** Comparisons between results on independent datasets or use of prior biological knowledge can help to confirm the biological relevance of the features extracted [186].

6.3.5.6 Heterogeneity of Data

Omics data are often obtained from differing technologies (e.g., RNA-Seq, microarray, MS) or populations (e.g., healthy and diseased patients) [187]. The platform is linked to bias, sensitivity, and specificity, and the amalgamation of these diverse data sources is a significant challenge [188].

Methods for Handling Data Heterogeneity:

- **Data harmonization:** Techniques like batch effect removal (e.g., ComBat) and standardization methods are applied to harmonize data across different sources [189].
- **Multi-platform integration:** Methods like canonical correlation analysis (CCA) or network-based approaches enable integration of diverse omics data to obtain more accurate feature extraction [190].

6.3.5.7 Biological Interpretation

Finally, even after feature extraction, it is challenging to interpret the features from a biological perspective. The extracted features must be mapped back to biological mechanisms, pathways, or disease, which requires in-depth knowledge of the biological system under study [191]:

- **Gene ontology and pathway enrichment:** Tools like gene ontology (GO) analysis or Kyoto Encyclopedia of Genes and Genomes (KEGG) pathway analysis can be utilized to position extracted features into a known context of biological function [192].
- **Interpretability in machine learning models:** Techniques like SHAP or LIME (Local Interpretable Model-agnostic Explanations) can be employed in interpreting feature contributions of specific features in ML models [193].

6.4 ADVANCED STATISTICAL METHODS FOR FEATURE EXTRACTION AND SELECTION IN OMICS DATA ANALYSIS

Feature selection and extraction are key steps in analyzing high-dimensional omics data that allow scientists to identify molecular signatures for specific diseases. Statistical methods are at the core of dimension reduction, reduction of noise, and simplification of complex biological data to be more interpretable [194].

6.4.1 Principal Component Analysis (PCA)

Definition: PCA is a method of dimension reduction that transforms data of high dimensions into a sequence of orthogonal components (principal components) such that most of the variance in the dataset is preserved. PCA is applied frequently in exploratory data visualization and analysis [195].

- **Related Case Study:** A triple-negative breast cancer (TNBC) analysis applied PCA to RNA-seq data to stratify different molecular subtypes. Researchers identified key gene expression patterns associated with immune response pathways and assisted in selecting targeted therapies [196].

6.4.2 ICA (Independent Component Analysis)

Definition: ICA is an unsupervised statistical method that separates mixed signals into statistically independent components, making it useful for uncovering hidden patterns in omics data [197].

- **Related Case Study (2024):** ICA was applied in a multi-omics study of Parkinson's disease, where it helped separate metabolic signatures from confounding biological noise. The study identified a distinct set of metabolites linked to early neurodegeneration, improving early detection of the disease [198].

6.4.3 SVD (Singular Value Decomposition)

Definition: SVD decomposes a matrix into three components (U, Σ, V), extracting meaningful patterns from high-dimensional omics datasets. It is widely used for noise reduction and feature extraction [199].

- **Related Case Study:** Researchers applied SVD to pancreatic cancer proteomics data to extract significant protein interaction modules. The study identified novel biomarkers involved in tumor progression, which were validated using patient-derived tissue samples (doi: 10.3390/proteomes11030024) [200].

6.4.4 Autoencoders

Definition: Autoencoders are neural networks that learn concise feature representations via compressing and reconstructing the input data and can identify intrinsic biological patterns from omics data [201].

- **Related case study:** An autoencoder model was used in glioblastoma transcriptomics to discover a sparse gene signature having predictive prognosis of patient survival. The strategy reduced the dimensionality with preserved key genomic characteristics, superior to conventional feature selection methods [202].

6.4.5 Convolutional Neural Networks (CNNs) and Recurrent Neural Networks (RNNs)

Definition: CNNs are DL models capable of feature extraction from structured data (e.g., histopathology images), whereas RNNs are capable of sequential data analysis (e.g., time-series gene expression data) [203].

- **Related case study:** CNNs were applied to colorectal cancer histopathological image analysis that automatically detected molecular features linked to genetic mutations. This resulted in a substantially enhanced diagnostic accuracy. RNNs were used to analyze longitudinal transcriptomic fluctuations in Type 1 Diabetes to detect predictive patterns of gene expression upon disease onset [204].

6.4.6 Decision Trees and Random Forests, Support Vector Machines (SVMs)

ML methods such as decision trees, random forests, SVMs, and DL models (CNNs, RNNs) are all frequently employed for feature selection and classification in high-dimensional omics data [205]

6.4.6.1 Decision Trees and Random Forests

Definition: Decision trees build hierarchical models by repeatedly splitting the dataset along feature importance, while random forests enhance robustness by aggregating multiple trees to avoid overfitting [206].

- **Example Case:** Researchers of a certain group had employed random forests to subtype breast cancer from multi-omics data and identified the significant metabolite and gene expression signatures [207].

6.4.6.2 Support Vector Machines (SVMs)

Definition: SVMs distinguish classes by using hyperplanes to maximize separation and are suitable for small, high-dimensional datasets such as transcriptomics and proteomics studies [208].

- **Example Case:** In a study based on Alzheimer's disease biomarkers, SVMs were applied to microRNA expression profiles, and high accuracy was identified in early stage cases [209].

6.4.7 Recursive Feature Elimination

Definition: RFE is a popular feature selection technique that iteratively deletes less important features from a dataset to enhance model performance. It is especially effective in high-dimensional omics data, as redundant or irrelevant variables can act as dampeners on predictive accuracy continues by learning a ML model, ranking features by importance, removing the least important ones, and doing this iteratively until an optimal subset remains [210].

- **Example Case:** For instance, RFE used on leukemia patient single-cell RNA sequencing (scRNA-seq) data helped to identify the most informative genes for separating disease types [211]. By removing a thousand irrelevant genes, the study was able to successfully lower the dimension of the dataset without sacrificing the cost of over 95% classification accuracy from an SVM model. The subset of selected genes was subsequently confirmed as candidate biomarkers for the early detection of leukemia [212].

The main strength of RFE is that it can reduce feature sets while maintaining predictive ability. It is, however, computationally expensive, particularly when used with large omics datasets, and sensitive to the selection of ML algorithms [213].

6.4.8 Sequential Feature Selection and Genetic Algorithms Methods

6.4.8.1 SFS

Definition: Sequential feature selection (SFS) is a model-improvement-based, iterative algorithm to choose the best features either through forward selection (incremental addition by model improvement) or backward elimination (initialization with all features and iterative removal of the least useful ones) [213]. SFS finds its best applications in model tuning in transcriptomics, proteomics, and metabolomics research [214].

- **Example case:** An SFS-based study was used to investigate proteomic evidence from Alzheimer patients. The procedure started with a null set and added proteins successively that would best enhance the performance of a predictive model. The final 120 significant proteins improved early diagnosis of Alzheimer's by 92% compared to conventional means [215].

The greatest benefit of SFS is its ability to be highly efficient in the reduction of overfitting since it only uses the most useful features. But it is also sensitive to the feature selection order and does not always identify the global optimum [216].

6.4.8.2 *Genetic Algorithm*

Definition: GA draw upon evolutionary principles and are used to maximize feature selection through an imitation of natural selection. GAs progress sets of features using crossover, mutation, and selection and ultimately retain the most predictive features. This is especially beneficial for multi-omics integration, where various omics layers are contributing to disease phenotyping [217].

- **Example case:** For instance, bioinformatics method used GA to metabolomics data for pancreatic cancer diagnosis. The approach first examined more than 7,000 metabolites and, by evolutionary optimization, converged to a group of 90 important metabolites, which reached 88% accuracy in discriminating between pancreatic cancer and benign diseases [218].

The key strength of GA is that it can efficiently search through various combinations of features, but it is computationally costly and needs sensitive parameter tuning to prevent local optima [219].

6.4.9 Wrapper (Lasso and Ridge Regression) and Embedded Methods

6.4.9.1 *Wrapper Method*

Definition: Wrapper techniques like Ridge regression (L2 regularization) and Lasso regression (L1 regularization) are significant for feature selection since they charge less important features, effectively reducing the model. Lasso regression sets some coefficients to zero to eliminate non-informative features, whereas Ridge regression sets coefficients to zero to avoid overfitting without elimination [220].

- **Example case:** Researchers in 2024 Nature Communications applied Lasso regression to interpret breast cancer patient transcriptomics profiles to discover a set of genes significantly correlated with response to treatment. This reduced the predictive models' complexity but kept them very accurate. Analogously, Ridge regression was applied in an IEEE Transactions on Bioinformatics study to consolidate multi-omics data for forecasting colorectal cancer prognosis and handled collinearity of thousands of features successfully [221].

6.4.9.2 *Embedded Methods*

Definition: Embedded methods like decision trees and DL models incorporate feature selection as an integral component of model training [222]. Random forests and XGBoost provide importance scores for features, permitting automatic selection of the most informative features. Convolutional neural networks (CNNs) and recurrent neural networks (RNNs) may also learn hierarchical representations of omics data, thus being effective for biomarker discovery [223].

- **Example case:** A recent publication in cell reports medicine applied CNNs to feature selection in spatial transcriptomics data, separating tumor microenvironment signatures with high accuracy [223]. Embedded techniques provide efficiency by combining feature selection with the learning process but need heavy training and hyperparameter optimization [224].

These sophisticated feature selection methods altogether advance biomarker discovery, enhance disease classification, and optimize predictive modelling in bioinformatics and precision medicine [225].

6.5 COMPUTATIONAL TOOLS AND SOFTWARE

6.5.1 Overview of Bioinformatics Tools for Omics Data Analysis

Feature extraction and selection methods are important in the identification of disease-related molecular signatures by avoiding dimensionality reduction, irrelevant variable removal, and model improvement. Genomics, transcriptomics, proteomics, and metabolomics data integration through bioinformatics methods contributes to improved classification of diseases, biomarker identification, and precision medicine. Below is an extensive compilation of tools and databases, Table 6.2, that facilitate feature extraction, selection, and bioinformatics analysis in omics research [226].

With the growing availability of high-throughput omics data, selecting relevant features and extracting meaningful molecular signatures is critical for disease research [227]. Feature selection tools like mRMR, Boruta, and SHAP help refine predictive models by eliminating irrelevant variables, while DL-driven methods such as DeepFeature Selection reveal intricate interactions [228]. Bioinformatics repositories like TCGA, GEO, and HMDB supply the necessary data resources that support integrative analysis. With further developments in AI-driven platforms like DeepOmics and BioBERT, feature extraction procedures will most probably be enriched by integrating knowledge graphs, NLP, and sophisticated ML strategies [229]. The next steps should concentrate on enhancing interpretability of DL models and more effectively integrating multi-omics data to push precision medicine [230].

6.6 IDENTIFYING MOLECULAR SIGNATURES FOR SPECIFIC DISEASES

Identification of molecular signatures is critical for defining disease mechanisms, facilitating early diagnosis, and devising targeted treatment [231]. A molecular signature consists of specific characteristics—such as genes, proteins, or metabolites—that unambiguously differentiate among states of disease [232]. Identification and validation of the signatures provide for:

- **Disease-specific biomarkers:** These are susceptible to application in diagnosis or prognosis and allow clinicians to assess risk and disease progression [233].
- **Predictive models:** ML models with molecular signatures can predict susceptibility to disease, treatment response, and outcomes for patients [234].

Multi-omics integration, combined with ML methodologies, has significantly improved the identification of disease-specific molecular signatures over the recent past. The following are case studies describing the applications of molecular signatures in cancer, cardiovascular disease, and neurological disorders [235].

TABLE 6.2 Tools and Databases for Feature Extraction and Selection in Omics Data Analysis

Category	Tool/Database/Algorithms	Description	Advantages	Disadvantages
Feature selection and machine learning tools	mRMR (minimum redundancy maximum relevance)	Selects relevant yet non-redundant features for classification tasks.	Reduces dimensionality while preserving relevant features; computationally efficient	May not capture nonlinear interactions
	Boruta	Wrapper algorithm built around random forests to select all relevant features	Robust against overfitting; identifies all relevant features	Computationally expensive for high-dimensional data
	Elastic net	Combines LASSO and Ridge regression to handle collinearity	Handles correlated features better than LASSO.	No hyperparameter tuning
	XGBoost feature importance	Uses gradient boosting for feature ranking and selection	Highly efficient and accurate; works well with structured omics data	Requires careful hyperparameter tuning
	Deep feature selection (DFS)	Applies deep learning-based feature selection techniques	Uncovers complex patterns in high-dimensional omics data.	Requires large datasets to avoid overfitting
	LASSO (Least Absolute Shrinkage and Selection Operator)	Performs feature selection by shrinking coefficients of less important variables	Shrinks less important features to zero, improving model interpretability	Struggles with highly correlated features
	SHAP (SHapley Additive exPlanations)	Explains the impact of each feature in AI-based models	Provides explainability to feature importance in AI models	Can be slow for large datasets
Transcriptomics databases and tools	GEO (gene expression omnibus)	Repository of gene expression datasets from various experiments.	Large dataset collection; supports meta-analysis	Data variability due to different experimental conditions
	TCGA (The Cancer Genome Atlas)	Provides multi-omics data for cancer research	Comprehensive cancer genomics resource	Mostly focused on cancer; limited non-cancer datasets
	GTEx (genotype-tissue expression project)	Gene expression database across multiple tissues	Valuable for studying gene regulation in normal tissues	Limited disease-specific datasets
Proteomics databases and tools	PRIDE (proteomics identifications database)	Stores proteomics data from mass spectrometry experiments	High-quality mass spectrometry-based proteomics data	Requires bioinformatics expertise to process raw data
	UniProtKB	Comprehensive database of protein sequences and functional annotations	Extensive curation of protein functions and structures	May lack experimental validation for some predicted annotations
	DeepProt	AI-driven tool for protein function prediction	Uses deep learning for accurate protein function annotation	Requires extensive computational resources for model training
Metabolomics databases and tools	HMDB (human metabolome database)	Extensive database of metabolites and disease associations	Provides metabolite structures, pathways, and disease links	Some entries lack experimental validation
	MetaboAnalyst	Web-based tool for metabolomics data analysis	User-friendly statistical analysis and visualization	Requires preprocessing of metabolomics data

(Continued)

TABLE 6.2 *(Continued)* Tools and Databases for Feature Extraction and Selection in Omics Data Analysis

Category	Tool/Database/Algorithms	Description	Advantages	Disadvantages
Microbiomics databases and tools	QIIME2	Workflow for microbiome analysis using 16S rRNA sequencing	Widely used for microbiome research; strong community support	Requires command-line knowledge
	MGnify	Integrates multiple metagenomic datasets for microbiome analysis	Provides metagenomic annotations across environments	Some datasets have inconsistent metadata
Pathway and network analysis tools	KEGG (Kyoto Encyclopedia of Genes and Genomes)	Curated database of metabolic and signaling pathways	Rich repository for pathway analysis and functional annotation.	Some pathways lack complete experimental validation
	Reactome	Database of molecular pathways involved in various diseases	Provides curated biological pathways with functional interactions	Focused on human data, limited non-human pathways
	STRING	Protein-protein interaction (PPI) network database	Identifies functional interactions between proteins	Includes predicted interactions that may not be experimentally validated
Multi-omics integration and data fusion tools	MOFA+ (multi-omics factor analysis)	Integrates multiple omics layers for unsupervised feature extraction	Captures complex interactions between omics data types	Computationally demanding for large datasets
	DIABLO	Identifies biomarkers across multiple omics layers	Enables multi-omics feature selection for classification	Requires careful selection of input parameters
AI-driven and deep learning tools	BioBERT (bidirectional encoder representations from transformers for biomedical texts)	Extracts features from biomedical literature using NLP	Identifies relevant omics features from unstructured text	Requires GPU support for large-scale applications
	DeepOmics	AI-based tool for multi-omics feature selection	Uses deep learning for biomarker discovery	Model interpretability can be challenging
	GraphOmics	Uses graph-based machine learning to integrate omics data	Captures network-level interactions between biomolecules	Requires expertise in graph neural networks

6.6.1 Case Studies in Disease Research

6.6.1.1 Cancer Biomarkers

Lung cancer remains a major global health problem, with most instances being non-small-cell lung cancer (NSCLC). Outcome is often hampered by late diagnosis, heterogeneity of tumor subtypes, and the complexity of underlying molecular alterations [236]:

- **Lung Adenocarcinoma (LUAD):** Current research has proven the role of molecular profiling in the detection of actionable mutations in lung cancer. A 2024 paper highlighted the heterogeneity of EGFR mutations in LUAD, with rare EGFR mutations tending to co-occur with KRAS and TP53 mutations, particularly among smokers. This depicts the utility of extensive molecular testing, e.g., NGS, in guiding treatment approaches [237].

6.6.1.2 Neurological Disorder Markers

The discovery of molecular signatures in neurodegenerative illnesses has tremendously propagated the research in biomarker discovery. Recent metabolomic and proteomic investigations have furnished fundamental knowledge regarding the pathophysiology of neurological disease [238]:

- **Alzheimer's Disease (AD):** Metabolomics has offered new insights into the discovery of diagnostic biomarkers for AD. High-throughput metabolite profiling in CSF and plasma has revealed metabolic changes associated with disease progression. A neurodegenerative-based study revealed substantial alterations in polyamine metabolism, markers of oxidative stress, and lipid metabolites in AD patients, proposing their use as early-stage biomarkers [239].
- **Lewy Body Dementia and Multiple Sclerosis:** Current proteomics research has been able to identify Lewy body-linked dementia and multiple sclerosis (MS) potential biomarkers. A certain study revealed that certain neuroinflammation and synaptic dysfunction-related protein markers are greatly disturbed in Lewy body dementia patients [240]. The discoveries help create diagnostic tools for discriminating among neurodegenerative diseases. Technological progress in biomarker studies has opened avenues for precision medicine strategies in neurologic disorders. The integration of multi-omics, including genomics, transcriptomics, proteomics, and metabolomics, is likely to enhance the discovery of disease-specific molecular signatures and facilitate the development of targeted treatments [241].

6.7 FUTURE TRENDS AND RESEARCH DIRECTIONS

Disease-specific molecular signatures from omics data are being uncovered at a rate of knots, with the help of improvements in feature extraction, feature selection, and bioinformatics tools [242]. The convergence of ML, DL, and multi-omics is revolutionizing biomarker discovery, leading to precision medicine and data-driven diagnostics. The next few sections highlight some of the most significant emerging trends and directions of research [243].

6.7.1 Advances in Machine Learning and AI

Feature extraction and selection are essential processes for dimensionality reduction and enhancing interpretability of omics data. Advances in AI-based methods in recent times have greatly improved the capacity to recognize pertinent molecular signatures [244].

- **Autoencoders for Nonlinear Feature Extraction:** DL-based autoencoders are increasingly being utilized to acquire latent high-dimensional omics data representation. For example, some researchers had employed the use of variational autoencoders (VAEs) in the application to transcriptomic data to subtype breast cancer and found new transcriptomic features correlated with tumor aggressiveness [245].
- **Explainable AI (XAI) for Feature Selection:** To provide more explainability to ML models in biomarker discovery, in recent times, there has been an interest in applying XAI approaches. SHAP and LIME (Local Interpretable Model-Agnostic Explanations) have been employed successfully combined with proteomics data, giving feature importance scores to disease classification in Alzheimer's disease [246].
- **Graph Neural Networks (GNNs) for Multi-Omics Feature Integration:** GNNs are increasingly being shown to be effective in learning molecular entity relationships (genes, proteins, metabolites). Another study illustrated the capability of graph-based feature extraction to merge genomics and metabolomics information for the discovery of metabolic dysregulations in type 2 diabetes [246].

Future work will target automated feature selection pipelines that include self-supervised learning, as well as developing AI systems capable of dynamically varying feature selection with disease-specific context [247].

6.7.2 Emerging Technologies in Omics Research

The convergence of genomics, transcriptomics, proteomics, metabolomics, and epigenomics has made new discoveries on disease-specific molecular signatures possible. State-of-the-art experimental and computational methods are fueling the next generation of biomarker discovery [248]:

- **Single-Cell Multi-Omics Analysis:** Single-cell tools allow research at the resolution of cells for molecular signatures. Single-cell RNA sequencing (scRNA-seq) and single-cell proteomics were recently successfully used to map immune subtypes in rheumatoid arthritis, discovering new immunotherapy targets [249].
- **Spatial Omics to Extract Features:** Spatial transcriptomics and spatial proteomics add a spatial dimension to molecular features within tissue. One study employed the use of spatial metabolomics in an investigation into glioblastoma tumor heterogeneity, thereby recognizing unique metabolic microenvironments correlated with therapy resistance [250].

- **AI-Aided Mass Spectrometry for Metabolomics:** AI-aided peak detection and feature extraction in MS has greatly enhanced the detection of metabolic biomarkers. DL models were recently used to deconvolute intricate MS spectra to identify rare lipid biomarkers for Parkinson's disease [251].

The future of omics-driven feature selection will be based on combining multi-modal data fusion, high-throughput AI-based feature extraction, and cloud-based bioinformatics platforms for large-scale data analysis [252].

6.7.3 Personalized Medicine and Precision Health

Personalized medicine seeks to customize therapies to the unique molecular profiles by the promise of omics-guided feature selection to enhance diagnostics and therapeutics [253].

- **Multi-Omics Biomarkers for Personalized Cancer Therapy:** For example, researchers showed that combining genomics, transcriptomics, and proteomics provided more personalized treatment regimens for TNBC, revealing patient-specific oncogenic pathways for drug repurposing [254].
- **N-of-1 Trials for Orphan Diseases:** Precision medicine is advancing with N-of-1 trials, in which interventions are individualized to a patient's own omics signature. A recent trial in Duchenne muscular dystrophy used patient-specific gene-editing strategies to show successful reconstitution of dystrophin expression [255].
- **Digital Twins for Predictive Medicine:** AI-driven digital twin models, modelling an individual's metabolism from real-time omics data, are being developed for diabetes and cardiovascular disease. One study employed a digital twin platform to predict personalized insulin responses, optimizing treatment strategies [256].

The future will witness the creation of customized disease models, artificial intelligence-based drug response predictors, and omics-driven clinical choice support systems to encourage the application of biomarker results to improve medical practice [257].

REFERENCES

1. Abbasi, E. Y., Deng, Z., Ali, Q., Khan, A., Shaikh, A., Reshan, M. S. A., Sulaiman, A., & Alshahrani, H. (2024). A machine learning and deep learning-based integrated multi-omics technique for leukemia prediction. *Heliyon, 10*(3), e25369. https://doi.org/10.1016/j.heliyon.2024.e25369
2. Abdelmoaty, M. M., Lu, E., Kadry, R., Foster, E. G., Bhattarai, S., Mosley, R. L., & Gendelman, H. E. (2023). Clinical biomarkers for Lewy body diseases. *Cell & Bioscience, 13*(1), 209. https://doi.org/10.1186/s13578-023-01152-x
3. Abidin, S. Z., Mat Pauzi, N. A., Mansor, N. I., Mohd Isa, N. I., & Hamid, A. A. (2023). A new perspective on Alzheimer's disease: microRNAs and circular RNAs. *Frontiers in Genetics, 14*, 1231486. https://doi.org/10.3389/fgene.2023.1231486
4. Adhit, K. K., Wanjari, A., Menon, S., & K, S. (2023). Liquid biopsy: An evolving paradigm for non-invasive disease diagnosis and monitoring in medicine. *Cureus*. https://doi.org/10.7759/cureus.50176

5. Ahmad, A., Imran, M., & Ahsan, H. (2023). Biomarkers as biomedical bioindicators: Approaches and techniques for the detection, analysis, and validation of novel biomarkers of diseases. *Pharmaceutics, 15*(6), 1630. https://doi.org/10.3390/pharmaceutics15061630
6. Ahuja, S., & Zaheer, S. (2025). Advancements in pathology: Digital transformation, precision medicine, and beyond. *Journal of Pathology Informatics, 16*, 100408. https://doi.org/10.1016/j.jpi.2024.100408
7. Al-Amrani, S., Al-Jabri, Z., Al-Zaabi, A., Alshekaili, J., & Al-Khabori, M. (2021). Proteomics: Concepts and applications in human medicine. *World Journal of Biological Chemistry, 12*(5), 57–69. https://doi.org/10.4331/wjbc.v12.i5.57
8. Alemu, R., Sharew, N. T., Arsano, Y. Y., Ahmed, M., Tekola-Ayele, F., Mersha, T. B., & Amare, A. T. (2025). Multi-omics approaches for understanding gene-environment interactions in noncommunicable diseases: Techniques, translation, and equity issues. *Human Genomics, 19*(1), 8. https://doi.org/10.1186/s40246-025-00718-9
9. Alexakis, K., Benekis, V., Kokkinakos, P., & Askounis, D. (2025). Genetic algorithm-based multi-objective optimisation for energy-efficient building retrofitting: A systematic review. *Energy and Buildings, 328*, 115216. https://doi.org/10.1016/j.enbuild.2024.115216
10. Alkhanbouli, R., Matar Abdulla Almadhaani, H., Alhosani, F., & Simsekler, M. C. E. (2025). The role of explainable artificial intelligence in disease prediction: A systematic literature review and future research directions. *BMC Medical Informatics and Decision Making, 25*(1), 110. https://doi.org/10.1186/s12911-025-02944-6
11. Allioui, H., & Mourdi, Y. (2023). Exploring the full potentials of IoT for better financial growth and stability: A comprehensive survey. *Sensors, 23*(19), 8015. https://doi.org/10.3390/s23198015
12. Alum, E. U. (2025). AI-driven biomarker discovery: Enhancing precision in cancer diagnosis and prognosis. *Discover Oncology, 16*(1), 313. https://doi.org/10.1007/s12672-025-02064-7
13. Anghel, S. A., Dinu-Pirvu, C.-E., Costache, M.-A., Voiculescu, A. M., Ghica, M. V., Anuța, V., & Popa, L. (2024). Receptor pharmacogenomics: Deciphering genetic influence on drug response. *International Journal of Molecular Sciences, 25*(17), 9371. https://doi.org/10.3390/ijms25179371
14. Arjmand, B., Hamidpour, S. K., Tayanloo-Beik, A., Goodarzi, P., Aghayan, H. R., Adibi, H., & Larijani, B. (2022). Machine learning: A new prospect in multi-omics data analysis of cancer. *Frontiers in Genetics, 13*, 824451. https://doi.org/10.3389/fgene.2022.824451
15. Athieniti, E., & Spyrou, G. M. (2023). A guide to multi-omics data collection and integration for translational medicine. *Computational and Structural Biotechnology Journal, 21*, 134–149. https://doi.org/10.1016/j.csbj.2022.11.050
16. Augustin, R. C., Cai, W. L., Luke, J. J., & Bao, R. (2024). Facts and hopes in using omics to advance combined immunotherapy strategies. *Clinical Cancer Research, 30*(9), 1724–1732. https://doi.org/10.1158/1078-0432.CCR-22-2241
17. Awad, M., & Fraihat, S. (2023). Recursive feature elimination with cross-validation with decision tree: feature selection method for machine learning-based intrusion detection systems. *Journal of Sensor and Actuator Networks, 12*(5), 67. https://doi.org/10.3390/jsan12050067
18. Azad, M., Nehal, T. H., & Moshkov, M. (2025). A novel ensemble learning method using majority based voting of multiple selective decision trees. *Computing, 107*(1), 42. https://doi.org/10.1007/s00607-024-01394-8
19. Babu, M., & Snyder, M. (2023). Multi-omics profiling for health. *Molecular & Cellular Proteomics, 22*(6), 100561. https://doi.org/10.1016/j.mcpro.2023.100561
20. Baldwin, E., Han, J., Luo, W., Zhou, J., An, L., Liu, J., Zhang, H. H., & Li, H. (2020). On fusion methods for knowledge discovery from multi-omics datasets. *Computational and Structural Biotechnology Journal, 18*, 509–517. https://doi.org/10.1016/j.csbj.2020.02.011
21. Baldwin, J. R., Pingault, J.-B., Schoeler, T., Sallis, H. M., & Munafò, M. R. (2022). Protecting against researcher bias in secondary data analysis: Challenges and potential solutions. *European Journal of Epidemiology, 37*(1), 1–10. https://doi.org/10.1007/s10654-021-00839-0

22. Ballard, J. L., Wang, Z., Li, W., Shen, L., & Long, Q. (2024). Deep learning-based approaches for multi-omics data integration and analysis. *BioData Mining, 17*(1), 38. https://doi.org/10.1186/s13040-024-00391-z
23. Beaulieu-Jones, B. K., Frau, F., Bozzi, S., Chandross, K. J., Peterschmitt, M. J., Cohen, C., Coulovrat, C., Kumar, D., Kruger, M. J., Lipnick, S. L., Fitzsimmons, L., Kohane, I. S., & Scherzer, C. R. (2024). Disease progression strikingly differs in research and real-world Parkinson's populations. *Npj Parkinson's Disease, 10*(1), 58. https://doi.org/10.1038/s41531-024-00667-5
24. Beck, A. G., Muhoberac, M., Randolph, C. E., Beveridge, C. H., Wijewardhane, P. R., Kenttämaa, H. I., & Chopra, G. (2024). Recent developments in machine learning for mass spectrometry. *ACS Measurement Science Au, 4*(3), 233–246. https://doi.org/10.1021/acsmeasuresciau.3c00060
25. Beger, R. D., Goodacre, R., Jones, C. M., Lippa, K. A., Mayboroda, O. A., O'Neill, D., Najdekr, L., Ntai, I., Wilson, I. D., & Dunn, W. B. (2024). Analysis types and quantification methods applied in UHPLC-MS metabolomics research: A tutorial. *Metabolomics, 20*(5), 95. https://doi.org/10.1007/s11306-024-02155-6
26. Bhadra, T., Mallik, S., Hasan, N., & Zhao, Z. (2022). Comparison of five supervised feature selection algorithms leading to top features and gene signatures from multi-omics data in cancer. *BMC Bioinformatics, 23*(S3), 153. https://doi.org/10.1186/s12859-022-04678-y
27. Bhinder, B., Gilvary, C., Madhukar, N. S., & Elemento, O. (2021). Artificial intelligence in cancer research and precision medicine. *Cancer Discovery, 11*(4), 900–915. https://doi.org/10.1158/2159-8290.CD-21-0090
28. Bodaghi, A., Fattahi, N., & Ramazani, A. (2023). Biomarkers: Promising and valuable tools towards diagnosis, prognosis and treatment of Covid-19 and other diseases. *Heliyon, 9*(2), e13323. https://doi.org/10.1016/j.heliyon.2023.e13323
29. Broeckling, C. D., Beger, R. D., Cheng, L. L., Cumeras, R., Cuthbertson, D. J., Dasari, S., Davis, W. C., Dunn, W. B., Evans, A. M., Fernández-Ochoa, A., Gika, H., Goodacre, R., Goodman, K. D., Gouveia, G. J., Hsu, P.-C., Kirwan, J. A., Kodra, D., Kuligowski, J., Lan, R. S.-L., … Mosley, J. D. (2023). Current practices in LC-MS untargeted metabolomics: A scoping review on the use of pooled quality control samples. *Analytical Chemistry, 95*(51), 18645–18654. https://doi.org/10.1021/acs.analchem.3c02924
30. Bustin, S. A., & Jellinger, K. A. (2023). Advances in molecular medicine: Unravelling disease complexity and pioneering precision healthcare. *International Journal of Molecular Sciences, 24*(18), 14168. https://doi.org/10.3390/ijms241814168
31. Caliskan, A., Dangwal, S., & Dandekar, T. (2023). Metadata integrity in bioinformatics: Bridging the gap between data and knowledge. *Computational and Structural Biotechnology Journal, 21*, 4895–4913. https://doi.org/10.1016/j.csbj.2023.10.006
32. Cao, Y., Xu, Y., Cao, M., Chen, N., Zeng, Q., Lai, M. K. P., Fan, D., Sethi, G., & Cao, Y. (2025). Fluid-based biomarkers for neurodegenerative diseases. *Ageing Research Reviews, 108*, 102739. https://doi.org/10.1016/j.arr.2025.102739
33. Carrasco-Zanini, J., Pietzner, M., Davitte, J., Surendran, P., Croteau-Chonka, D. C., Robins, C., Torralbo, A., Tomlinson, C., Grünschläger, F., Fitzpatrick, N., Ytsma, C., Kanno, T., Gade, S., Freitag, D., Ziebell, F., Haas, S., Denaxas, S., Betts, J. C., Wareham, N. J., … Langenberg, C. (2024). Proteomic signatures improve risk prediction for common and rare diseases. *Nature Medicine, 30*(9), 2489–2498. https://doi.org/10.1038/s41591-024-03142-z
34. Cascianelli, S., Molineris, I., Isella, C., Masseroli, M., & Medico, E. (2020). Machine learning for RNA sequencing-based intrinsic subtyping of breast cancer. *Scientific Reports, 10*(1), 14071. https://doi.org/10.1038/s41598-020-70832-2
35. Chen, L., Hu, Y., Li, Y., Zhang, B., Wang, J., Deng, M., Zhang, J., Zhu, W., Gu, H., & Zhang, L. (2024). Integrated multiomics analysis identified comprehensive crosstalk between diverse programmed cell death patterns and novel molecular subtypes in hepatocellular carcinoma. *Scientific Reports, 14*(1), 27529. https://doi.org/10.1038/s41598-024-78911-4

36. Chen, Q., Meng, Z., & Su, R. (2020). WERFE: A Gene selection algorithm based on recursive feature elimination and ensemble strategy. *Frontiers in Bioengineering and Biotechnology, 8*, 496. https://doi.org/10.3389/fbioe.2020.00496
37. Chen, T., Tao, Y., Wang, Q., Pei, Y., Zhao, Z., Yang, W., & Lu, Y. (2024). Utilizing an integrated bioinformatics and machine learning approach to uncover biomarkers linking ulcerative colitis to purine metabolism-related genes. *Heliyon, 10*(21), e38403. https://doi.org/10.1016/j.heliyon.2024.e38403
38. Chiesa, M., Maioli, G., Colombo, G. I., & Piacentini, L. (2020). GARS: Genetic algorithm for the identification of a robust subset of features in high-dimensional datasets. *BMC Bioinformatics, 21*(1), 54. https://doi.org/10.1186/s12859-020-3400-6
39. Collin, C. B., Gebhardt, T., Golebiewski, M., Karaderi, T., Hillemanns, M., Khan, F. M., Salehzadeh-Yazdi, A., Kirschner, M., Krobitsch, S., Kuepfer, L., & EU-STANDS4PM Consortium (2022). Computational models for clinical applications in personalized medicine—Guidelines and recommendations for data integration and model validation. *Journal of Personalized Medicine, 12*(2), 166. https://doi.org/10.3390/jpm12020166
40. Conte, L., Caruso, G., Philip, A. K., Cucci, F., De Nunzio, G., Cascio, D., & Caffo, M. (2025). Artificial intelligence-assisted drug and biomarker discovery for glioblastoma: A scoping review of the literature. *Cancers, 17*(4), 571. https://doi.org/10.3390/cancers17040571
41. Corsaro, C., Vasi, S., Neri, F., Mezzasalma, A. M., Neri, G., & Fazio, E. (2022). NMR in metabolomics: From conventional statistics to machine learning and neural network approaches. *Applied Sciences, 12*(6), 2824. https://doi.org/10.3390/app12062824
42. Cruz, Y. J., Villalonga, A., Castaño, F., Rivas, M., & Haber, R. E. (2024). Automated machine learning methodology for optimizing production processes in small and medium-sized enterprises. *Operations Research Perspectives, 12*, 100308. https://doi.org/10.1016/j.orp.2024.100308
43. Cui, J.-W., Li, Y., Yang, Y., Yang, H.-K., Dong, J.-M., Xiao, Z.-H., He, X., Guo, J.-H., Wang, R.-Q., Dai, B., & Zhou, Z.-L. (2024). Tumor immunotherapy resistance: Revealing the mechanism of PD-1 / PD-L1-mediated tumor immune escape. *Biomedicine & Pharmacotherapy, 171*, 116203. https://doi.org/10.1016/j.biopha.2024.116203
44. Cui, X., Wu, R., Liu, Y., Chen, P., Chang, Q., Liang, P., & He, C. (2025). scSMD: A deep learning method for accurate clustering of single cells based on auto-encoder. *BMC Bioinformatics, 26*(1), 33. https://doi.org/10.1186/s12859-025-06047-x
45. Curti, N., Levi, G., Giampieri, E., Castellani, G., & Remondini, D. (2022). A network approach for low dimensional signatures from high throughput data. *Scientific Reports, 12*(1), 22253. https://doi.org/10.1038/s41598-022-25549-9
46. Dai, X., & Shen, L. (2022). Advances and trends in omics technology development. *Frontiers in Medicine, 9*, 911861. https://doi.org/10.3389/fmed.2022.911861
47. Dai, Y., Wang, W., & Zhou, X. (2023). Development of a CD8+ T cell-based molecular classification for predicting prognosis and heterogeneity in triple-negative breast cancer by integrated analysis of single-cell and bulk RNA-sequencing. *Heliyon, 9*(9), e19798. https://doi.org/10.1016/j.heliyon.2023.e19798
48. Dalal, N., Jalandra, R., Sharma, M., Prakash, H., Makharia, G. K., Solanki, P. R., Singh, R., & Kumar, A. (2020). Omics technologies for improved diagnosis and treatment of colorectal cancer: Technical advancement and major perspectives. *Biomedicine & Pharmacotherapy, 131*, 110648. https://doi.org/10.1016/j.biopha.2020.110648
49. Dauda, K. A., & Lamidi, R. K. (2025). Consistency and stability in feature selection for high-dimensional microarray survival data in diffuse large B-cell lymphoma cancer. *Data, 10*(2), 26. https://doi.org/10.3390/data10020026
50. deAndrés-Galiana, E. J., Fernández-Martínez, J. L., Álvarez-Machancoses, Ó, Bea, G., Galmarini, C. M., & Kloczkowski, A. (2022). Analysis of transcriptomic responses to SARS-CoV-2 reveals plausible defective pathways responsible for increased susceptibility to infection and complications and helps to develop fast-track repositioning of drugs against COVID-19. *Computers in Biology and Medicine, 149*, 106029. https://doi.org/10.1016/j.compbiomed.2022.106029

51. Dhinakaran, D., Srinivasan, L., Edwin Raja, S., Valarmathi, K., & Gomathy Nayagam, M. (2025). Synergistic feature selection and distributed classification framework for high-dimensional medical data analysis. *MethodsX*, *14*, 103219. https://doi.org/10.1016/j.mex.2025.103219
52. Di Minno, A., Gelzo, M., Stornaiuolo, M., Ruoppolo, M., & Castaldo, G. (2021). The evolving landscape of untargeted metabolomics. *Nutrition, Metabolism and Cardiovascular Diseases*, *31*(6), 1645–1652. https://doi.org/10.1016/j.numecd.2021.01.008
53. Ding, X., Yang, F., & Ma, F. (2022). An efficient model selection for linear discriminant function-based recursive feature elimination. *Journal of Biomedical Informatics*, *129*, 104070. https://doi.org/10.1016/j.jbi.2022.104070
54. Dressler, F. F., Brägelmann, J., Reischl, M., & Perner, S. (2022). Normics: Proteomic normalization by variance and data-inherent correlation structure. *Molecular & Cellular Proteomics*, *21*(9), 100269. https://doi.org/10.1016/j.mcpro.2022.100269
55. Du, P., Fan, R., Zhang, N., Wu, C., & Zhang, Y. (2024). Advances in integrated multi-omics analysis for drug-target identification. *Biomolecules*, *14*(6), 692. https://doi.org/10.3390/biom14060692
56. Duan, X.-P., Qin, B.-D., Jiao, X.-D., Liu, K., Wang, Z., & Zang, Y.-S. (2024). New clinical trial design in precision medicine: Discovery, development and direction. *Signal Transduction and Targeted Therapy*, *9*(1), 57. https://doi.org/10.1038/s41392-024-01760-0
57. Elgammal, Y. M., Zahran, M. A., & Abdelsalam, M. M. (2022). A new strategy for the early detection of Alzheimer disease stages using multifractal geometry analysis based on K-nearest neighbor algorithm. *Scientific Reports*, *12*(1), 22381. https://doi.org/10.1038/s41598-022-26958-6
58. Elrashedy, A., Mousa, W., Nayel, M., Salama, A., Zaghawa, A., Elsify, A., & Hasan, M. E. (2025). Advances in bioinformatics and multi-omics integration: Transforming viral infectious disease research in veterinary medicine. *Virology Journal*, *22*(1), 22. https://doi.org/10.1186/s12985-025-02640-x
59. Fadhel, M. A., Duhaim, A. M., Saihood, A., Sewify, A., Al-Hamadani, M. N. A., Albahri, A. S., Alzubaidi, L., Gupta, A., Mirjalili, S., & Gu, Y. (2024). Comprehensive systematic review of information fusion methods in smart cities and urban environments. *Information Fusion*, *107*, 102317. https://doi.org/10.1016/j.inffus.2024.102317
60. Federico, A., Serra, A., Ha, M. K., Kohonen, P., Choi, J.-S., Liampa, I., Nymark, P., Sanabria, N., Cattelani, L., Fratello, M., Kinaret, P. A. S., Jagiello, K., Puzyn, T., Melagraki, G., Gulumian, M., Afantitis, A., Sarimveis, H., Yoon, T.-H., Grafström, R., & Greco, D. (2020). Transcriptomics in toxicogenomics, Part II: Preprocessing and differential expression analysis for high quality data. *Nanomaterials*, *10*(5), 903. https://doi.org/10.3390/nano10050903
61. Feldner-Busztin, D., Firbas Nisantzis, P., Edmunds, S. J., Boza, G., Racimo, F., Gopalakrishnan, S., Limborg, M. T., Lahti, L., & De Polavieja, G. G. (2023). Dealing with dimensionality: The application of machine learning to multi-omics data. *Bioinformatics*, *39*(2), btad021. https://doi.org/10.1093/bioinformatics/btad021
62. Felsky, D., Cannitelli, A., & Pipitone, J. (2023). Whole person modeling: A transdisciplinary approach to mental health research. *Discover Mental Health*, *3*(1), 16. https://doi.org/10.1007/s44192-023-00041-6
63. Fitz, V., Panzenboeck, L., Schoeny, H., Foels, E., & Koellensperger, G. (2024). Isotope dilution with isotopically labeled biomass: An effective alternative for quantitative metabolomics. *Analytica Chimica Acta*, *1318*, 342909. https://doi.org/10.1016/j.aca.2024.342909
64. Flores, J. E., Claborne, D. M., Weller, Z. D., Webb-Robertson, B.-J. M., Waters, K. M., & Bramer, L. M. (2023). Missing data in multi-omics integration: Recent advances through artificial intelligence. *Frontiers in Artificial Intelligence*, *6*, 1098308. https://doi.org/10.3389/frai.2023.1098308
65. Fu, C., & Chen, Q. (2025). The future of pharmaceuticals: Artificial intelligence in drug discovery and development. *Journal of Pharmaceutical Analysis*, 101248. https://doi.org/10.1016/j.jpha.2025.101248

66. Galal, A., Talal, M., & Moustafa, A. (2022). Applications of machine learning in metabolomics: Disease modeling and classification. *Frontiers in Genetics, 13*, 1017340. https://doi.org/10.3389/fgene.2022.1017340
67. Garg, P., Singhal, S., Kulkarni, P., Horne, D., Malhotra, J., Salgia, R., & Singhal, S. S. (2024). Advances in non-small cell lung cancer: Current insights and future directions. *Journal of Clinical Medicine, 13*(14), 4189. https://doi.org/10.3390/jcm13144189
68. Ghodeshwar, G. K., Dube, A., & Khobragade, D. (2023). Impact of lifestyle modifications on cardiovascular health: A narrative review. *Cureus*. https://doi.org/10.7759/cureus.42616
69. Ghosh, S., Zhao, X., Alim, M., Brudno, M., & Bhat, M. (2025). Artificial intelligence applied to 'omics data in liver disease: Towards a personalised approach for diagnosis, prognosis and treatment. *Gut, 74*(2), 295–311. https://doi.org/10.1136/gutjnl-2023-331740
70. Ginhoux, F., Yalin, A., Dutertre, C. A., & Amit, I. (2022). Single-cell immunology: Past, present, and future. *Immunity, 55*(3), 393–404. https://doi.org/10.1016/j.immuni.2022.02.006
71. Gobena, S., Admassu, B., Kinde, M. Z., & Gessese, A. T. (2024). Proteomics and its current application in biomedical area: Concise review. *The Scientific World Journal, 2024*, 1–13. https://doi.org/10.1155/2024/4454744
72. Goh, W. W. B., & Wong, L. (2019). Advanced bioinformatics methods for practical applications in proteomics. *Briefings in Bioinformatics, 20*(1), 347–355. https://doi.org/10.1093/bib/bbx128
73. Gong, T.-T., Guo, S., Liu, F.-H., Huo, Y.-L., Zhang, M., Yan, S., Zhou, H.-X., Pan, X., Wang, X.-Y., Xu, H.-L., Kang, Y., Li, Y.-Z., Qin, X., Xiao, Q., Huang, D.-H., Li, X.-Y., Zhao, Y.-Y., Zhao, X.-X., Wang, Y.-L., … Wu, Q.-J. (2023). Proteomic characterization of epithelial ovarian cancer delineates molecular signatures and therapeutic targets in distinct histological subtypes. *Nature Communications, 14*(1), 7802. https://doi.org/10.1038/s41467-023-43282-3
74. Gotsmy, M., Brunmair, J., Büschl, C., Gerner, C., & Zanghellini, J. (2022). Probabilistic quotient's work and pharmacokinetics' contribution: Countering size effect in metabolic time series measurements. *BMC Bioinformatics, 23*(1), 379. https://doi.org/10.1186/s12859-022-04918-1
75. Gowran, A., Brioschi, M., Rovina, D., Chiesa, M., Piacentini, L., Mallia, S., Banfi, C., Pompilio, G., & Santoro, R. (2021). Multiomic approaches to uncover the complexities of dystrophin-associated cardiomyopathy. *International Journal of Molecular Sciences, 22*(16), 8954. https://doi.org/10.3390/ijms22168954
76. Graw, S., Chappell, K., Washam, C. L., Gies, A., Bird, J., Robeson, M. S., & Byrum, S. D. (2021). Multi-omics data integration considerations and study design for biological systems and disease. *Molecular Omics, 17*(2), 170–185. https://doi.org/10.1039/D0MO00041H
77. Guido, R., Ferrisi, S., Lofaro, D., & Conforti, D. (2024). An overview on the advancements of support vector machine models in healthcare applications: A review. *Information, 15*(4), 235. https://doi.org/10.3390/info15040235
78. Hachem, S., Yehya, A., El Masri, J., Mavingire, N., Johnson, J. R., Dwead, A. M., Kattour, N., Bouchi, Y., Kobeissy, F., Rais-Bahrami, S., Mechref, Y., Abou-Kheir, W., & Woods-Burnham, L. (2024). Contemporary update on clinical and experimental prostate cancer biomarkers: A multi-omics-focused approach to detection and risk stratification. *Biology, 13*(10), 762. https://doi.org/10.3390/biology13100762
79. Hamdia, K. M., Zhuang, X., & Rabczuk, T. (2021). An efficient optimization approach for designing machine learning models based on genetic algorithm. *Neural Computing and Applications, 33*(6), 1923–1933. https://doi.org/10.1007/s00521-020-05035-x
80. Han, Y., Wang, D., Peng, L., Huang, T., He, X., Wang, J., & Ou, C. (2022). Single-cell sequencing: A promising approach for uncovering the mechanisms of tumor metastasis. *Journal of Hematology & Oncology, 15*(1), 59. https://doi.org/10.1186/s13045-022-01280-w
81. Harris, L., Fondrie, W. E., Oh, S., & Noble, W. S. (2023). Evaluating proteomics imputation methods with improved criteria. *Journal of Proteome Research, 22*(11), 3427–3438. https://doi.org/10.1021/acs.jproteome.3c00205

82. Hayes, C. N., Nakahara, H., Ono, A., Tsuge, M., & Oka, S. (2024). From omics to multi-omics: A review of advantages and tradeoffs. *Genes*, *15*(12), 1551. https://doi.org/10.3390/genes15121551
83. He, X., Liu, X., Zuo, F., Shi, H., & Jing, J. (2023). Artificial intelligence-based multi-omics analysis fuels cancer precision medicine. *Seminars in Cancer Biology*, *88*, 187–200. https://doi.org/10.1016/j.semcancer.2022.12.009
84. Hiransuchalert, R., Poarsa, C., Yocawibun, P., Amparyup, P., Taranart, T., Wachirachaikarn, A., Wongphayak, S., Kondo, H., & Hirono, I. (2024). Transcriptome analysis and identification of differentially expressed genes between early and mature ovarian stages in the female mantis shrimp (Harpiosquilla raphidea) using RNA-Seq. *Aquaculture Reports*, *34*, 101910. https://doi.org/10.1016/j.aqrep.2023.101910
85. Ho, D., Quake, S. R., McCabe, E. R. B., Chng, W. J., Chow, E. K., Ding, X., Gelb, B. D., Ginsburg, G. S., Hassenstab, J., Ho, C.-M., Mobley, W. C., Nolan, G. P., Rosen, S. T., Tan, P., Yen, Y., & Zarrinpar, A. (2020). Enabling technologies for personalized and precision medicine. *Trends in Biotechnology*, *38*(5), 497–518. https://doi.org/10.1016/j.tibtech.2019.12.021
86. Hohlfeld, A., Kredo, T., & Clarke, M. (2024). A scoping review of activities intended to reduce publication bias in randomised trials. *Systematic Reviews*, *13*(1), 310. https://doi.org/10.1186/s13643-024-02728-5
87. Homeyer, A., Geißler, C., Schwen, L. O., Zakrzewski, F., Evans, T., Strohmenger, K., Westphal, M., Bülow, R. D., Kargl, M., Karjauv, A., Munné-Bertran, I., Retzlaff, C. O., Romero-López, A., Sołtysiński, T., Plass, M., Carvalho, R., Steinbach, P., Lan, Y.-C., Bouteldja, N., … Zerbe, N. (2022). Recommendations on compiling test datasets for evaluating artificial intelligence solutions in pathology. *Modern Pathology*, *35*(12), 1759–1769. https://doi.org/10.1038/s41379-022-01147-y
88. Huang, L., Song, M., Shen, H., Hong, H., Gong, P., Deng, H.-W., & Zhang, C. (2023). Deep learning methods for omics data imputation. *Biology*, *12*(10), 1313. https://doi.org/10.3390/biology12101313
89. Huang, T., Yin, H., & Huang, X. (2024). Improved genetic algorithm for multi-threshold optimization in digital pathology image segmentation. *Scientific Reports*, *14*(1), 22454. https://doi.org/10.1038/s41598-024-73335-6
90. Huang, Y., Sulek, K., Stinson, S. E., Holm, L. A., Kim, M., Trost, K., Hooshmand, K., Lund, M. A. V., Fonvig, C. E., Juel, H. B., Nielsen, T., Ängquist, L., Rossing, P., Thiele, M., Krag, A., Holm, J.-C., Legido-Quigley, C., & Hansen, T. (2025). Lipid profiling identifies modifiable signatures of cardiometabolic risk in children and adolescents with obesity. *Nature Medicine*, *31*(1), 294–305. https://doi.org/10.1038/s41591-024-03279-x
91. Hui, H. W. H., Kong, W., & Goh, W. W. B. (2024). Thinking points for effective batch correction on biomedical data. *Briefings in Bioinformatics*, *25*(6), bbae515. https://doi.org/10.1093/bib/bbae515
92. Islam, M. K., Islam, M. R., Rahman, M. H., Islam, M. Z., Hasan, M. M., Mamun, M. M. I., & Moni, M. A. (2023). Integrated bioinformatics and statistical approach to identify the common molecular mechanisms of obesity that are linked to the development of two psychiatric disorders: Schizophrenia and major depressive disorder. *PLOS ONE*, *18*(7), e0276820. https://doi.org/10.1371/journal.pone.0276820
93. Islam, T., Hill, E., Abrahamson, E. E., Servaes, S., Smirnov, D. S., Zeng, X., Sehrawat, A., Chen, Y., Kac, P. R., Kvartsberg, H., Olsson, M., Sjons, E., Gonzalez-Ortiz, F., Therriault, J., Tissot, C., Del Popolo, I., Rahmouni, N., Richardson, A., Mitchell, V., … Karikari, T. K. (2025). Phospho-tau serine-262 and serine-356 as biomarkers of pre-tangle soluble tau assemblies in Alzheimer's disease. *Nature Medicine*, *31*(2), 574–588. https://doi.org/10.1038/s41591-024-03400-0
94. Jia, W., Sun, M., Lian, J., & Hou, S. (2022). Feature dimensionality reduction: A review. *Complex & Intelligent Systems*, *8*(3), 2663–2693. https://doi.org/10.1007/s40747-021-00637-x

95. Jia, Z.-C., Yang, X., Wu, Y.-K., Li, M., Das, D., Chen, M.-X., & Wu, J. (2024). The art of finding the right drug target: Emerging methods and strategies. *Pharmacological Reviews, 76*(5), 896–914. https://doi.org/10.1124/pharmrev.123.001028
96. Jiang, K., Dong, M., Li, C., & Sheng, J. (2021). Unraveling heterogeneity of tumor cells and microenvironment and its clinical implications for triple negative breast cancer. *Frontiers in Oncology, 11*, 557477. https://doi.org/10.3389/fonc.2021.557477
97. Jin, B., Cheng, X., Fei, G., Sang, S., & Zhong, C. (2023). Identification of diagnostic biomarkers in Alzheimer's disease by integrated bioinformatic analysis and machine learning strategies. *Frontiers in Aging Neuroscience, 15*, 1169620. https://doi.org/10.3389/fnagi.2023.1169620
98. Joshi, C. J., Ke, W., Drangowska-Way, A., O'Rourke, E. J., & Lewis, N. E. (2022). What are housekeeping genes? *PLOS Computational Biology, 18*(7), e1010295. https://doi.org/10.1371/journal.pcbi.1010295
99. Kellman, L. N., Neela, P. H., Srinivasan, S., Siprashvili, Z., Shanderson, R. L., Hong, A. W., Rao, D., Porter, D. F., Reynolds, D. L., Meyers, R. M., Guo, M. G., Yang, X., Zhao, Y., Wozniak, G. G., Donohue, L. K. H., Shenoy, R., Ko, L. A., Nguyen, D. T., Mondal, S., … Khavari, P. A. (2025). Functional analysis of cancer-associated germline risk variants. *Nature Genetics, 57*(3), 718–728. https://doi.org/10.1038/s41588-024-02070-5
100. Khalid, T., Cutress, R. I., Remer, M., & Copson, E. R. (2025). Clinical impact of somatic genomic testing on breast cancer care. *Clinical Oncology, 37*, 103665. https://doi.org/10.1016/j.clon.2024.10.037
101. Khosroabadi, Z., Azaryar, S., Dianat-Moghadam, H., Amoozgar, Z., & Sharifi, M. (2025). Single cell RNA sequencing improves the next generation of approaches to AML treatment: Challenges and perspectives. *Molecular Medicine, 31*(1), 33. https://doi.org/10.1186/s10020-025-01085-w
102. Ko, S., Li, G. X., Choi, H., & Won, J.-H. (2021). Computationally scalable regression modeling for ultrahigh-dimensional omics data with ParProx. *Briefings in Bioinformatics, 22*(6), bbab256. https://doi.org/10.1093/bib/bbab256
103. Koh, H. W. L., Fermin, D., Vogel, C., Choi, K. P., Ewing, R. M., & Choi, H. (2019). iOmicsPASS: Network-based integration of multiomics data for predictive subnetwork discovery. *Npj Systems Biology and Applications, 5*(1), 22. https://doi.org/10.1038/s41540-019-0099-y
104. Krassowski, M., Das, V., Sahu, S. K., & Misra, B. B. (2020). State of the field in multi-omics research: From computational needs to data mining and sharing. *Frontiers in Genetics, 11*, 610798. https://doi.org/10.3389/fgene.2020.610798
105. Labory, J., Njomgue-Fotso, E., & Bottini, S. (2024). Benchmarking feature selection and feature extraction methods to improve the performances of machine-learning algorithms for patient classification using metabolomics biomedical data. *Computational and Structural Biotechnology Journal, 23*, 1274–1287. https://doi.org/10.1016/j.csbj.2024.03.016
106. Lamane, H., Mouhir, L., Moussadek, R., Baghdad, B., Kisi, O., & El Bilali, A. (2025). Interpreting machine learning models based on SHAP values in predicting suspended sediment concentration. *International Journal of Sediment Research, 40*(1), 91–107. https://doi.org/10.1016/j.ijsrc.2024.10.002
107. Lan, W., Liao, H., Chen, Q., Zhu, L., Pan, Y., & Chen, Y.-P. P. (2024). DeepKEGG: A multi-omics data integration framework with biological insights for cancer recurrence prediction and biomarker discovery. *Briefings in Bioinformatics, 25*(3), bbae185. https://doi.org/10.1093/bib/bbae185
108. Lawson, C. E., Martí, J. M., Radivojevic, T., Jonnalagadda, S. V. R., Gentz, R., Hillson, N. J., Peisert, S., Kim, J., Simmons, B. A., Petzold, C. J., Singer, S. W., Mukhopadhyay, A., Tanjore, D., Dunn, J. G., & Garcia Martin, H. (2021). Machine learning for metabolic engineering: A review. *Metabolic Engineering, 63*, 34–60. https://doi.org/10.1016/j.ymben.2020.10.005
109. Leipzig, J., Nüst, D., Hoyt, C. T., Ram, K., & Greenberg, J. (2021). The role of metadata in reproducible computational research. *Patterns, 2*(9), 100322. https://doi.org/10.1016/j.patter.2021.100322

110. Li, C., Gao, Z., Su, B., Xu, G., & Lin, X. (2022). Data analysis methods for defining biomarkers from omics data. *Analytical and Bioanalytical Chemistry, 414*(1), 235–250. https://doi.org/10.1007/s00216-021-03813-7
111. Li, G., Wu, J., & Wang, X. (2024). Predicting functional UTR variants by integrating region-specific features. *Briefings in Bioinformatics, 25*(4), bbae248. https://doi.org/10.1093/bib/bbae248
112. Li, S.-Z., Sun, H.-Y., Tian, Y., Zhou, L.-Q., & Zhou, T. (2024). Machine-learning derived identification of prognostic signature to forecast head and neck squamous cell carcinoma prognosis and drug response. *Frontiers in Immunology, 15*, 1469895. https://doi.org/10.3389/fimmu.2024.1469895
113. Li, X., Wang, S., Zeng, S., Wu, Y., & Yang, Y. (2024). A survey on LLM-based multi-agent systems: Workflow, infrastructure, and challenges. *Vicinagearth, 1*(1), 9. https://doi.org/10.1007/s44336-024-00009-2
114. Li, Y., Mansmann, U., Du, S., & Hornung, R. (2022). Benchmark study of feature selection strategies for multi-omics data. *BMC Bioinformatics, 23*(1), 412. https://doi.org/10.1186/s12859-022-04962-x
115. Li, Y., Wu, X., Fang, D., & Luo, Y. (2024). Informing immunotherapy with multi-omics driven machine learning. *Npj Digital Medicine, 7*(1), 67. https://doi.org/10.1038/s41746-024-01043-6
116. Lilley, L. M., Sanche, S., Moore, S. C., Salemi, M. R., Vu, D., Iyer, S., Hengartner, N. W., & Mukundan, H. (2022). Methods to capture proteomic and metabolomic signatures from cerebrospinal fluid and serum of healthy individuals. *Scientific Reports, 12*(1), 13339. https://doi.org/10.1038/s41598-022-16598-1
117. Limeta, A., Gatto, F., Herrgård, M. J., Ji, B., & Nielsen, J. (2023). Leveraging high-resolution omics data for predicting responses and adverse events to immune checkpoint inhibitors. *Computational and Structural Biotechnology Journal, 21*, 3912–3919. https://doi.org/10.1016/j.csbj.2023.07.032
118. Lin, X., Kang, K., Chen, P., Zeng, Z., Li, G., Xiong, W., Yi, M., & Xiang, B. (2024). Regulatory mechanisms of PD-1/PD-L1 in cancers. *Molecular Cancer, 23*(1), 108. https://doi.org/10.1186/s12943-024-02023-w
119. Linden, A. H., & Hönekopp, J. (2021). Heterogeneity of research results: A new perspective from which to assess and promote progress in psychological science. *Perspectives on Psychological Science, 16*(2), 358–376. https://doi.org/10.1177/1745691620964193
120. Lippa, K. A., Aristizabal-Henao, J. J., Beger, R. D., Bowden, J. A., Broeckling, C., Beecher, C., Clay Davis, W., Dunn, W. B., Flores, R., Goodacre, R., Gouveia, G. J., Harms, A. C., Hartung, T., Jones, C. M., Lewis, M. R., Ntai, I., Percy, A. J., Raftery, D., Schock, T. B., … Ubhi, B. K. (2022). Reference materials for MS-based untargeted metabolomics and lipidomics: A review by the metabolomics quality assurance and quality control consortium (mQACC). *Metabolomics, 18*(4), 24. https://doi.org/10.1007/s11306-021-01848-6
121. Liu, X., Shi, J., Jiao, Y., An, J., Tian, J., Yang, Y., & Zhuo, L. (2024). Integrated multi-omics with machine learning to uncover the intricacies of kidney disease. *Briefings in Bioinformatics, 25*(5), bbae364. https://doi.org/10.1093/bib/bbae364
122. Liu, X.-H., Wang, G.-R., Zhong, N.-N., Wang, W.-Y., Liu, B., Li, Z., & Bu, L.-L. (2025). Multi-omics in immunotherapy research for HNSCC: Present situation and future perspectives. *Npj Precision Oncology, 9*(1), 93. https://doi.org/10.1038/s41698-025-00886-w
123. Liu, Z., Liu, J., Liu, X., Wang, X., Xie, Q., Zhang, X., Kong, X., He, M., Yang, Y., Deng, X., Yang, L., Qi, Y., Li, J., Liu, Y., Yuan, L., Diao, L., He, F., & Li, D. (2022). CTR-DB, an omnibus for patient-derived gene expression signatures correlated with cancer drug response. *Nucleic Acids Research, 50*(D1), D1184–D1199. https://doi.org/10.1093/nar/gkab860
124. Lucena-Padros, H., Bravo-Gil, N., Tous, C., Rojano, E., Seoane-Zonjic, P., Fernández, R. M., Ranea, J. A. G., Antiñolo, G., & Borrego, S. (2024). Bioinformatics prediction for network-based integrative multi-omics expression data analysis in hirschsprung disease. *Biomolecules, 14*(2), 164. https://doi.org/10.3390/biom14020164

125. Łukaszuk, T., Krawczuk, J., Żyła, K., & Kęsik, J. (2024). Stability of feature selection in multi-omics data analysis. *Applied Sciences, 14*(23), 11103. https://doi.org/10.3390/app142311103
126. Machado, M. P., Tomaz, J. P., Lorand-Metze, I., Souza, C. A. D., Vigorito, A. C., Delamain, M. T., Bendit, I., Pereira, N. F., & Pagnano, K. B. B. (2011). Monitoring of BCR-ABL levels in chronic myeloid leukemia patients treated with imatinib in the chronic phase: The importance of a major molecular response. *Revista Brasileira de Hematologia e Hemoterapia, 33*(3), 211–215. https://doi.org/10.5581/1516-8484.20110056
127. Madhumita, & Paul, S. (2022). Capturing the latent space of an autoencoder for multi-omics integration and cancer subtyping. *Computers in Biology and Medicine, 148*, 105832. https://doi.org/10.1016/j.compbiomed.2022.105832
128. Mahbub, N. I., Hasan, M. I., Rahman, M. H., Naznin, F., Islam, M. Z., & Moni, M. A. (2022). Identifying molecular signatures and pathways shared between Alzheimer's and Huntington's disorders: A bioinformatics and systems biology approach. *Informatics in Medicine Unlocked, 30*, 100888. https://doi.org/10.1016/j.imu.2022.100888
129. Mallick, P., Maity, S., Chakrabarti, O., & Chakrabarti, S. (2022). Role of systems biology and multi-omics analyses in delineating spatial interconnectivity and temporal dynamicity of ER stress mediated cellular responses. *Biochimica et Biophysica Acta (BBA) - Molecular Cell Research, 1869*(4), 119210. https://doi.org/10.1016/j.bbamcr.2022.119210
130. Mammoliti, A., Smirnov, P., Nakano, M., Safikhani, Z., Eeles, C., Seo, H., Nair, S. K., Mer, A. S., Smith, I., Ho, C., Beri, G., Kusko, R., Massive Analysis Quality Control (MAQC) Society Board of Directors, Shraddha, T., Sansone, S.-A., Tong, W., Wolfinger, R. D., Mason, C. E., Jones, W., … Haibe-Kains, B. (2021). Orchestrating and sharing large multimodal data for transparent and reproducible research. *Nature Communications, 12*(1), 5797. https://doi.org/10.1038/s41467-021-25974-w
131. Marques, L., Costa, B., Pereira, M., Silva, A., Santos, J., Saldanha, L., Silva, I., Magalhães, P., Schmidt, S., & Vale, N. (2024). Advancing precision medicine: A review of innovative in silico approaches for drug development, clinical pharmacology and personalized healthcare. *Pharmaceutics, 16*(3), 332. https://doi.org/10.3390/pharmaceutics16030332
132. Maseno, E. M., & Wang, Z. (2024). Hybrid wrapper feature selection method based on genetic algorithm and extreme learning machine for intrusion detection. *Journal of Big Data, 11*(1), 24. https://doi.org/10.1186/s40537-024-00887-9
133. Mathew, D. E., Ebem, D. U., Ikegwu, A. C., Ukeoma, P. E., & Dibiaezue, N. F. (2025). Recent emerging techniques in explainable artificial intelligence to enhance the interpretable and understanding of AI models for human. *Neural Processing Letters, 57*(1), 16. https://doi.org/10.1007/s11063-025-11732-2
134. Mattes, R. D., Rowe, S. B., Ohlhorst, S. D., Brown, A. W., Hoffman, D. J., Liska, D. J., Feskens, E. J. M., Dhillon, J., Tucker, K. L., Epstein, L. H., Neufeld, L. M., Kelley, M., Fukagawa, N. K., Sunde, R. A., Zeisel, S. H., Basile, A. J., Borth, L. E., & Jackson, E. (2022). Valuing the diversity of research methods to advance nutrition science. *Advances in Nutrition, 13*(4), 1324–1393. https://doi.org/10.1093/advances/nmac043
135. Mazidi, M., Wright, N., Yao, P., Kartsonaki, C., Millwood, I. Y., Fry, H., Said, S., Pozarickij, A., Pei, P., Chen, Y., Avery, D., Du, H., Schmidt, D. V., Yang, L., Lv, J., Yu, C., Chen, J., Hill, M., Holmes, M. V., … Zhang, X. (2023). Plasma proteomics to identify drug targets for ischemic heart disease. *Journal of the American College of Cardiology, 82*(20), 1906–1920. https://doi.org/10.1016/j.jacc.2023.09.804
136. Mehdary, A., Chehri, A., Jakimi, A., & Saadane, R. (2024). Hyperparameter optimization with genetic algorithms and XGBoost: A step forward in smart grid fraud detection. *Sensors, 24*(4), 1230. https://doi.org/10.3390/s24041230
137. Menyhárt, O., & Győrffy, B. (2021). Multi-omics approaches in cancer research with applications in tumor subtyping, prognosis, and diagnosis. *Computational and Structural Biotechnology Journal, 19*, 949–960. https://doi.org/10.1016/j.csbj.2021.01.009

138. Mienye, I. D., & Swart, T. G. (2025). Deep autoencoder neural networks: A comprehensive review and new perspectives. *Archives of Computational Methods in Engineering*. https://doi.org/10.1007/s11831-025-10260-5
139. Miller, F. W. (2023). The increasing prevalence of autoimmunity and autoimmune diseases: An urgent call to action for improved understanding, diagnosis, treatment, and prevention. *Current Opinion in Immunology, 80*, 102266. https://doi.org/10.1016/j.coi.2022.102266
140. Mishra, P., Pandey, C., Singh, U., Keshri, A., & Sabaretnam, M. (2019). Selection of appropriate statistical methods for data analysis. *Annals of Cardiac Anaesthesia, 22*(3), 297. https://doi.org/10.4103/aca.ACA_248_18
141. Mohammadi-Shemirani, P., Sood, T., & Paré, G. (2023). From 'Omics to multi-omics technologies: The discovery of novel causal mediators. *Current Atherosclerosis Reports, 25*(2), 55–65. https://doi.org/10.1007/s11883-022-01078-8
142. Mohseni, P., & Ghorbani, A. (2024). Exploring the synergy of artificial intelligence in microbiology: Advancements, challenges, and future prospects. *Computational and Structural Biotechnology Reports, 1*, 100005. https://doi.org/10.1016/j.csbr.2024.100005
143. Mohsenirad, S., & Triantis, K. (2025). Testing for heterogeneity in data envelopment analysis. *Annals of Operations Research*. https://doi.org/10.1007/s10479-024-06460-0
144. Momeni, Z., Hassanzadeh, E., Saniee Abadeh, M., & Bellazzi, R. (2020). A survey on single and multi-omics data mining methods in cancer data classification. *Journal of Biomedical Informatics, 107*, 103466. https://doi.org/10.1016/j.jbi.2020.103466
145. Moradi, N., Habibirad, A., & Panahi, H. (2024). Leveraging neural networks for robust survival estimation in cox proportional hazards model with censored data. *Communications in Statistics: Case Studies, Data Analysis and Applications, 10*(3–4), 349–364. https://doi.org/10.1080/23737484.2024.2439847
146. Morrison, F. M. M., Rezaei, N., Arero, A. G., Graklanov, V., Iritsyan, S., Ivanovska, M., Makuku, R., Marquez, L. P., Minakova, K., Mmema, L. P., Rzymski, P., & Zavolodko, G. (2023). Maintaining scientific integrity and high research standards against the backdrop of rising artificial intelligence use across fields. *Journal of Medical Artificial Intelligence, 6*, 24–24. https://doi.org/10.21037/jmai-23-63
147. Mumuni, A., & Mumuni, F. (2025). Automated data processing and feature engineering for deep learning and big data applications: A survey. *Journal of Information and Intelligence, 3*(2), 113–153. https://doi.org/10.1016/j.jiixd.2024.01.002
148. Munquad, S., & Das, A. B. (2023). DeepAutoGlioma: A deep learning autoencoder-based multi-omics data integration and classification tools for glioma subtyping. *BioData Mining, 16*(1), 32. https://doi.org/10.1186/s13040-023-00349-7
149. Musib, L., Coletti, R., Lopes, M. B., Mouriño, H., & Carrasquinha, E. (2024). Priority-elastic net for binary disease outcome prediction based on multi-omics data. *BioData Mining, 17*(1), 45. https://doi.org/10.1186/s13040-024-00401-0
150. Nasir, M., Summerfield, N. S., Carreiro, S., Berlowitz, D., & Oztekin, A. (2025). A machine learning approach for diagnostic and prognostic predictions, key risk factors and interactions. *Health Services and Outcomes Research Methodology, 25*(1), 1–28. https://doi.org/10.1007/s10742-024-00324-7
151. Nedeljković, M., Vuletić, A., & Mirjačić Martinović, K. (2025). Divide and conquer—Targeted therapy for triple-negative breast cancer. *International Journal of Molecular Sciences, 26*(4), 1396. https://doi.org/10.3390/ijms26041396
152. Nooreldeen, R., & Bach, H. (2021). Current and future development in lung cancer diagnosis. *International Journal of Molecular Sciences, 22*(16), 8661. https://doi.org/10.3390/ijms22168661
153. Nssibi, M., Manita, G., Chhabra, A., Mirjalili, S., & Korbaa, O. (2024). Gene Selection for high dimensional biological datasets using hybrid island binary artificial bee colony with chaos game optimization. *Artificial Intelligence Review, 57*(3), 51. https://doi.org/10.1007/s10462-023-10675-1

154. Nunes, S., Bastos, R., Marinho, A. I., Vieira, R., Benício, I., De Noronha, M. A., Lírio, S., Brodskyn, C., & Tavares, N. M. (2025). Recent advances in the development and clinical application of miRNAs in infectious diseases. *Non-Coding RNA Research, 10*, 41–54. https://doi.org/10.1016/j.ncrna.2024.09.005
155. Obaido, G., Mienye, I. D., Egbelowo, O. F., Emmanuel, I. D., Ogunleye, A., Ogbuokiri, B., Mienye, P., & Aruleba, K. (2024). Supervised machine learning in drug discovery and development: Algorithms, applications, challenges, and prospects. *Machine Learning with Applications, 17*, 100576. https://doi.org/10.1016/j.mlwa.2024.100576
156. O'Connor, L. M., O'Connor, B. A., Lim, S. B., Zeng, J., & Lo, C. H. (2023). Integrative multi-omics and systems bioinformatics in translational neuroscience: A data mining perspective. *Journal of Pharmaceutical Analysis, 13*(8), 836–850. https://doi.org/10.1016/j.jpha.2023.06.011
157. Oliver, J., Nair, N., Orozco, G., Smith, S., Hyrich, K. L., Morgan, A., Isaacs, J., Wilson, A. G., BRAGGSS, Barton, A., & Plant, D. (2021). Transcriptome-wide study of TNF-inhibitor therapy in rheumatoid arthritis reveals early signature of successful treatment. *Arthritis Research & Therapy, 23*(1), 80. https://doi.org/10.1186/s13075-021-02451-9
158. Oropeza-Valdez, J. J., Padron-Manrique, C., Vázquez-Jiménez, A., Soberon, X., & Resendis-Antonio, O. (2024). Exploring metabolic anomalies in COVID-19 and post-COVID-19: A machine learning approach with explainable artificial intelligence. *Frontiers in Molecular Biosciences, 11*, 1429281. https://doi.org/10.3389/fmolb.2024.1429281
159. Ostrom, Q. T., Cioffi, G., Waite, K., Kruchko, C., & Barnholtz-Sloan, J. S. (2021). CBTRUS statistical report: Primary brain and other Central nervous system tumors diagnosed in the United States in 2014–2018. *Neuro-Oncology, 23*(Supplement_3), iii1–iii105. https://doi.org/10.1093/neuonc/noab200
160. Ouadfel, S., & Abd Elaziz, M. (2022). Efficient high-dimension feature selection based on enhanced equilibrium optimizer. *Expert Systems with Applications, 187*, 115882. https://doi.org/10.1016/j.eswa.2021.115882
161. Panahi, B., Hosseinzadeh Gharajeh, N., Mohammadzadeh Jalaly, H., & Golkari, S. (2024). Leveraging multi-omics and machine learning approaches in malting barley research: From farm cultivation to the final products. *Current Plant Biology, 39*, 100362. https://doi.org/10.1016/j.cpb.2024.100362
162. Papadaki, E., Kakkos, I., Vlamos, P., Petropoulou, O., Miloulis, S. T., Palamas, S., & Vrahatis, A. G. (2024). Recent web platforms for multi-omics integration unlocking biological complexity. *Applied Sciences, 15*(1), 329. https://doi.org/10.3390/app15010329
163. Peña-Enríquez, R., Bermejo, B., Pollán, M., Díaz-Chacón, A., Jerez Gilarranz, Y., Ponce Lorenzo, J. J., Fernández Aramburo, A., Cantos Sánchez De Ibargüen, B., Santaballa Bertrán, A., Galve-Calvo, E., Jiménez-Arranz, Á, Fernández, Y., Pérez, M. E., De La Cruz, S., Anton-Torres, A., Moreno, F., Vidal-Losada, M. J., López-Ceballos, M. H., Blancas, I., … De La Haba-Rodríguez, J. (2025). Molecular characterization of pregnancy-associated breast cancer and insights on timing from GEICAM-EMBARCAM study. *Npj Breast Cancer, 11*(1), 12. https://doi.org/10.1038/s41523-025-00718-x
164. Perazzoli, G., García-Valdeavero, O. M., Peña, M., Prados, J., Melguizo, C., & Jiménez-Luna, C. (2023). Evaluating metabolite-based biomarkers for early diagnosis of pancreatic cancer: A systematic review. *Metabolites, 13*(7), 872. https://doi.org/10.3390/metabo13070872
165. Picard, M., Scott-Boyer, M.-P., Bodein, A., Périn, O., & Droit, A. (2021). Integration strategies of multi-omics data for machine learning analysis. *Computational and Structural Biotechnology Journal, 19*, 3735–3746. https://doi.org/10.1016/j.csbj.2021.06.030
166. Pradat, Y., Viot, J., Yurchenko, A. A., Gunbin, K., Cerbone, L., Deloger, M., Grisay, G., Verlingue, L., Scott, V., Padioleau, I., Panunzi, L., Michiels, S., Hollebecque, A., Jules-Clément, G., Mezquita, L., Lainé, A., Loriot, Y., Besse, B., Friboulet, L., … Nikolaev, S. I. (2023). Integrative pan-cancer genomic and transcriptomic analyses of refractory metastatic cancer. *Cancer Discovery, 13*(5), 1116–1143. https://doi.org/10.1158/2159-8290.CD-22-0966

167. Prianichnikov, N., Koch, H., Koch, S., Lubeck, M., Heilig, R., Brehmer, S., Fischer, R., & Cox, J. (2020). MaxQuant software for ion mobility enhanced shotgun proteomics. *Molecular & Cellular Proteomics, 19*(6), 1058–1069. https://doi.org/10.1074/mcp.TIR119.001720
168. Priyatno, A. M., & Widiyaningtyas, T. (2024). A systematic literature review: Recursive feature elimination algorithms. *JITK (Jurnal Ilmu Pengetahuan Dan Teknologi Komputer*(2), 196–207. https://doi.org/10.33480/jitk.v9i2.5015
169. Pudjihartono, N., Fadason, T., Kempa-Liehr, A. W., & O'Sullivan, J. M. (2022). A review of feature selection methods for machine learning-based disease risk prediction. *Frontiers in Bioinformatics, 2*, 927312. https://doi.org/10.3389/fbinf.2022.927312
170. Pulsipher, M. A., Han, X., Maude, S. L., Laetsch, T. W., Qayed, M., Rives, S., Boyer, M. W., Hiramatsu, H., Yanik, G. A., Driscoll, T., Myers, G. D., Bader, P., Baruchel, A., Buechner, J., Stefanski, H. E., Kalfoglou, C., Nguyen, K., Waldron, E. R., Thudium Mueller, K., … Grupp, S. A. (2022). Next-generation sequencing of minimal residual disease for predicting relapse after tisagenlecleucel in children and young adults with acute lymphoblastic leukemia. *Blood Cancer Discovery, 3*(1), 66–81. https://doi.org/10.1158/2643-3230.BCD-21-0095
171. Rahnenführer, J., De Bin, R., Benner, A., Ambrogi, F., Lusa, L., Boulesteix, A.-L., Migliavacca, E., Binder, H., Michiels, S., Sauerbrei, W., McShane, L., & for topic group "High-dimensional data" (TG9) of the STRATOS initiative (2023). Statistical analysis of high-dimensional biomedical data: A gentle introduction to analytical goals, common approaches and challenges. *BMC Medicine, 21*(1), 182. https://doi.org/10.1186/s12916-023-02858-y
172. Rakusanova, S., & Cajka, T. (2024). Tips and tricks for LC–MS-based metabolomics and lipidomics analysis. *TrAC Trends in Analytical Chemistry, 180*, 117940. https://doi.org/10.1016/j.trac.2024.117940
173. Ran, X., Suyaroj, N., Tepsan, W., Ma, J., Zhou, X., & Deng, W. (2024). A hybrid genetic-fuzzy ant colony optimization algorithm for automatic K-means clustering in urban global positioning system. *Engineering Applications of Artificial Intelligence, 137*, 109237. https://doi.org/10.1016/j.engappai.2024.109237
174. Rao, A. M., Popper, S. J., Gupta, S., Davong, V., Vaidya, K., Chanthongthip, A., Dittrich, S., Robinson, M. T., Vongsouvath, M., Mayxay, M., Nawtaisong, P., Karmacharya, B., Thair, S. A., Bogoch, I., Sweeney, T. E., Newton, P. N., Andrews, J. R., Relman, D. A., & Khatri, P. (2022). A robust host-response-based signature distinguishes bacterial and viral infections across diverse global populations. *Cell Reports Medicine, 3*(12), 100842. https://doi.org/10.1016/j.xcrm.2022.100842
175. Rashid, M. M., & Selvarajoo, K. (2024). Advancing drug-response prediction using multimodal and -omics machine learning integration (MOMLIN): A case study on breast cancer clinical data. *Briefings in Bioinformatics, 25*(4), bbae300. https://doi.org/10.1093/bib/bbae300
176. Rehman, A. U., Li, M., Wu, B., Ali, Y., Rasheed, S., Shaheen, S., Liu, X., Luo, R., & Zhang, J. (2024). Role of artificial intelligence in revolutionizing drug discovery. *Fundamental Research*, S266732582400205X. https://doi.org/10.1016/j.fmre.2024.04.021
177. Rosati, D., Palmieri, M., Brunelli, G., Morrione, A., Iannelli, F., Frullanti, E., & Giordano, A. (2024). Differential gene expression analysis pipelines and bioinformatic tools for the identification of specific biomarkers: A review. *Computational and Structural Biotechnology Journal, 23*, 1154–1168. https://doi.org/10.1016/j.csbj.2024.02.018
178. Rosenblatt, M., Tejavibulya, L., Jiang, R., Noble, S., & Scheinost, D. (2024). Data leakage inflates prediction performance in connectome-based machine learning models. *Nature Communications, 15*(1), 1829. https://doi.org/10.1038/s41467-024-46150-w
179. Ruano-Ordás, D. (2024). Machine learning-based feature extraction and selection. *Applied Sciences, 14*(15), 6567. https://doi.org/10.3390/app14156567
180. S Band, S., Yarahmadi, A., Hsu, C.-C., Biyari, M., Sookhak, M., Ameri, R., Dehzangi, I., Chronopoulos, A. T., & Liang, H.-W. (2023). Application of explainable artificial intelligence in medical health: A systematic review of interpretability methods. *Informatics in Medicine Unlocked, 40*, 101286. https://doi.org/10.1016/j.imu.2023.101286

181. Saadh, M. J., Ahmed, H. H., Kareem, R. A., Yadav, A., Ganesan, S., Shankhyan, A., Sharma, G. C., Naidu, K. S., Rakhmatullaev, A., Sameer, H. N., Yaseen, A., Athab, Z. H., Adil, M., & Farhood, B. (2025). Advanced machine learning framework for enhancing breast cancer diagnostics through transcriptomic profiling. *Discover Oncology, 16*(1), 334. https://doi.org/10.1007/s12672-025-02111-3
182. Sabit, H., Arneth, B., Pawlik, T. M., Abdel-Ghany, S., Ghazy, A., Abdelazeem, R. M., Alqosaibi, A., Al-Dhuayan, I. S., Almulhim, J., Alrabiah, N. A., & Hashash, A. (2025). Leveraging single-cell multi-omics to decode tumor microenvironment diversity and therapeutic resistance. *Pharmaceuticals, 18*(1), 75. https://doi.org/10.3390/ph18010075
183. Sadee, W., Wang, D., Hartmann, K., & Toland, A. E. (2023). Pharmacogenomics: Driving personalized medicine. *Pharmacological Reviews, 75*(4), 789–814. https://doi.org/10.1124/pharmrev.122.000810
184. Sadeghi, Z., Alizadehsani, R., Cifci, M. A., Kausar, S., Rehman, R., Mahanta, P., Bora, P. K., Almasri, A., Alkhawaldeh, R. S., Hussain, S., Alatas, B., Shoeibi, A., Moosaei, H., Hladík, M., Nahavandi, S., & Pardalos, P. M. (2024). A review of explainable artificial intelligence in healthcare. *Computers and Electrical Engineering, 118*, 109370. https://doi.org/10.1016/j.compeleceng.2024.109370
185. Salih, S., Elliyanti, A., Alkatheeri, A., AlYafei, F., Almarri, B., & Khan, H. (2023). The role of molecular imaging in personalized medicine. *Journal of Personalized Medicine, 13*(2), 369. https://doi.org/10.3390/jpm13020369
186. Santos, M. R., Guedes, A., & Sanchez-Gendriz, I. (2024). SHAPley additive exPlanations (SHAP) for efficient feature selection in rolling bearing fault diagnosis. *Machine Learning and Knowledge Extraction, 6*(1), 316–341. https://doi.org/10.3390/make6010016
187. Sarker, A., Aziz, M. A., Hossen, M. B., Mollah, M. M. H., Al-Amin, & Mollah, M. N. H. (2024). Discovery of key molecular signatures for diagnosis and therapies of glioblastoma by combining supervised and unsupervised learning approaches. *Scientific Reports, 14*(1), 27545. https://doi.org/10.1038/s41598-024-79391-2
188. Serdar, C. C., Cihan, M., Yücel, D., & Serdar, M. A. (2021). Sample size, power and effect size revisited: Simplified and practical approaches in pre-clinical, clinical and laboratory studies. *Biochemia Medica, 31*(1), 27–53. https://doi.org/10.11613/BM.2021.010502
189. Shantaraman, A., Dammer, E. B., Ugochukwu, O., Duong, D. M., Yin, L., Carter, E. K., Gearing, M., Chen-Plotkin, A., Lee, E. B., Trojanowski, J. Q., Bennett, D. A., Lah, J. J., Levey, A. I., Seyfried, N. T., & Higginbotham, L. (2024). *Network Proteomics of the Lewy Body Dementia Brain Reveals Presynaptic Signatures Distinct from Alzheimer's Disease.* https://doi.org/10.1101/2024.01.23.576728
190. Sharma, A., Debik, J., Naume, B., Ohnstad, H. O., Oslo Breast Cancer Consortium (OSBREAC), Sahlber, K. K., Borgen, E., Børresen-Dale, A.-L., Engebråten, O., Fritzman, B., Garred, Ø, Geisler, J., Geitvik, G. A., Hofvind, S., Kristensen, V. N., Kåresen, R., Langerød, A., Lingjærde, O. C., Mælandsmo, G. M., … Giskeødegård, G. F.(2024). Comprehensive multi-omics analysis of breast cancer reveals distinct long-term prognostic subtypes. *Oncogenesis, 13*(1), 22. https://doi.org/10.1038/s41389-024-00521-6
191. Sharma, A., Lysenko, A., Jia, S., Boroevich, K. A., & Tsunoda, T. (2024). Advances in AI and machine learning for predictive medicine. *Journal of Human Genetics, 69*(10), 487–497. https://doi.org/10.1038/s10038-024-01231-y
192. Shen, S., Wang, X., Zhu, X., Rasam, S., Ma, M., Huo, S., Qian, S., Zhang, M., Qu, M., Hu, C., Jin, L., Tian, Y., Sethi, S., Poulsen, D., Wang, J., Tu, C., & Qu, J. (2023). High-quality and robust protein quantification in large clinical/pharmaceutical cohorts with IonStar proteomics investigation. *Nature Protocols, 18*(3), 700–731. https://doi.org/10.1038/s41596-022-00780-w
193. Shi, Y., Yu, Y., Zhao, J., Huang, L., Wang, Q., Sun, Q., Liu, L., & Sun, C. (2025). Development of a prognostic model based on four genes related to exhausted CD8+ T cell in triple-negative breast cancer patients: A comprehensive analysis integrating scRNA-seq and bulk RNA-seq. *Discover Oncology, 16*(1), 114. https://doi.org/10.1007/s12672-025-01812-z

194. Sidak, D., Schwarzerová, J., Weckwerth, W., & Waldherr, S. (2022). Interpretable machine learning methods for predictions in systems biology from omics data. *Frontiers in Molecular Biosciences*, *9*, 926623. https://doi.org/10.3389/fmolb.2022.926623
195. Siedlecki-Wullich, D., Miñano-Molina, A. J., & Rodríguez-Álvarez, J. (2021). microRNAs as early biomarkers of Alzheimer's disease: A synaptic perspective. *Cells*, *10*(1), 113. https://doi.org/10.3390/cells10010113
196. Spendlove, S. J., Bondhus, L., Lluri, G., Sul, J. H., & Arboleda, V. A. (2022). Polygenic risk scores of endo-phenotypes identify the effect of genetic background in congenital heart disease. *Human Genetics and Genomics Advances*, *3*(3), 100112. https://doi.org/10.1016/j.xhgg.2022.100112
197. Stroggilos, R., Tserga, A., Zoidakis, J., Vlahou, A., & Makridakis, M. (2024). Tissue proteomics repositories for data reanalysis. *Mass Spectrometry Reviews*, *43*(6), 1270–1284. https://doi.org/10.1002/mas.21860
198. Su, J., Yang, L., Sun, Z., & Zhan, X. (2024). Personalized drug therapy: Innovative concept guided with proteoformics. *Molecular & Cellular Proteomics*, *23*(3), 100737. https://doi.org/10.1016/j.mcpro.2024.100737
199. Subramanian, I., Verma, S., Kumar, S., Jere, A., & Anamika, K. (2020). Multi-omics data integration, interpretation, and its application. *Bioinformatics and Biology Insights*, *14*, 117793221989905. https://doi.org/10.1177/1177932219899051
200. Sun, F., Li, H., Sun, D., Fu, S., Gu, L., Shao, X., Wang, Q., Dong, X., Duan, B., Xing, F., Wu, J., Xiao, M., Zhao, F., Han, J.-D. J., Liu, Q., Fan, X., Li, C., Wang, C., & Shi, T. (2025). Single-cell omics: Experimental workflow, data analyses and applications. *Science China Life Sciences*, *68*(1), 5–102. https://doi.org/10.1007/s11427-023-2561-0
201. Sun, J., & Xia, Y. (2024). Pretreating and normalizing metabolomics data for statistical analysis. *Genes & Diseases*, *11*(3), 100979. https://doi.org/10.1016/j.gendis.2023.04.018
202. Sun, L., Fan, X., Zhao, Y., Zhang, Q., & Jiang, M. (2024). Deep learning-based metabolomics data study of prostate cancer. *BMC Bioinformatics*, *25*(1), 391. https://doi.org/10.1186/s12859-024-06016-w
203. Sun, R., Ge, W., Zhu, Y., Sayad, A., Luna, A., Lyu, M., Liang, S., Tobalina, L., Rajapakse, V. N., Yu, C., Zhang, H., Fang, J., Wu, F., Xie, H., Saez-Rodriguez, J., Ying, H., Reinhold, W. C., Sander, C., Pommier, Y., … Guo, T. (2023). Proteomic dynamics of breast cancer cell lines identifies potential therapeutic protein targets. *Molecular & Cellular Proteomics*, *22*(8), 100602. https://doi.org/10.1016/j.mcpro.2023.100602
204. Taherdoost, H., & Ghofrani, A. (2024). AI's role in revolutionizing personalized medicine by reshaping pharmacogenomics and drug therapy. *Intelligent Pharmacy*, *2*(5), 643–650. https://doi.org/10.1016/j.ipha.2024.08.005
205. Tandon, R., Levey, A. I., Lah, J. J., Seyfried, N. T., & Mitchell, C. S. (2023). Machine learning selection of most predictive brain proteins suggests role of sugar metabolism in Alzheimer's disease. *Journal of Alzheimer's Disease*, *92*(2), 411–424. https://doi.org/10.3233/JAD-220683
206. Tarazona, S., Balzano-Nogueira, L., Gómez-Cabrero, D., Schmidt, A., Imhof, A., Hankemeier, T., Tegnér, J., Westerhuis, J. A., & Conesa, A. (2020). Harmonization of quality metrics and power calculation in multi-omic studies. *Nature Communications*, *11*(1), 3092. https://doi.org/10.1038/s41467-020-16937-8
207. Tenchov, R., Sapra, A. K., Sasso, J., Ralhan, K., Tummala, A., Azoulay, N., & Zhou, Q. A. (2024). Biomarkers for early cancer detection: A landscape view of recent advancements, spotlighting pancreatic and liver cancers. *ACS Pharmacology & Translational Science*, *7*(3), 586–613. https://doi.org/10.1021/acsptsci.3c00346
208. Thompson, A., & Suomalainen, J. (2025). GAOR: Genetic algorithm-based optimization for machine learning robustness in communication networks. *Network*, *5*(1), 6. https://doi.org/10.3390/network5010006
209. Tolani, P., Gupta, S., Yadav, K., Aggarwal, S., & Yadav, A. K. (2021). Big data, integrative omics and network biology. In *Advances in Protein Chemistry and Structural Biology* (Vol. 127, pp. 127–160). Elsevier. https://doi.org/10.1016/bs.apcsb.2021.03.006

210. Torres-Martos, Á, Bustos-Aibar, M., Ramírez-Mena, A., Cámara-Sánchez, S., Anguita-Ruiz, A., Alcalá, R., Aguilera, C. M., & Alcalá-Fdez, J. (2023). Omics data preprocessing for machine learning: A case study in childhood obesity. *Genes*, *14*(2), 248. https://doi.org/10.3390/genes14020248
211. Toussaint, P. A., Leiser, F., Thiebes, S., Schlesner, M., Brors, B., & Sunyaev, A. (2023). Explainable artificial intelligence for omics data: A systematic mapping study. *Briefings in Bioinformatics*, *25*(1), bbad453. https://doi.org/10.1093/bib/bbad453
212. Tran, K. A., Kondrashova, O., Bradley, A., Williams, E. D., Pearson, J. V., & Waddell, N. (2021). Deep learning in cancer diagnosis, prognosis and treatment selection. *Genome Medicine*, *13*(1), 152. https://doi.org/10.1186/s13073-021-00968-x
213. Tsantilas, K. A., Merrihew, G. E., Robbins, J. E., Johnson, R. S., Park, J., Plubell, D. L., Canterbury, J. D., Huang, E., Riffle, M., Sharma, V., MacLean, B. X., Eckels, J., Wu, C. C., Bereman, M. S., Spencer, S. E., Hoofnagle, A. N., & MacCoss, M. J. (2024). A framework for quality control in quantitative proteomics. *Journal of Proteome Research*, *23*(10), 4392–4408. https://doi.org/10.1021/acs.jproteome.4c00363
214. Tsimenidis, S., Vrochidou, E., & Papakostas, G. A. (2022). Omics data and data representations for deep learning-based predictive modeling. *International Journal of Molecular Sciences*, *23*(20), 12272. https://doi.org/10.3390/ijms232012272
215. Tuly, K., Hossen, M., Islam, M., Kibria, M., Alam, M., Harun-Or-Roshid, M., Begum, A., Hasan, S., Mahumud, R., & Mollah, M. (2023). Robust identification of differential Gene expression patterns from multiple transcriptomics datasets for early diagnosis, prognosis, and therapies for breast cancer. *Medicina*, *59*(10), 1705. https://doi.org/10.3390/medicina59101705
216. Tyagi, N., Mehla, K., & Gupta, D. (2023). Deciphering novel common gene signatures for rheumatoid arthritis and systemic lupus erythematosus by integrative analysis of transcriptomic profiles. *PLOS ONE*, *18*(3), e0281637. https://doi.org/10.1371/journal.pone.0281637
217. Vacher, M., Canovas, R., Laws, S. M., & Doecke, J. D. (2024). A comprehensive multi-omics analysis reveals unique signatures to predict Alzheimer's disease. *Frontiers in Bioinformatics*, *4*, 1390607. https://doi.org/10.3389/fbinf.2024.1390607
218. Vahabi, N., & Michailidis, G. (2022). Unsupervised multi-omics data integration methods: A comprehensive review. *Frontiers in Genetics*, *13*, 854752. https://doi.org/10.3389/fgene.2022.854752
219. Valous, N. A., Popp, F., Zörnig, I., Jäger, D., & Charoentong, P. (2024). Graph machine learning for integrated multi-omics analysis. *British Journal of Cancer*, *131*(2), 205–211. https://doi.org/10.1038/s41416-024-02706-7
220. Van De Sande, B., Lee, J. S., Mutasa-Gottgens, E., Naughton, B., Bacon, W., Manning, J., Wang, Y., Pollard, J., Mendez, M., Hill, J., Kumar, N., Cao, X., Chen, X., Khaladkar, M., Wen, J., Leach, A., & Ferran, E. (2023). Applications of single-cell RNA sequencing in drug discovery and development. *Nature Reviews Drug Discovery*, *22*(6), 496–520. https://doi.org/10.1038/s41573-023-00688-4
221. Vihinen, M. (2020). Strategy for disease diagnosis, progression prediction, risk group stratification and treatment—Case of COVID-19. *Frontiers in Medicine*, *7*, 294. https://doi.org/10.3389/fmed.2020.00294
222. Virág, D., Dalmadi-Kiss, B., Vékey, K., Drahos, L., Klebovich, I., Antal, I., & Ludányi, K. (2020). Current trends in the analysis of post-translational modifications. *Chromatographia*, *83*(1), 1–10. https://doi.org/10.1007/s10337-019-03796-9
223. Vitorino, R. (2024). Transforming clinical research: The power of high-throughput omics integration. *Proteomes*, *12*(3), 25. https://doi.org/10.3390/proteomes12030025
224. Wang, J., Zeng, Z., Li, Z., Liu, G., Zhang, S., Luo, C., Hu, S., Wan, S., & Zhao, L. (2025). The clinical application of artificial intelligence in cancer precision treatment. *Journal of Translational Medicine*, *23*(1), 120. https://doi.org/10.1186/s12967-025-06139-5

225. Wang, Q., Gümüş, Z. H., Colarossi, C., Memeo, L., Wang, X., Kong, C. Y., & Boffetta, P. (2023). SCLC: Epidemiology, risk factors, genetic susceptibility, molecular pathology, screening, and early detection. *Journal of Thoracic Oncology, 18*(1), 31–46. https://doi.org/10.1016/j.jtho.2022.10.002
226. Wang, X., & Chen, H. (2022). Prognosis prediction through an integrated analysis of single-cell and bulk RNA-sequencing data in triple-negative breast cancer. *Frontiers in Genetics, 13*, 928175. https://doi.org/10.3389/fgene.2022.928175
227. Watson, E. R., Taherian Fard, A., & Mar, J. C. (2022). Computational methods for single-cell imaging and omics data integration. *Frontiers in Molecular Biosciences, 8*, 768106. https://doi.org/10.3389/fmolb.2021.768106
228. Wei, L., Niraula, D., Gates, E. D. H., Fu, J., Luo, Y., Nyflot, M. J., Bowen, S. R., El Naqa, I. M., & Cui, S. (2023). Artificial intelligence (AI) and machine learning (ML) in precision oncology: A review on enhancing discoverability through multiomics integration. *The British Journal of Radiology, 96*(1150), 20230211. https://doi.org/10.1259/bjr.20230211
229. Werling, D. M., Pochareddy, S., Choi, J., An, J.-Y., Sheppard, B., Peng, M., Li, Z., Dastmalchi, C., Santpere, G., Sousa, A. M. M., Tebbenkamp, A. T. N., Kaur, N., Gulden, F. O., Breen, M. S., Liang, L., Gilson, M. C., Zhao, X., Dong, S., Klei, L., … Sestan, N. (2020). Whole-genome and RNA sequencing reveal variation and transcriptomic coordination in the developing human prefrontal cortex. *Cell Reports, 31*(1), 107489. https://doi.org/10.1016/j.celrep.2020.03.053
230. Woo, V., & Alenghat, T. (2022). Epigenetic regulation by gut microbiota. *Gut Microbes, 14*(1), 2022407. https://doi.org/10.1080/19490976.2021.2022407
231. Wu, C.-H., Zhou, X., & Chen, M. (2025). Exploring and mitigating shortcomings in single-cell differential expression analysis with a new statistical paradigm. *Genome Biology, 26*(1), 58. https://doi.org/10.1186/s13059-025-03525-6
232. Wu, T., Hu, E., Xu, S., Chen, M., Guo, P., Dai, Z., Feng, T., Zhou, L., Tang, W., Zhan, L., Fu, X., Liu, S., Bo, X., & Yu, G. (2021). clusterProfiler 4.0: A universal enrichment tool for interpreting omics data. *The Innovation, 2*(3), 100141. https://doi.org/10.1016/j.xinn.2021.100141
233. Wu, X., Yang, X., Dai, Y., Zhao, Z., Zhu, J., Guo, H., & Yang, R. (2024). Single-cell sequencing to multi-omics: Technologies and applications. *Biomarker Research, 12*(1), 110. https://doi.org/10.1186/s40364-024-00643-4
234. Xie, L., He, B., Varathan, P., Nho, K., Risacher, S. L., Saykin, A. J., Salama, P., & Yan, J. (2021). Integrative-omics for discovery of network-level disease biomarkers: A case study in Alzheimer's disease. *Briefings in Bioinformatics, 22*(6), bbab121. https://doi.org/10.1093/bib/bbab121
235. Xiong, Q., Huang, S., Li, Y., Lv, N., Lv, C., Ding, Y., Liu, W., Wang, L., Chen, Y., Sun, L., Zhao, Y., Liao, S., Zhang, M., Zhu, B., & Yu, L. (2020). Single-cell RNA sequencing of t(8;21) acute myeloid leukemia for risk prediction. *Oncology Reports*. https://doi.org/10.3892/or.2020.7507
236. Xu, K., Berthiller, F., Metzler-Zebeli, B. U., & Schwartz-Zimmermann, H. E. (2025). Development and validation of targeted metabolomics methods using liquid Chromatography–Tandem mass spectrometry (LC-MS/MS) for the quantification of 235 plasma metabolites. *Molecules, 30*(3), 706. https://doi.org/10.3390/molecules30030706
237. Yao, K., Tong, C.-Y., & Cheng, C. (2022). A framework to predict the applicability of oncotype DX, MammaPrint, and E2F4 Gene signatures for improving breast cancer prognostic prediction. *Scientific Reports, 12*(1), 2211. https://doi.org/10.1038/s41598-022-06230-7
238. Yaqoob, A., Musheer Aziz, R., & Verma, N. K. (2023). Applications and techniques of machine learning in cancer classification: A systematic review. *Human-Centric Intelligent Systems, 3*(4), 588–615. https://doi.org/10.1007/s44230-023-00041-3
239. Yaqoob, A., Verma, N. K., Aziz, R. M., & Shah, M. A. (2024). RNA-seq analysis for breast cancer detection: A study on paired tissue samples using hybrid optimization and deep learning techniques. *Journal of Cancer Research and Clinical Oncology, 150*(10), 455. https://doi.org/10.1007/s00432-024-05968-z

240. Yaqoob, A., Verma, N. K., Mir, M. A., Tejani, G. G., Eisa, N. H. B., Mamoun Hussien Osman, H., & Shah, M. A. (2025). SGA-driven feature selection and random forest classification for enhanced breast cancer diagnosis: A comparative study. *Scientific Reports, 15*(1), 10944. https://doi.org/10.1038/s41598-025-95786-1
241. Yen, N. T. H., Tien, N. T. N., Thu, N. Q., Ducatez, F., Mauhin, W., Lidove, O., Bekri, S., Tebani, A., & Long, N. P. (2025). A multi-omics-empowered framework for precision diagnosis and treatment of lysosomal diseases. *Journal of Pharmaceutical Analysis*, 101274. https://doi.org/10.1016/j.jpha.2025.101274
242. Yin, L., Duan, J.-J., Bian, X.-W., & Yu, S. (2020). Triple-negative breast cancer molecular subtyping and treatment progress. *Breast Cancer Research, 22*(1), 61. https://doi.org/10.1186/s13058-020-01296-5
243. You, Y., Dong, X., Wee, Y. K., Maxwell, M. J., Alhamdoosh, M., Smyth, G. K., Hickey, P. F., Ritchie, M. E., & Law, C. W. (2023). Modeling group heteroscedasticity in single-cell RNA-seq pseudo-bulk data. *Genome Biology, 24*(1), 107. https://doi.org/10.1186/s13059-023-02949-2
244. Yousef, M., Kumar, A., & Bakir-Gungor, B. (2020). Application of biological domain knowledge based feature selection on Gene expression data. *Entropy, 23*(1), 2. https://doi.org/10.3390/e23010002
245. Yu, S.-H., Chen, S.-C., Wu, P.-S., Kuo, P.-I., Chen, T.-A., Lee, H.-Y., & Lin, M.-H. (2024). Quantification quality control emerges as a crucial factor to enhance single-cell proteomics data analysis. *Molecular & Cellular Proteomics, 23*(5), 100768. https://doi.org/10.1016/j.mcpro.2024.100768
246. Yu, Y., Mai, Y., Zheng, Y., & Shi, L. (2024). Assessing and mitigating batch effects in large-scale omics studies. *Genome Biology, 25*(1), 254. https://doi.org/10.1186/s13059-024-03401-9
247. Yu, Y., Zhang, N., Mai, Y., Ren, L., Chen, Q., Cao, Z., Chen, Q., Liu, Y., Hou, W., Yang, J., Hong, H., Xu, J., Tong, W., Dong, L., Shi, L., Fang, X., & Zheng, Y. (2023). Correcting batch effects in large-scale multiomics studies using a reference-material-based ratio method. *Genome Biology, 24*(1), 201. https://doi.org/10.1186/s13059-023-03047-z
248. Zacharias, H. U., Kaleta, C., Cossais, F., Schaeffer, E., Berndt, H., Best, L., Dost, T., Glüsing, S., Groussin, M., Poyet, M., Heinzel, S., Bang, C., Siebert, L., Demetrowitsch, T., Leypoldt, F., Adelung, R., Bartsch, T., Bosy-Westphal, A., Schwarz, K., & Berg, D. (2022). Microbiome and metabolome insights into the role of the gastrointestinal–brain axis in Parkinson's and Alzheimer's disease: Unveiling potential therapeutic targets. *Metabolites, 12*(12), 1222. https://doi.org/10.3390/metabo12121222
249. Zhai, Z., Lin, Z., Meng, X., Zheng, X., Du, Y., Li, Z., Zhang, X., Liu, C., Zhou, L., Zhang, X., Tian, Z., Ma, Q., Li, J., Li, Q., & Pan, J. (2024). DiSignAtlas: An atlas of human and mouse disease signatures based on bulk and single-cell transcriptomics. *Nucleic Acids Research, 52*(D1), D1236–D1245. https://doi.org/10.1093/nar/gkad961
250. Zhang, F., Petersen, M., Johnson, L., Hall, J., & O'Bryant, S. E. (2021). Recursive support vector machine biomarker selection for Alzheimer's disease. *Journal of Alzheimer's Disease, 79*(4), 1691–1700. https://doi.org/10.3233/JAD-201254
251. Zhang, X. (2025). Highly effective batch effect correction method for RNA-seq count data. *Computational and Structural Biotechnology Journal, 27*, 58–64. https://doi.org/10.1016/j.csbj.2024.12.010
252. Zhao, H., Zhou, X., Wang, G., Yu, Y., Li, Y., Chen, Z., Song, W., Zhao, L., Wang, L., Wang, X., Cao, X., & Tian, Y. (2024). Integrating bulk and single-cell RNA-seq to construct a macrophage-related prognostic model for prognostic stratification in triple-negative breast cancer. *Journal of Cancer, 15*(18), 6002–6015. https://doi.org/10.7150/jca.101042
253. Zheng, Y., Liu, Y., Yang, J., Dong, L., Zhang, R., Tian, S., Yu, Y., Ren, L., Hou, W., Zhu, F., Mai, Y., Han, J., Zhang, L., Jiang, H., Lin, L., Lou, J., Li, R., Lin, J., Liu, H., … Shi, L. (2024). Multi-omics data integration using ratio-based quantitative profiling with quartet reference materials. *Nature Biotechnology, 42*(7), 1133–1149. https://doi.org/10.1038/s41587-023-01934-1

254. Zhou, J., & Troyanskaya, O. G. (2021). An analytical framework for interpretable and generalizable single-cell data analysis. *Nature Methods*, *18*(11), 1317–1321. https://doi.org/10.1038/s41592-021-01286-1
255. Zhou, Y., Geng, P., Zhang, S., Xiao, F., Cai, G., Chen, L., For the Alzheimer's Disease Neuroimaging Initiative, & Lu, Q. (2024). Multimodal functional deep learning for multiomics data. *Briefings in Bioinformatics*, *25*(5), bbae448. https://doi.org/10.1093/bib/bbae448
256. Zhu, A., Ibrahim, J. G., & Love, M. I. (2019). Heavy-tailed prior distributions for sequence count data: Removing the noise and preserving large differences. *Bioinformatics*, *35*(12), 2084–2092. https://doi.org/10.1093/bioinformatics/bty895
257. Zoran, T., Seelbinder, B., White, P., Price, J., Kraus, S., Kurzai, O., Linde, J., Häder, A., Loeffler, C., Grigoleit, G., Einsele, H., Panagiotou, G., Loeffler, J., & Schäuble, S. (2022). Molecular profiling reveals characteristic and decisive signatures in patients after allogeneic stem cell transplantation suffering from invasive pulmonary aspergillosis. *Journal of Fungi*, *8*(2), 171. https://doi.org/10.3390/jof8020171

Feature Extraction and Selection Methods Outperform Machine Learning and Deep Learning Techniques

Tuward Jade Dweh and Selorm Adablanu

7.1 INTRODUCTION

In the age of data-driven biology, feature extraction and feature selection (FE/FS) methods serve as foundational tools in improving the performance and interpretability of machine learning (ML) and deep learning (DL) models. These techniques are especially crucial when dealing with complex biological datasets, such as those encountered in omics-driven research, where high dimensionality, small sample size, and data noise often hinder accurate predictions and model generalization (Alghamdi et al., 2022; Abiodun et al., 2023).

The role of FE/FS has evolved significantly with the rise of data-intensive biomedical challenges. Classical ML models, often reliant on engineered or filtered features, benefit from careful variable selection that improves classification accuracy, model stability, and training efficiency (Alghamdi et al., 2022; Narkhede et al., 2021). Even in comparison with modern end-to-end DL systems, where raw data is passed directly through deep architectures, studies have shown that well-chosen features can often outperform or rival deep networks in domains such as disease detection, signal analysis, and image classification (Alzubaidi et al., 2023; Houssem et al., 2017; Elshawi et al., 2025). For instance, in Alzheimer's disease detection, integrating FE techniques into ML/DL pipelines significantly improved diagnostic performance over raw DL alone (Abiodun et al., 2023).

In the context of multi-omics integration, FE/FS remains especially relevant. Omics data—such as genomics, transcriptomics, proteomics, and metabolomics—introduce massive, noisy, and heterogeneous variables. DL models, although powerful, are often

DOI: 10.1201/9781003591658-7

opaque and require enormous training data, which is rarely available in biomedical studies. As such, FS techniques play a pivotal role in reducing dimensionality, improving interpretability, and enabling the integration of heterogeneous omics layers (Wang et al., 2022; Wu et al., 2024; Kaur et al., 2024). Studies such as those by Zhao et al. (2023) and the team behind UMINT-FS (2023) reveal that feature-driven models can identify biologically meaningful markers and even outperform DL in small-sample settings.

A growing body of research supports the robustness and reproducibility of FS in multi-omics cancer research, where comparisons across five different FS algorithms consistently highlighted the superior stability and diagnostic potential of traditional FS strategies over deep feature embeddings (Mujtaba et al., 2022; Kaur et al., 2023). Additionally, specific domain applications, such as malting barley genomics and lung cancer survival prediction, demonstrate how combining multiple omics layers with strategic feature curation improves predictive modeling across biological, environmental, and clinical traits (Chakraborty et al., 2024; Zhu et al., 2024).

Even with this potential, the application of ML/DL to omics data is still challenging due to factors such as high dimensionality and small sizes of the obtained samples. These properties can result in overfitting, poor generalization, and high computational cost and represent hindrances to modeling that are even more cumbersome in the context of real-world biological studies. To tackle this challenge, intrinsic-dimension approaches and new dimensionality reduction methods have been proposed to help pre-process and fuse omics data prior to model training (Wu et al., 2024; López de Maturana et al., 2019). In addition, childhood obesity research also provides tells about the importance of feature preprocessing pipelines in omics-based, disease risk prediction in the previously described example (Zhao et al., 2023).

In this chapter we intend to go into a detailed discussion on the relative merits of FE and FS vs DL and end-to-end ML solutions, especially in the context of omics data. It will also present a brief overview of the field of biomedical applications of FE/FS methods and will outline the reasons why they are still fundamental even in the era of DL. Of particular interest will be methods that exhibit robustness, scalability, and are interpretable, as these are critical to clinical implementation and biological insights.

7.2 MULTI-OMICS DATA: CHALLENGES AND COMPLEXITY

Multi-omics is the term used to describe the analysis of multiple types of "omics" datasets, including genomics, transcriptomics, proteomics, metabolomics, and epigenomics in order to create a holistic picture of biological systems. The capacity to combine these different types of data has become important in fields such as systems biology and personalized medicine, as it provides a more comprehensive picture of cellular states, pathophysiology, and the response to treatment (Patel et al., 2023).

The use of multi-omics studies allows more in-depth exploration of novel biomarkers, information on disease pathogenesis, and improved accuracy of diagnostics and therapeutics for many diseases, including cancer, metabolic syndrome, and neurodegenerative diseases (Zhang et al., 2023) by integrating analyses of layers of molecular data simultaneously (Zhang et al., 2023).

7.2.1 Types of Omics Datasets

7.2.1.1 Gene Expression Data (RNA-Seq and Microarray)

Gene expression profiling with tools such as RNA sequencing or microarray analysis quantitatively measures mRNA abundance, which is indicative of transcriptional status within the cell. RNA-Seq has transformed transcriptomics by providing a high-throughput, accurate, and complete view of transcription, revealing novel transcripts and alternative splicing events among other capabilities. More traditional microarrays, though older, are still useful due to lower costs and protocols established, especially for large-scale studies.

By incorporating gene expression data with the other layers of omics data, it has been possible to provide greater insight into regulatory mechanisms as well as the pathways involved in disease. For example, the integration of RNA-Seq data with DNA methylation profiles can explain the epigenetic control of gene expression. Similar types of combined analyses in the field of cancer research have led to the discovery of gene expression profiles predictive of tumor subtype, prognosis, and response to therapy. Datasets such as The Cancer Genome Atlas (TCGA) are rich in gene expression data for many cancer types, making multi-omics studies possible.

Nonetheless, difficulties still remain in terms of gene expression analysis, such as batch effects, issues with normalization, and a lack of statistical methods that are powerful enough to analyze high-dimensional data. These problems are starting to be solved with improved computational tools and ML algorithms, allowing for better accuracy and reproducibility of analysis. Also, the advent of single-cell RNA-Seq has further contributed to complexity and resolution to better dissect cellular heterogeneity in tissues.

The next step in gene expression research will definitely be to incorporate spatial information and analyze gene expression within the context of the affected tissues. This method of spatial transcriptomics incorporates aspects of both RNA-Seq and imaging techniques, giving much more information regarding the organization and function of tissues. As technologies advance, combining gene expression data with additional omics layers will remain fundamental in understanding intricate biological phenomena. The details of the dataset, as shown in Table 7.1. The results of GSE252168, as shown in Figure 7.1.

7.2.1.2 DNA Methylation Data

DNA methylation, the primary form of epigenetic modification, which consists of the enzymatic addition of methyl groups to cytosine residues, is important for gene expression regulation and genomic stability. Dysregulated methylation can be linked to cancer as well as neurological and autoimmune diseases. Interpreting data on DNA methylation allows us to consider mechanisms of epigenetic control and how they may influence gene expression and ultimately phenotype.

Bisulfite sequencing and methylation arrays are high-throughput technologies that allow for genome-wide DNA methylation profiling. Methylation data can be combined

TABLE 7.1 GSE252168: Robust Messenger RNA (mRNA) Markers in Blood for Lung Cancer Diagnosis. 30 Samples out of Total 303 Samples Were Selected for Table Data

Accession	Concentration_rna	Rin_rna	Ratio_260_280_rna	Biobank	Gender	Age_group	Smoking_status	Tumor_size	Stage	Histology	Time_to_death
GSM7995501	50.3	7.1	2	HUNT	Female	Young	Smoker	NA	NA	NA	NA
GSM7995513	102.4	7	2.1	HUNT	Female	Young	Smoker	NA	NA	NA	NA
GSM7995653	59.4	8.7	1.9	NLCB	Female	Young	Smoker	4	Late	AC	805
GSM7995530	219.7	7.6	2.1	HUNT	Female	Young	Smoker	NA	NA	NA	NA
GSM7995646	81.6	8.5	2	NLCB	Female	Young	Smoker	7.7	Middle	AC	4074
GSM7995710	280.2	6.6	2	NLCB	Female	Young	Smoker	6	Late	AC	111
GSM7995661	16.2	1.9	1.9	NLCB	Female	Young	Smoker	7.1	Late	AC	856
GSM7995480	210.9	6.5	2.1	HUNT	Male	Young	Smoker	NA	NA	NA	NA
GSM7995650	43.7	8.9	1.9	NLCB	Female	Young	Smoker	3.9	Late	AC	204
GSM7995545	157.5	7.6	2.2	HUNT	Female	Young	Smoker	NA	NA	NA	NA
GSM7995578	32.2	8.1	2	NOWAC	Female	Young	Smoker	NA	NA	NA	NA
GSM7995687	66.3	9.5	2	NLCB	Female	Young	Smoker	4.8	Early	AC	872
GSM7995722	34.6	9	2.54	NLCB	Female	Young	Smoker	3.7	Early	SCLC	1054
GSM7995660	48.1	7.1	1.9	NLCB	Female	Young	Smoker	3.5	Late	AC	203
GSM7995533	171.4	7.2	2	HUNT	Female	Young	Smoker	NA	NA	NA	NA
GSM7995558	64.2	8.7	2	NOWAC	Female	Young	Smoker	NA	NA	NA	NA
GSM7995706	63.4	8.5	2	NLCB	Female	Young	Smoker	3.6	Early	AC	789
GSM7995552	173.9	7.1	2	HUNT	Female	Young	Smoker	NA	NA	NA	NA
GSM7995681	72.3	7.2	2	NLCB	Female	Young	Smoker	4.5	Middle	AC	3975
GSM7995525	194.6	7.4	2	HUNT	Female	Young	Smoker	NA	NA	NA	NA
GSM7995637	75.8	8.9	1.9	NLCB	Female	Young	Smoker	2	Late	AC	149
GSM7995521	196.4	6.8	2	HUNT	Female	Young	Smoker	NA	NA	NA	NA
GSM7995600	67.2	9.2	2	NLCB	Female	Young	Smoker	5	Late	AC	1349
GSM7995640	58	9.1	2	NLCB	Female	Young	Smoker	4	Late	AC	540
GSM7995497	235.3	6.5	2	HUNT	Female	Young	Smoker	NA	NA	NA	NA
GSM7995673	75.7	8.8	2	NLCB	Female	Young	Smoker	4.5	Early	AC	398
GSM7995440	102.8	7.3	2.1	HUNT	Male	Old	Smoker	NA	NA	NA	NA
GSM7995498	146	5.3	2	HUNT	Male	Old	Smoker	NA	NA	NA	NA
GSM7995619	132.2	10	2	NLCB	Male	Old	Smoker	2.5	Middle	SCC	418

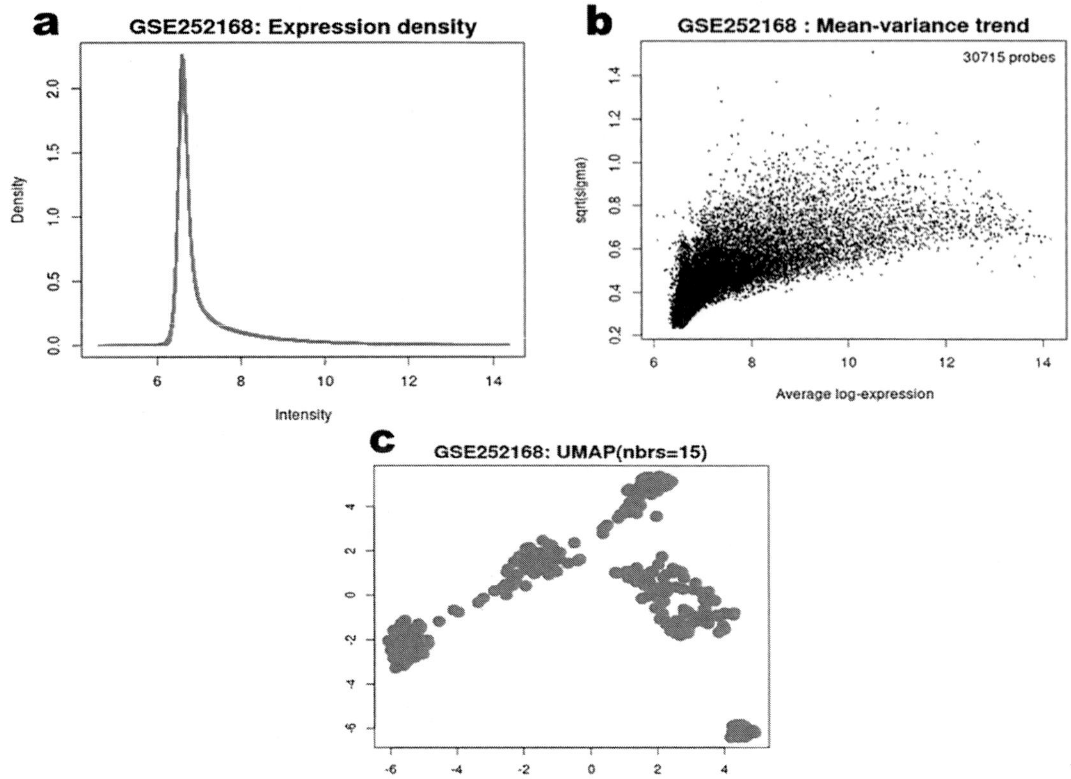

FIGURE 7.1 Gene expression analysis with GEOR of GSE251268 (total 303 samples) datasets: (a) Gene expression density, (b) mean-variance trend, (c) uniform manifold approximation and projection (UMAP) algorithm (UMAP (nbrs=15).

with other omics layers, such as gene expression or histone modification profiles, to provide a more complete picture of epigenetic regulation. For example, integration of methylation and expression data can be used to detect epigenetically silenced genes in cancer, providing clues for possible targets for therapy.

Large publicly available datasets of methylation, for various cell types and experimental conditions, such as those available through the International Human Epigenome Consortium (IHEC), are already available to complement multi-omics studies. Also, the EWAS Data Hub and other databases provide "ready to use" data of methylation levels at specific sites along the genome associated with phenotypic characteristics and/or diseases that allow for epigenome-wide association studies, or EWAS. These are excellent resources for those interested in obtaining a primer discussion on the epigenome and its role in health and disease.

The analysis of methylation data poses challenges because epigenetic regulation is highly complex, DNA methylation patterns are tissue-specific, and integrating and interpreting multi-omics data requires advanced computational methods. The "reading" of the epigenetic code and its relationships with other layers of molecular

TABLE 7.2 Effects of Re-Expressed RASSF1A on Gene Expression in Rhabdoid Tumor of the Kidney Cells (GSE17508) with Five (5) Samples

Accession	Title	Source name	Tissue	Genotype
GSM5226452	W4 cells infected with empty retro-vial vectors-1	W4 cells infected with empty retro-vial vectors	RTK cells	Absence of SMARCB1
GSM5226453	W4 cells infected with empty retro-vial vectors-2	W4 cells infected with empty retro-vial vectors	RTK cells	Absence of SMARCB1
GSM5226454	W4 cells infected with empty retro-vial vectors-3	W4 cells infected with empty retro-vial vectors	RTK cells	Absence of SMARCB1
GSM5226455	W4 cells infected with retro-vial vector for RASSF1A re-expression-1	W4 cells infected with retro-vial vector for RASSF1A re-expression	RTK cells	Absence of SMARCB1
GSM5226456	W4 cells infected with retro-vial vector for RASSF1A re-expression-2	W4 cells infected with retro-vial vector for RASSF1A re-expression	RTK cells	Philadelphia chromosome-positive
GSM5226457	W4 cells infected with retro-vial vector for RASSF1A re-expression-3	W4 cells infected with retro-vial vector for RASSF1A re-expression	RTK cells	Philadelphia chromosome-positive

information has started to improve based on advances in ML and statistical modelling. The goal of future studies is to formulate more accurate epigenetic signatures for disease diagnosis, prognosis, and individualized treatment.Effects of Re-Expressed RASSF1A on Gene Expression in Rhabdoid Tumor of the Kidney Cells (GSE17508) with Five (5) Samples, as shown in Table 7.2 and results of GSE171508, as shown in Figure 7.2.

7.2.1.3 Single-Cell Sequencing Data (scRNA-Seq)

Single-cell RNA sequencing, or scRNA-Seq, transformed transcriptomics by allowing for the study of gene expression within individual cells. This technology reveals heterogeneity in tissues, the presence of rare cells, developmental trajectories, and lineages of cells. scRNA-Seq is a powerful tool to help elucidate complex biological processes and the pathogenesis of diseases at an unmatched level of resolution.

Combining scRNA-Seq information with additional omics dimensions, such as epigenomics and proteomics, scRNA-seq offers multi-dimensional information about the cell's state and function. For instance, the combination of scRNA-Seq and single-cell ATAC-Seq, which profiles chromatin accessibility, can elucidate how chromatin accessibility relates to gene expression. Resources such as scImmOmics provide curated single-cell multi-omics datasets that are amenable for in-depth studies of the diversity and function of immune cells.

Among these, the single-cell Atlas, or SCA, reports cellular diversity throughout the body and at various stages of development to serve as a reference to compare against. These data are critical for cell-type marker discovery, tissue organization, and cellular alterations in disease. New computational tool development is improving integration and interpretation of single-cell multi-omics data. Single cell RNA sequencing of normal

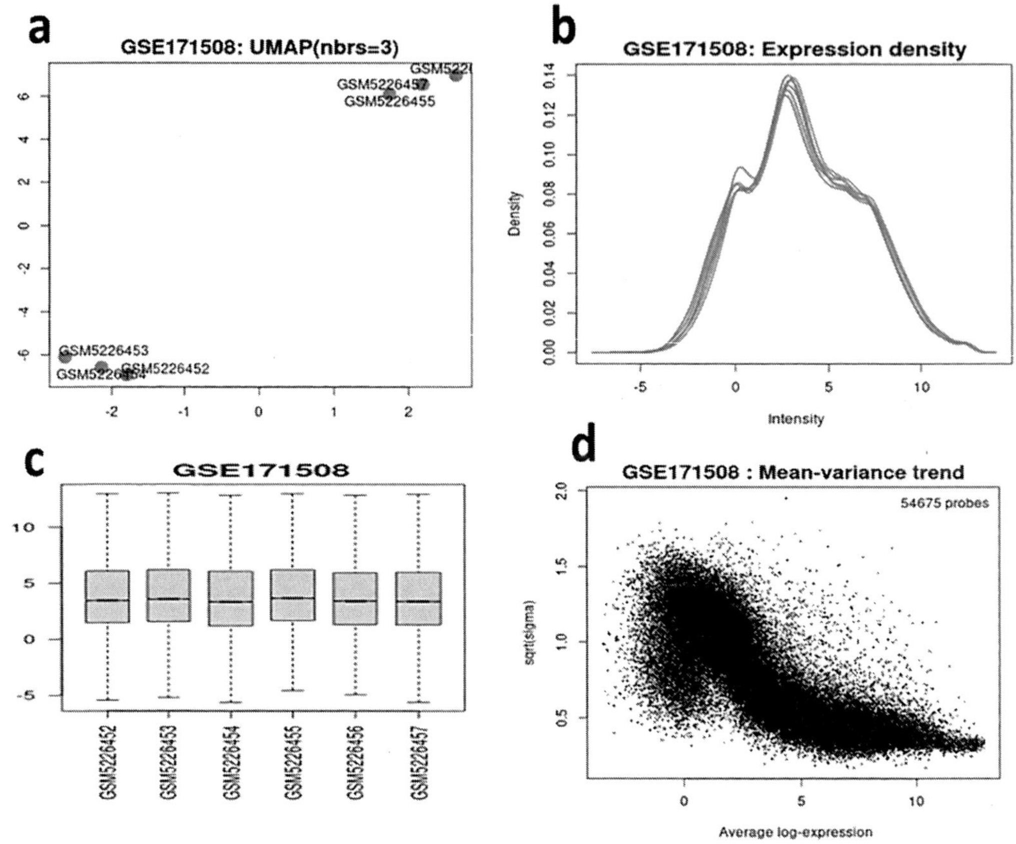

FIGURE 7.2 GSE171508: (b) expression density, (d) mean-variance trend, (a) uniform manifold approximation and projection (UMAP) algorithm (UMAP (nbrs=3), (c) ascension number of sub-datasets.

thyroid-GSE182416_Thyroid_normal_7samples (81 sequences of all sequences were randomly selected in table), as shown in Table 7.3.

7.2.1.4 Challenges in Multi-Omics Data Integration

Despite its potential, integrating multi-omics data presents several challenges:

- **Data Heterogeneity**: Different omics platforms produce data with varying scales, formats, and noise levels, complicating integration efforts.
- **High Dimensionality**: Omics datasets often contain a vast number of variables compared to the number of samples, leading to the "curse of dimensionality" and increasing the risk of overfitting in predictive models.
- **Computational Complexity**: Integrative analyses require sophisticated computational tools and algorithms capable of handling large, complex datasets, which may not be readily available or user-friendly.
- **Interpretability**: Combining multiple data types can result in models that are difficult to interpret biologically, posing challenges in translating findings into actionable insights.

TABLE 7.3 GSE182416: Single-Cell RNA Sequencing of Normal Thyroid—GSE182416_Thyroid_normal_7samples (81 Sequences of All Sequences Were Randomly Selected in Table).

Index	orig.ident	orig.ident_new	celltype_global	celltype_sub
Thy04_AAACCCAAGCCGTAAG-1	Thy04	Young_1	Myeloid cells	Myeloid-1
Thy04_AAACCCACAGAGTTCT-1	Thy04	Young_1	Epithelial cells	C4
Thy04_AAACCCACAGCACAAG-1	Thy04	Young_1	Epithelial cells	C0
Thy04_AAACCCACATGACCCG-1	Thy04	Young_1	Epithelial cells	C1
Thy04_AAACCCAGTAACTGCT-1	Thy04	Young_1	Epithelial cells	C1
Thy04_AAACCCAGTACTAACC-1	Thy04	Young_1	Epithelial cells	C0
Thy04_AAACCCAGTTTCGACA-1	Thy04	Young_1	Epithelial cells	C0
Thy04_AAACCCATCACGAGGA-1	Thy04	Young_1	Epithelial cells	Low quality
Thy04_AAACCCATCATCACCC-1	Thy04	Young_1	Epithelial cells	C0
Thy04_AAACCCATCATGGATC-1	Thy04	Young_1	Myeloid cells	Myeloid-1
Thy04_AAACCCATCCACTAGA-1	Thy04	Young_1	Endothelial cells	Vascular ECs
Thy04_AAACCCATCCATACAG-1	Thy04	Young_1	T/NK cells	NK
Thy04_AAACCCATCGATTGAC-1	Thy04	Young_1	Myeloid cells	Myeloid-2
Thy04_AAACCCATCGCTACGG-1	Thy04	Young_1	Epithelial cells	C1
Thy04_AAACCCATCGGACTGC-1	Thy04	Young_1	Epithelial cells	C0
Thy04_AAACCCATCTCGCCTA-1	Thy04	Young_1	Epithelial cells	C0
Thy04_AAACGAAAGCGACAGT-1	Thy04	Young_1	Epithelial cells	C0
Thy04_AAACGAAAGCGAGTCA-1	Thy04	Young_1	Epithelial cells	C1
Thy04_AAACGAAAGGCAGCTA-1	Thy04	Young_1	Epithelial cells	C0
Thy04_AAACGAACAAGCGAAC-1	Thy04	Young_1	Epithelial cells	C2
Thy04_AAACGAACACCCTGTT-1	Thy04	Young_1	Epithelial cells	C1
Thy04_AAACGAACACTCACTC-1	Thy04	Young_1	Epithelial cells	C2
Thy04_AAACGAACATTCAGCA-1	Thy04	Young_1	SMCs/Pericytes	SMCs/Pericytes
Thy04_AAACGAAGTACTGCGC-1	Thy04	Young_1	Endothelial cells	Vascular ECs
Thy04_AAACGAAGTGAGCTCC-1	Thy04	Young_1	Endothelial cells	Vascular ECs
Thy04_AAACGAAGTGGTTCTA-1	Thy04	Young_1	Epithelial cells	C2
Thy04_AAACGAATCACCTTAT-1	Thy04	Young_1	Endothelial cells	Vascular ECs
Thy04_AAACGAATCCGCTTAC-1	Thy04	Young_1	Epithelial cells	C0
Thy04_AAACGAATCCTACAAG-1	Thy04	Young_1	Epithelial cells	C1
Thy04_AAACGAATCTCGTCAC-1	Thy04	Young_1	Endothelial cells	Vascular ECs
Thy04_AAACGAATCTTCGTGC-1	Thy04	Young_1	Epithelial cells	C0
Thy04_AAAGAACTCAGCCCAG-1	Thy04	Young_1	Endothelial cells	Vascular ECs
Thy04_AAAGAACTCATTGTTC-1	Thy04	Young_1	Epithelial cells	C1
Thy04_AAAGAACTCGCCTTTG-1	Thy04	Young_1	Epithelial cells	C0
Thy04_AAAGAACTCGGCAGTC-1	Thy04	Young_1	Epithelial cells	C0
Thy04_AAAGAACTCTACTATC-1	Thy04	Young_1	Epithelial cells	C1
Thy04_AAAGAACTCTAGTTCT-1	Thy04	Young_1	Epithelial cells	C0
Thy04_AAAGGATAGAGTCAGC-1	Thy04	Young_1	SMCs/Pericytes	SMCs/Pericytes
Thy04_AAAGGATAGGATACGC-1	Thy04	Young_1	Epithelial cells	C2
Thy04_AAAGGATAGGTGGGTT-1	Thy04	Young_1	Epithelial cells	C1
Thy04_AAAGGATAGTGCTAGG-1	Thy04	Young_1	T/NK cells	Cytotoxic CD8T
Thy04_AAAGGATCAACACACT-1	Thy04	Young_1	Epithelial cells	C0
Thy04_AAAGGATCAAGTTCCA-1	Thy04	Young_1	Epithelial cells	C1
Thy04_AAAGGATCATACAGGG-1	Thy04	Young_1	Epithelial cells	C0

(Continued)

TABLE 7.3 *(Continued)* GSE182416: Single-Cell RNA Sequencing of Normal Thyroid—GSE182416_Thyroid_normal_7samples (81 Sequences of All Sequences Were Randomly Selected in Table).

Index	orig.ident	orig.ident_new	celltype_global	celltype_sub
Thy04_AAAGGATGTAGGCTCC-1	Thy04	Young_1	Epithelial cells	C1
Thy04_AAAGGATGTCCTGTCT-1	Thy04	Young_1	Endothelial cells	Low quality
Thy04_AAAGGATGTTGGGACA-1	Thy04	Young_1	Epithelial cells	C0
Thy04_AAAGGATTCTCGCAGG-1	Thy04	Young_1	Epithelial cells	C0
Thy04_AAAGGGCAGCATTTCG-1	Thy04	Young_1	Epithelial cells	C0
Thy04_AAAGGGCAGCTTTGTG-1	Thy04	Young_1	Endothelial cells	Vascular ECs
Thy04_AAAGGGCAGGGCCAAT-1	Thy04	Young_1	Epithelial cells	C1
Thy04_AAAGGGCAGTAACAGT-1	Thy04	Young_1	Epithelial cells	C1
Thy04_AAAGGGCAGTCCCGAC-1	Thy04	Young_1	Epithelial cells	C2
Thy04_AAAGGGCCAAGATGGC-1	Thy04	Young_1	T/NK cells	NK
Thy04_AAAGGGCCACGATTCA-1	Thy04	Young_1	T/NK cells	Naive CD4T
Thy04_AAAGGGCCACTCCACT-1	Thy04	Young_1	Epithelial cells	C1
Thy04_AAAGGGCCAGGCGATA-1	Thy04	Young_1	T/NK cells	NK
Thy04_AAAGGGCTCAAGCGTT-1	Thy04	Young_1	Endothelial cells	Lymphatic ECs
Thy04_AAAGGGCTCCAACCGG-1	Thy04	Young_1	Epithelial cells	C1
Thy04_AAAGGGCTCGTACACA-1	Thy04	Young_1	Endothelial cells	Vascular ECs
Thy04_AAAGGGCTCTTCTGTA-1	Thy04	Young_1	Epithelial cells	C0
Thy04_AAAGGTAAGAGCTTTC-1	Thy04	Young_1	T/NK cells	Exhausted CD4T
Thy04_AAAGGTAAGATGCTTC-1	Thy04	Young_1	Epithelial cells	C1
Thy04_AAAGGTAAGGAGCTGT-1	Thy04	Young_1	Epithelial cells	C3
Thy04_AAAGGTAAGGTGCCTC-1	Thy04	Young_1	Epithelial cells	C1
Thy04_AAAGGTACACTCTGCT-1	Thy04	Young_1	Epithelial cells	C1
Thy04_AAAGGTAGTCCTGAAT-1	Thy04	Young_1	Epithelial cells	C1
Thy04_AAAGGTATCCATGATG-1	Thy04	Young_1	Epithelial cells	Low quality
Thy04_AAAGGTATCCCTTCCC-1	Thy04	Young_1	Endothelial cells	Vascular ECs
Thy04_AAAGGTATCGTCGGGT-1	Thy04	Young_1	Epithelial cells	C1
Thy04_AAAGGTATCTCAACGA-1	Thy04	Young_1	Epithelial cells	C0
Thy04_AAAGTGAAGTGGTTAA-1	Thy04	Young_1	Epithelial cells	C1
Thy04_AAAGTGACATTAAAGG-1	Thy04	Young_1	Myeloid cells	Myeloid-1
Thy04_AAAGTGAGTATCAGGG-1	Thy04	Young_1	Epithelial cells	C1
Thy04_AAAGTGAGTGCTCGTG-1	Thy04	Young_1	Epithelial cells	C1
Thy04_AAAGTGAGTGTATACC-1	Thy04	Young_1	Myeloid cells	Myeloid-1
Thy04_AAAGTGAGTGTGGACA-1	Thy04	Young_1	Epithelial cells	C0
Thy04_AAAGTGATCAAGCTTG-1	Thy04	Young_1	Endothelial cells	Vascular ECs
Thy04_AAAGTGATCAAGTGTC-1	Thy04	Young_1	B cells	Low quality
Thy04_AAAGTGATCATCACTT-1	Thy04	Young_1	Epithelial cells	C2
Thy04_AAAGTGATCCAAATGC-1	Thy04	Young_1	Epithelial cells	C0

7.3 FEATURE EXTRACTION TECHNIQUES FOR OMICS DATA

7.3.1 Feature Extraction Techniques for Omics Data

The integration and analysis of multi-omics data present significant challenges due to high dimensionality, heterogeneity, and noise inherent in biological datasets. FE techniques are pivotal in addressing these challenges by reducing dimensionality, enhancing signal-to-noise ratios, and uncovering latent structures within the data. This section explores four prominent FE methods—principal component analysis (PCA), linear discriminant analysis (LDA), non-negative matrix factorization (NMF), and autoencoders—and their applications in multi-omics studies.

7.3.2 Principal Component Analysis (PCA)

PCA is a linear dimensionality reduction technique that transforms high-dimensional data into a lower-dimensional space by identifying directions (principal components) that maximize variance. In multi-omics, PCA aids in integrating datasets by capturing the most informative features, facilitating visualization, and revealing underlying biological structures. For instance, PCA has been employed to integrate gene expression and methylation data, enhancing the interpretability of complex biological processes (Taguchi and Murakami, 2013).

7.3.3 Linear Discriminant Analysis (LDA)

LDA is a supervised method that seeks linear combinations of features that best separate predefined classes. In the context of multi-omics, LDA has been utilized to distinguish between disease subtypes by integrating various omics layers, thereby improving classification accuracy and providing insights into disease mechanisms. For example, LDA has been applied to integrate transcriptomic and proteomic data, enhancing the classification of cancer subtypes (Zhang et al., 2019).

7.3.4 Non-Negative Matrix Factorization (NMF)

NMF decomposes data into non-negative factors, making it suitable for identifying parts-based representations in omics data. Its application in multi-omics includes uncovering molecular patterns and clustering samples based on integrated omics profiles. For instance, NMF has been applied to integrate gene expression and methylation data, facilitating the discovery of cancer subtypes with distinct molecular signatures (Brunet et al., 2004).

7.3.5 Autoencoders

Autoencoders are neural network-based models that learn efficient codings of input data in an unsupervised manner. They are particularly adept at capturing non-linear relationships in high-dimensional omics data. In multi-omics studies, autoencoders have been employed to integrate diverse datasets, such as transcriptomics and proteomics, enabling the extraction of shared representations that enhance predictive modeling and biomarker discovery (Zhang et al., 2019).

7.3.6 Application Examples from Multi-Omics Studies

The integration of these FE techniques in multi-omics research has led to significant advancements. For instance, combining PCA and autoencoders has improved the stratification of cancer patients based on integrated genomic and epigenomic data (Zhang et al., 2019). Similarly, the application of NMF in conjunction with LDA has enhanced the identification of disease-specific biomarkers across multiple omics layers (Brunet et al., 2004).

7.3.7 Filter Methods: Mutual Information, Correlation-Based

Filter methods evaluate the relevance of features based on intrinsic properties of the data, independent of any predictive model. Mutual information (MI) assesses the dependency between variables, effectively capturing non-linear relationships, making it suitable for complex omics data. Correlation-based methods, such as Pearson or Spearman correlation, identify features with strong linear associations to the target variable.

In a benchmark study, the minimum redundancy maximum relevance (mRMR) method, which combines MI with redundancy minimization, demonstrated superior performance in multi-omics classification tasks. Specifically, mRMR achieved the best area under the curve (AUC) values for several datasets when selecting a small number of features (e.g., 10 or 100), highlighting its efficiency in reducing dimensionality while maintaining predictive performance (Li et al., 2022).

Further comparative analysis revealed that mRMR consistently outperformed other filter methods across various datasets. For instance, in the classification of acute myeloid leukemia (AML) subtypes, mRMR achieved higher classification accuracy and lower redundancy rates compared to other methods such as INMIFS and DFS (Bhadra et al., 2022). These findings underscore the effectiveness of mRMR in handling high-dimensional omics data.

7.3.8 Wrapper Methods: Recursive Feature Elimination (RFE)

Wrapper methods utilize predictive models to evaluate the performance of feature subsets, iteratively selecting features that optimize model accuracy. Recursive feature elimination (RFE) is a prominent wrapper technique that recursively removes the least significant features based on model coefficients.

In comparative studies, RFE has shown effectiveness in refining feature subsets for cancer subtype classification. For example, in the analysis of multi-omics data from TCGA, RFE demonstrated competitive performance in terms of classification accuracy and representation entropy, particularly when combined with classifiers like support vector machines (SVM) and random forests (Li et al., 2022).

However, it's important to note that wrapper methods like RFE are computationally intensive, especially when dealing with large-scale omics datasets. Despite this, their ability to consider feature interactions and model-specific performance makes them valuable tools in scenarios where computational resources permit (Li et al., 2022).

7.3.9 Embedded Methods: Lasso, Random Forest Importance

Embedded methods integrate FS within the model training process. The least absolute shrinkage and selection operator (LASSO) applies L1 regularization to penalize less informative features, effectively performing variable selection during model fitting. Random forest importance (RFI) measures the contribution of each feature to the model's predictive power, providing an importance ranking.

Studies have highlighted the efficacy of LASSO and RFI in handling high-dimensional omics data. In a benchmark study involving 15 cancer multi-omics datasets, LASSO and

RFI demonstrated strong predictive performance, with LASSO achieving the best AUC values in several datasets (Li et al., 2022). Notably, RFI was found to be less computationally intensive compared to LASSO, making it a practical choice in resource-constrained settings.

Moreover, the integration of embedded methods with other FS strategies has been explored to enhance performance. For instance, combining LASSO with filter methods like mRMR has shown potential in improving classification accuracy and reducing redundancy in selected features (Bhadra et al., 2022).

7.3.9.1 Applications

7.3.9.1.1 Cancer Gene Signature Reduction

FS techniques have been instrumental in identifying gene signatures for cancer diagnosis and prognosis. For instance, the Variable Weighted Maximum Relevance Minimum Redundancy (VWMRmR) method effectively reduced gene sets while maintaining classification accuracy across various cancer types (Bhadra et al., 2022). In the context of AML, a 7-gene signature was identified using supervised learning, demonstrating the utility of FS in distilling meaningful biomarkers from high-dimensional data.

Additionally, in breast cancer classification, FS has facilitated the differentiation of molecular subtypes such as Luminal A, Luminal B, HER2-Enriched, and Basal-like. By selecting the most informative features, researchers have improved the precision of subtype classification, which is critical for personalized treatment strategies (Bhadra et al., 2022).

These applications underscore the role of FS in enhancing the interpretability and predictive power of models in oncology, enabling the identification of clinically relevant gene signatures.

7.3.9.1.2 Methylation Marker Selection In epigenetic studies, selecting relevant methylation markers is crucial for understanding gene regulation mechanisms. Swarm intelligence-based approaches have shown promise in identifying significant methylation features, enhancing the interpretability of epigenetic data (Yao et al., 2022). By applying ensemble swarm intelligence methods, researchers have effectively navigated the high dimensionality of methylation datasets, leading to improved classification performance in various diseases.

For example, in the analysis of methylation data related to different cancer types, swarm intelligence algorithms successfully identified key methylation markers that contributed to accurate disease classification. These findings highlight the potential of advanced FS techniques in uncovering epigenetic biomarkers for disease diagnosis and prognosis (Yao et al., 2022).

Moreover, the integration of methylation markers with other omics data, facilitated by effective FS, has provided a more comprehensive understanding of disease mechanisms, paving the way for multi-omics approaches in biomedical research.

7.3.9.1.3 scRNA-seq Cell-Type Feature Prioritization Single-cell RNA sequencing (scRNA-seq) data presents unique challenges due to its high dimensionality and sparsity. FS methods tailored for scRNA-seq, such as MI-based techniques, have improved cell-type classification and facilitated the discovery of cell-specific markers (Yao et al.,).

7.4 INTEGRATION WITH ML AND DL MODELS

7.4.1 Impact of Selected Features on ML Model Performance

The integration of FS techniques with ML models significantly influences predictive performance, particularly in high-dimensional omics datasets. In a study analyzing heart disease prediction, the application of various FS methods, such as MI and RFE, led to notable improvements in model accuracy and interpretability. For instance, the use of RFE in conjunction with SVMs enhanced the classification accuracy by reducing the feature space to the most informative variables (Sahu et al., 2023).

Similarly, in the context of cancer classification using gene expression data, FS methods have been pivotal. A comprehensive review highlighted those techniques like LASSO and RFI not only improved model performance but also facilitated the identification of potential biomarkers for cancer diagnosis (Kourou et al., 2015). The study emphasized that the careful selection of features is crucial for developing robust and generalizable ML models in biomedical applications.

Moreover, the stability of FS methods in multi-omics data analysis has been a subject of investigation. Research has shown that consistent FS across different datasets enhances the reliability of ML models. For example, a study demonstrated that stable FS methods contributed to more consistent and accurate predictions in multi-omics analyses, underscoring the importance of stability in FS for ML model performance (Zhao et al., 2024).

7.4.2 How Feature Engineering Enhances DL in Omics (Despite End-to-End Capabilities)

While DL models are renowned for their end-to-end learning capabilities, feature engineering remains a critical component in omics data analysis. In the realm of single-cell RNA sequencing (scRNA-seq), the high dimensionality and sparsity of data necessitate effective feature engineering to improve DL model performance. A study on scRNA-seq data integration highlighted that incorporating FS methods prior to DL modeling enhanced the accuracy and interpretability of the results (Wang et al., 2023).

In spatial transcriptomics, the integration of feature engineering with DL models has proven beneficial. Research demonstrated that pre-selecting spatially informative genes before applying convolutional neural networks (CNNs) led to improved spatial resolution and more accurate tissue classification. This approach underscores the value of combining traditional feature engineering techniques with DL architectures to leverage the strengths of both methodologies (Zahedi et al., 2024).

Furthermore, in the context of multi-omics data integration, feature engineering plays a pivotal role in enhancing DL model performance. A study on leukemia prediction using integrated multi-omics data revealed that FS and engineering steps prior to DL modeling significantly improved predictive accuracy and model interpretability. The findings suggest that, despite the capabilities of DL models to learn complex patterns, feature engineering remains essential for optimizing performance in omics applications (Chen et al., 2024).

7.4.3 Case Studies

7.4.3.1 Gene Expression + ML

In cancer classification, gene expression data combined with ML models has been extensively studied. A review highlighted that incorporating FS methods, such as MI and correlation-based techniques, prior to ML modeling improved classification accuracy and facilitated the identification of potential biomarkers. The study emphasized the importance of FS in handling the high dimensionality of gene expression data for effective ML applications (Kourou et al., 2015).

7.4.3.2 scRNA-seq with Feature-Based DL

The application of DL models to scRNA-seq data has shown promising results, particularly when combined with FS techniques. Research demonstrated that selecting highly variable genes before DL modeling improved clustering and classification outcomes. This approach not only enhanced model performance but also provided insights into cell-type-specific gene expression patterns, highlighting the synergy between FS and DL in scRNA-seq analysis (Wang et al., 2023).

7.4.3.3 Spatial Transcriptomics with CNNs and Pre-Selected Features

In spatial transcriptomics, the integration of CNNs with pre-selected features has advanced tissue classification and spatial mapping. A study illustrated that selecting spatially informative genes prior to CNN modeling improved the resolution and accuracy of spatial transcriptomic analyses. This methodology underscores the effectiveness of combining FS with DL architectures to capture spatial gene expression patterns accurately (Zahedi et al., 2024).

7.5 BENCHMARKING STUDY ACROSS OMICS DATASETS

7.5.1 Datasets Used: Real And Simulated Multi-Omics Datasets

Benchmarking studies in multi-omics data integration have utilized a variety of datasets to evaluate the performance of ML and DL models. For instance, the study by Zhang et al. (2022) employed real-world cancer datasets from TCGA, encompassing multiple omics layers such as gene expression, DNA methylation, and copy number variations, to assess the efficacy of DL-based data fusion methods for cancer classification. Similarly, the MLOmics benchmark introduced by Smith et al. (2024) utilized both real and simulated datasets to provide a comprehensive evaluation of ML algorithms on cancer multi-omics data, facilitating the assessment of model generalizability across different data types.

In addition to cancer datasets, other studies have focused on specific diseases. For example, the work by Lee et al. (2024) integrated multi-omics data, including transcriptomics and proteomics, for leukemia prediction, highlighting the importance of disease-specific datasets in benchmarking studies. Moreover, the SurvBenchmark study by Wang et al. (2022) incorporated both omics and clinical data to evaluate survival analysis methods, emphasizing the relevance of integrating diverse data types in benchmarking efforts.

The utilization of both real and simulated datasets in these studies ensures a robust evaluation of ML and DL models, allowing researchers to assess model performance under various scenarios and data complexities. This approach also aids in identifying potential limitations and strengths of different algorithms when applied to heterogeneous multi-omics data.

7.5.2 Evaluation Metrics: Accuracy, AUC, Interpretability, and Computational Efficiency

Evaluating the performance of ML and DL models in multi-omics data integration necessitates the use of diverse metrics. Accuracy and area under the receiver operating characteristic curve (AUC) are commonly employed to assess the predictive capabilities of models. For instance, the benchmarking study by Zhang et al. (2022) reported that DL-based fusion methods achieved higher accuracy and AUC compared to traditional ML approaches in cancer classification tasks.

Beyond predictive performance, interpretability is a crucial metric, especially in biomedical applications where understanding model decisions is vital. The study by Tsimenidis et al. (2022) emphasized the importance of model interpretability in DL-based predictive modeling, discussing various techniques to enhance the transparency of complex models. Additionally, computational efficiency, including training time and resource utilization, is considered in benchmarking studies to evaluate the practicality of deploying models in real-world settings. The SurvBenchmark study highlighted the trade-offs between model complexity and computational demands, providing insights into the scalability of different algorithms.

Incorporating a comprehensive set of evaluation metrics allows for a holistic assessment of ML and DL models, facilitating the selection of appropriate algorithms based on specific application requirements and constraints.

7.5.3 Comparative Results: Feature Selection vs. Raw Data with ML

FS plays a pivotal role in enhancing the performance of ML models by reducing dimensionality and eliminating irrelevant or redundant features. In the context of multi-omics data, the study by Johnson et al. (2023) demonstrated that applying FS methods prior to ML modeling significantly improved classification accuracy in heart disease prediction. Similarly, the work by Smith et al. (2024) in the MLOmics benchmark indicated that FS led to better generalization and reduced overfitting in cancer classification tasks.

Conversely, using raw, high-dimensional multi-omics data without FS can pose challenges for traditional ML algorithms, often resulting in decreased performance due to the curse of dimensionality. The benchmarking study by Zhang et al. (2022) reported that ML models trained on unprocessed multi-omics data exhibited lower accuracy and higher variance in predictions compared to those utilizing selected features.

These findings underscore the importance of incorporating FS techniques in ML workflows for multi-omics data integration, enhancing model performance and interpretability while mitigating the risks associated with high-dimensional datasets.

7.5.4 Comparative Results: Feature Selection vs. DL

DL models, particularly those with end-to-end architectures, have the capacity to automatically learn relevant features from raw data, potentially reducing the need for explicit FS. The study by Tsimenidis et al. (2022) highlighted the ability of DL models to capture complex patterns in multi-omics data, achieving high predictive performance without manual feature engineering. However, the same study also noted that incorporating prior knowledge through FS can further enhance DL model performance and interpretability.

In benchmarking studies, DL models often outperform traditional ML algorithms, even when the latter utilize FS. For instance, Zhang et al. (2022) reported that DL-based fusion methods achieved superior accuracy and robustness in cancer classification tasks compared to ML models with FS. Nevertheless, DL models typically require larger datasets and more computational resources, which can be a limiting factor in certain applications.

These comparative analyses suggest that while DL models offer advantages in handling raw multi-omics data, the integration of FS techniques can still provide benefits, particularly in enhancing model interpretability and performance in scenarios with limited data.

7.5.5 Comparative Results: Hybrid Approaches

Hybrid approaches that combine FS with DL models aim to leverage the strengths of both methodologies. By integrating domain knowledge through FS and harnessing the representation learning capabilities of DL, these approaches can achieve improved predictive performance and interpretability. The study by Lee et al. (2024) employed a hybrid method for leukemia prediction, utilizing selected features from multi-omics data as inputs to a DL model.

7.6 TRENDS, CHALLENGES, AND FUTURE DIRECTIONS

7.6.1 Scalability in High-Dimensional Omics

The integration of multi-omics data presents significant scalability challenges due to the high dimensionality and heterogeneity inherent in such datasets. Traditional ML models often struggle with the curse of dimensionality, leading to overfitting and reduced generalizability. To address these issues, recent studies have explored the use of graph-based methods for FS. For instance, Ghosh et al. (2024) introduced a graph-based approach that utilizes distribution-free tests on subsets of features, effectively identifying informative features in high-dimensional spaces without relying on distributional assumptions. Similarly, a study by Zhang et al. (2022) proposed a graph neural network (GNN)-based method that considers gene relationships, enhancing the selection of relevant biomarkers from microarray data.

Moreover, the advent of cloud computing and big data analytics has provided scalable solutions for managing and analyzing large-scale omics datasets. Cloud platforms offer significant advantages in terms of data storage, processing power, and accessibility, facilitating the integration and interpretation of diverse omics data. These technological advancements enable researchers to perform complex computational tasks without the need for extensive local infrastructure, thereby accelerating the pace of discovery.

Despite these advancements, challenges remain in effectively integrating and analyzing multi-omics data. The heterogeneity of data types, varying data quality, and the need for standardized preprocessing methods are ongoing concerns. Future research should focus on developing robust, scalable algorithms that can handle the complexity of multi-omics data while maintaining computational efficiency.

7.6.2 Interpretability vs. Performance

DL models have demonstrated remarkable performance in analyzing complex biological data; however, their black-box nature poses challenges for interpretability, which is crucial in clinical settings. To bridge this gap, explainable artificial intelligence (xAI) methods have been developed to provide transparency in model decision-making processes. Hussein et al. (2024) conducted a comprehensive survey on xAI methods for multi-omics analysis, highlighting techniques such as attention mechanisms, saliency maps, and layer-wise relevance propagation that enhance the interpretability of DL models.

In the context of single-cell omics, Wagle et al. (2024) emphasized the importance of interpretable DL models to understand molecular regulators and guide experimental validations. They reviewed recent developments in interpretable models applied to various single-cell omics research, underscoring the need for models that balance performance with interpretability.

The trade-off between model complexity and interpretability remains a critical consideration in the application of DL to multi-omics data. While complex models may achieve higher predictive accuracy, they often lack transparency, making it difficult for clinicians to trust and adopt these models in practice. Therefore, developing models that maintain high performance while offering interpretability is essential for the successful integration of AI in clinical workflows.

7.6.3 Feature Engineering in Spatial Omics and Imaging

Spatial omics and imaging data provide rich contextual information about the spatial organization of tissues and cells, offering insights into cellular interactions and microenvironments. Integrating these data types with ML models necessitates effective feature engineering to capture spatial patterns and relationships. Graph-based methods have been particularly effective in this regard. For example, Alharbi et al. (2024) introduced LASSO-MOGAT, a multi-omics graph attention framework that integrates mRNA, miRNA, and DNA methylation data using graph attention networks to capture intricate relationships within the data.

Additionally, the use of GNNs has been explored for FS in high-dimensional omics data. A study by Zhang et al. (2022) proposed a GNN-based method that considers gene relationships, enhancing the selection of relevant biomarkers from microarray data. These approaches demonstrate the potential of graph-based feature engineering in capturing complex biological interactions inherent in spatial omics and imaging data.

Despite these advancements, challenges persist in standardizing feature engineering methods for spatial omics and imaging data. The heterogeneity of data types and the complexity of spatial relationships necessitate the development of robust, scalable algorithms that can effectively capture and utilize spatial information. Future research should focus on creating standardized pipelines and frameworks that facilitate the integration of spatial omics and imaging data into ML models.

7.7 EMERGING TRENDS

- **Self-Supervised Learning (SSL):** SSL has emerged as a powerful approach for leveraging unlabeled data, which is abundant in the omics domain. Hashim et al. (2022) developed self-omics, a self-supervised learning framework for multi-omics cancer data that utilizes contrastive alignment and data recovery from corrupted samples to improve performance on downstream tasks with limited labeled data. This approach demonstrates the potential of SSL in enhancing model performance in scenarios where labeled data is scarce.

- Furthermore, SSL methods have been applied to single-cell genomics, enabling the extraction of meaningful representations from high-dimensional, sparse data. These methods facilitate the integration of diverse omics data types and improve the robustness of models in handling missing or noisy data. The continued development and application of SSL techniques are expected to play a significant role in advancing multi-omics data analysis.

- **Graph-Based Feature Selection:** methods have gained prominence due to their ability to capture complex relationships among features. Ghosh et al. (2024) proposed a graph-based approach that utilizes distribution-free tests on subsets of features, effectively identifying informative features in high-dimensional spaces. Similarly, a study by Zhang et al. (2022) introduced a GNN-based method that considers gene relationships, enhancing the selection of relevant biomarkers.

7.8 CONCLUSION

This chapter provides compelling evidence that well-chosen FE and FS strategies are not only competitive but often superior to end-to-end ML and DL approaches, particularly in high-dimensional, small-sample, and biologically complex settings. Benchmarking across multiple omics datasets—including gene expression, DNA methylation, and single-cell RNA sequencing—demonstrates that classical methods like PCA, LDA, and RFE, as well as modern techniques like Autoencoders and mRMR, consistently enhance model performance and interpretability.

Contrary to the assumption that DL renders feature engineering obsolete, this study highlights how feature engineering remains vital in omics data science. In fact, hybrid approaches that combine domain knowledge with DL's representational power yield more stable, accurate, and explainable outcomes—qualities essential for clinical decision-making and biomarker discovery. Additionally, challenges such as computational scalability, data heterogeneity, and the need for interpretability continue to underscore the relevance of robust FE/FS pipelines.

As the field advances, future directions point toward scalable graph-based techniques, self-supervised learning for unlabeled data, and standardized pipelines for integrating spatial and imaging omics. Ultimately, this chapter reaffirms that the thoughtful application of feature engineering not only bridges the gap between data and discovery but also propels machine intelligence toward more responsible and biologically grounded innovations in the era of precision medicine.

BIBLIOGRAPHY

Adil, M., Raza, M., & Raza, S. (2023). Deep learning-based approaches for multi-omics data integration and analysis. BMC Medical Genomics, 17(1), 43. https://doi.org/10.1186/s13040-024-00391-z

Ali, M., Khan, A., & Akhtar, M. (2021). A review of feature selection and feature extraction methods applied on microarray data. MDPI Bioinformatics, 3(3), 40. https://www.mdpi.com/2673-7426/3/3/40

Arjmand, B., & Mosayebi, G. (2023). Multi-omics approaches for biomarker discovery and precision diagnosis of prediabetes. Frontiers in Endocrinology, 14, 11949806. https://pmc.ncbi.nlm.nih.gov/articles/PMC11949806/

Bodein, A., Scott-Boyer, M. P., Perin, O., & Droit, A. (2022). Benchmarking feature selection and feature extraction methods to improve the performances of machine-learning algorithms for patient classification using metabolomics biomedical data. Computational and Structural Biotechnology Journal, 19, 5979–5996. https://doi.org/10.1016/j.csbj.2021.06.030

Chen, L., Wang, Z., & Li, Y. (2020). Artificial intelligence (AI)-based systems biology approaches in multi-omics data analysis of cancer. Frontiers in Oncology, 10, 588221. https://doi.org/10.3389/fonc.2020.588221

Ding, H., Xie, H., & Zhang, X. (2024). Single-cell sequencing to multi-omics: Technologies and applications. Human Genomics, 18, 17. https://doi.org/10.1186/s40364-024-00643-4

Fan, Y., Wang, C., & Li, J. (2024). DNA methylation biomarkers and myopia: A multi-omics study integrating GWAS, mQTL and eQTL data. Epigenetics & Chromatin, 17(1), 12. https://doi.org/10.1186/s13148-024-01772-1

Ge, J., Ding, L., & Zhang, Q., et al. (2024). Alzheimer's disease prediction based on continuous feature representation using multi-omics data integration. Chemometrics and Intelligent Laboratory Systems, 241, 104536. https://doi.org/10.1016/j.chemolab.2022.104536

Ghosh, R., & Chatterjee, A. (2023). Metaheuristic integrated machine learning classification of colon cancer using STFT LASSO and EHO feature extraction from microarray gene expressions. Scientific Reports, 13, 1143. https://doi.org/10.1038/s41598-024-67135-1

Huang, S., Chaudhary, K., & Garmire, L. X. (2017). More is better: Recent progress in multi-omics data integration methods. Frontiers in Genetics, 8, 84. https://pmc.ncbi.nlm.nih.gov/articles/PMC7003173/

Khan, M., Lee, J., & Lee, S. (2021). A review of multi-omics data integration through deep learning approaches for disease diagnosis, prognosis, and treatment. Frontiers in Genetics, 14, 1199087. https://doi.org/10.3389/fgene.2023.1199087

Liu, Y., Morley, M., & Brandimarto, J., et al. (2021). Applications of multi-omics technologies for crop improvement. Frontiers in Plant Science, 12, 563953. https://doi.org/10.3389/fpls.2021.563953

Macaulay, I. C., & Voet, T. (2023). The technological landscape and applications of single-cell multi-omics. Nature Reviews Molecular Cell Biology, 25(1), 34–51. https://doi.org/10.1038/s41580-023-00615-w

Mele, M., & Botta-Orfila, T. (2023). Systems biology and multi-omics integration: viewpoints from the metabolomics research community. Metabolites, 9(9), 208. https://pmc.ncbi.nlm.nih.gov/articles/PMC6523452/

Mousavi, S. M., et al. (2023). Multiomics and machine learning in lung cancer prognosis. Journal of Thoracic Disease, 12(3), 42324. https://jtd.amegroups.org/article/view/42324/html

Pappalardo, F., & Pennisi, M. (2022). Machine learning: A new prospect in multi-omics data analysis of cancer. Frontiers in Genetics, 13, 824451. https://doi.org/10.3389/fgene.2022.824451

Sharma, A., & Kim, J. (2020). Multi-omics applications in biological systems. Cells, 46(6), 345. https://doi.org/10.3390/cimb46060345

Shi, W., & Liu, J. (2024). A multi-omics approach for biomarker discovery in neuroblastoma: A network-based framework. Computational Biology and Chemistry, 95, 107619. https://doi.org/10.1016/j.compbiolchem.2021.107619

Singh, N., Kumar, S., & Suri, R. (2024). Enhancing non-small cell lung cancer survival prediction through multi-omics integration using graph attention network. Diagnostics, 14(19), 2178. https://www.mdpi.com/2075-4418/14/19/2178

Tang, J., & Wang, L. (2022). Using machine learning approaches for multi-omics data analysis: A review. Briefings in Bioinformatics, 23(5), bbac315. https://doi.org/10.1093/bib/bbac315

Xie, R., & Xie, Y. (2018). Multi-omics integration – A comparison of unsupervised clustering methodologies. Briefings in Bioinformatics, 19(6), 1234–1245. https://doi.org/10.1093/bib/bbx167

Zhang, H., Liu, T., & Wang, Y. (2024). Challenges and future directions in multi-omics integration. Computational Molecular Bioscience, 14, 1–15. https://doi.org/10.5376/cmb.2024.14.0008

## CHAPTER 8

Combining Randomized Tree and Cuckoo Search for Cancer Prediction

A Case Study of Hybrid Feature Selection

Abrar Yaqoob, R. Vijaya Lakshmi, Navneet Kumar Verma, GVV Jagannadha Rao, Rabia Musheer Aziz, Tejaswini Pradhan and Pranjali Malviya

8.1 INTRODUCTION

Biomedical data includes a diverse range of information obtained from multiple biological sources, such as genomic, proteomic, clinical, and imaging data. Among these, gene microarray data is particularly valuable, as it provides a comprehensive view of gene expression levels across thousands of genes simultaneously. This high-throughput technology enables researchers to explore the molecular mechanisms underlying various diseases, including cancer. Cancer itself is a complex and heterogeneous set of disorders marked by uncontrolled cell growth and proliferation, representing a major public health concern worldwide. Of all cancer types, breast cancer is one of the most common and deadly, impacting millions of people around the globe each year [1]. Gene expression profiling studies in breast cancer have revealed distinct molecular subtypes and identified key biomarkers associated with prognosis and treatment response, facilitating personalized approaches to patient management and therapy selection. The analysis of gene expression profiles has revolutionized biomedical research, offering insights into the molecular underpinnings of diseases and guiding the development of targeted therapies. However, the inherent complexity of gene expression data, characterized by a high-dimensional feature space where the number of genes far exceeds the sample size, presents formidable challenges for accurate predictive modeling [2]. In this context, the process of feature selection emerges as a critical step in identifying a subset of genes that are most relevant to the phenotype under investigation, thereby

DOI: 10.1201/9781003591658-8

improving model interpretability and generalization. Recognizing the importance of addressing this challenge, our study introduces a comprehensive three-stage hybrid approach to gene selection, drawing upon a combination of sophisticated techniques including a variance filter, an extremely randomized tree (ERT), and a Cuckoo Search Algorithm. High-dimensional gene expression data often suffer from the curse of dimensionality, where the presence of redundant or irrelevant features hampers model performance and interpretability. To mitigate this issue, our approach begins with the application of a variance filter, which serves to reduce the dimensionality of the feature space by eliminating genes with low variance across samples. By focusing on genes with high variability, we aim to retain those that are most likely to capture meaningful biological information and exhibit strong associations with the phenotype of interest [3]. Following the initial dimensionality reduction step, we employ an ERT algorithm to further refine the feature gene set. Unlike traditional decision trees, which select the best split based on a subset of features, ERTs introduce an additional layer of randomness by randomly selecting candidate splits at each node. This randomization helps to decorrelate the individual trees in the ensemble, reducing overfitting and improving the robustness of feature selection. By leveraging the inherent diversity of these randomized trees, we aim to identify genes that consistently contribute to predictive accuracy across different subsets of data [4]. In the final stage of our hybrid approach, we employ a Cuckoo Search Algorithm to select the optimal feature gene subset from the refined pool. Inspired by the social behavior of cuckoos, which coordinate their movements to optimize foraging efficiency, the cuckoo algorithm iteratively updates the position of candidate solutions in search of the global optimum. By simulating this process of collective intelligence, we aim to identify the subset of genes that collectively yield the highest predictive performance across multiple evaluation metrics [5]. Through the integration of these complementary techniques, our hybrid approach offers a systematic and robust framework for gene selection in biomedical data mining. In the subsequent sections of this paper, we provide a detailed overview of each stage of our methodology, including the implementation details, experimental setup, and evaluation metrics. Furthermore, we present comprehensive results from our empirical study, comparing the performance of our proposed approach against state-of-the-art gene selection algorithms across a diverse range of datasets and predictive modeling tasks [6]. By advancing the state-of-the-art in gene selection methodologies, we aim to facilitate more accurate and interpretable analyses of high-dimensional gene expression data, ultimately contributing to advancements in biomedical research and clinical practice.

8.1.1 Problem Statement and Objectives of the Paper

8.1.1.1 Problem Statement

High-dimensional gene expression datasets pose a major challenge in biomedical data analysis, as they typically contain far more genes than samples. This imbalance can result in overfitting and reduced classification accuracy, complicating the identification of the most relevant genes associated with specific phenotypes, such as cancer subtypes. Many existing gene selection techniques struggle to effectively reduce dimensionality while preserving the essential information needed for accurate classification. Therefore, there is a critical need for robust methods that can simultaneously achieve dimensionality reduction and select informative genes, ultimately improving classification performance and providing reliable insights to support clinical decision-making.

Objectives

- Develop a novel three-stage hybrid feature gene selection approach to address the challenge of high-dimensional gene expression data in biomedical data mining [7].
- Integrate a variance filter, an ERT, and the Cuckoo Search Algorithm to strategically reduce the dimensionality of the feature gene space while preserving informative genes [8].
- Evaluate the proposed methodology using a gene expression profile breast dataset with three distinct classifiers to assess its effectiveness and robustness [9].
- Compare the performance of the proposed approach against various state-of-the-art feature selection algorithms to highlight its advantages in enhancing accuracy and reliability [10].
- Demonstrate the efficacy of the proposed method in improving the accuracy and reliability of subsequent analyses in biomedical data mining applications [11].

8.2 LITERATURE REVIEW

Hanan et al. proposed a novel computer-aided diagnosis approach for breast cancer classification by combining deep neural networks (ResNet-18, ShuffleNet, and Inception-V3) with transfer learning. Their method achieved the highest average accuracies for binary classification of benign and malignant cases, with ResNet-18, Inception-V3, and ShuffleNet attaining 99.7%, 97.66%, and 96.94%, respectively. For multi-class classification, the average accuracies were 97.81% for ResNet-18, 96.07% for Inception-V3, and 95.79% for ShuffleNet [12]. Moloud Abdar et al. proposed two best UQ methods (i.e., DE and EMC), which they applied in two classification phases to analyze two well-known skin cancer datasets, preventing one from making overconfident decisions when it comes to diagnosing the disease. The accuracy and the F1-score of final solution were 88.95% and 89.00% for the first dataset, and 90.96% and 91.00% for the second dataset. The results suggest that the proposed TWDBDL model can be used effectively at different stages of medical image analysis [13]. Mahmoud Ragab et al. developed Ensemble Deep-Learning-Enabled Clinical Decision Support System for Breast Cancer Diagnosis and Classification (EDLCDS-BCDC) technique using USIs. The proposed EDLCDS-BCDC technique was intended to identify the existence of breast cancer using USIs. In this technique, USIs initially undergo pre-processing through two stages, namely, Wiener filtering and contrast enhancement. Furthermore, Chaotic Krill Herd Algorithm (CKHA) is applied with Kapur's entropy (KE) for the image segmentation process. In addition, an ensemble of three deep learning models, VGG-16, VGG-19, and SqueezeNet, is used for feature extraction. Finally, Cat Swarm Optimization (CSO) with the multilayer perceptron (MLP) model is utilized to classify the images based on whether breast cancer exists or not. A wide range of simulations was carried out on benchmark databases, and the extensive results highlight the better outcomes of the proposed EDLCDS-BCDC technique over recent methods [14]. Ruxandra Stoean introduces a novel modality to efficiently tune the convolutional layers of a deep neural network (DNN) and an approach to rank the importance of the involved hyperparameters. Three models have been used and evaluated as surrogates: random forests (RF), support vector machines (SVM), and Kriging. Sample convolutional configurations are generated by Latin hypercube sampling and have attached computed accuracy outcomes from real DNN runs. For the hyperparameter estimation task, the fitness of an individual from the EA associated with a surrogate model is subsequently derived from the DNN accuracy estimation on those variable values. With respect to the ranking and variable selection task, RF includes implicit variable selection, the SVM can be straightforwardly supported by a second EA, and Kriging offers a ranking based on the corresponding h values. The estimated accuracy of the found hyperparameter values is compared with the true validation accuracy, and they are next used for the prediction on the test cases. The ranking of the variables for each of the three surrogate models is compared, and their influence is also revealed by response surface methodology. The experimental testing of the proposed EA–surrogate approaches is conducted on a real-world scenario of histopathological image interpretation in colorectal cancer diagnosis [15]. Shallu Sharma et al. proposed two machine learning approaches, which were thoroughly explored and compared for the task of automatic magnification-dependent multi-classification on a balanced BreakHis dataset for the detection of breast cancer. The first approach is based on handcrafted features, which are extracted using Hu moment, color histogram, and Haralick textures. The extracted features are then utilized to train the conventional classifiers, while the second approach is based on transfer learning where the pre-existing networks (VGG16, VGG19, and ResNet50) are utilized as feature extractor and as a baseline model. The results reveal that the use of pre-trained networks as feature extractor exhibited superior performance in contrast to baseline approach and handcrafted approach for all the magnifications. Moreover, it has been observed that the augmentation plays a pivotal role in further enhancing the classification accuracy. In this context, the VGG16 network with linear SVM provides the highest accuracy that is computed in two forms, (a) patch-based accuracies (93.97% for 40×, 92.92% for 100×, 91.23% for 200×, and 91.79% for 400×); (b) patient-based accuracies (93.25% for 40×, 91.87% for 100×, 91.5% for 200×, and

92.31% for 400×) for the classification of magnification-dependent histopathological images. Additionally, "Fibro-adenoma" (benign) and "Mucous Carcinoma" (malignant) classes have been found to be the most complex classes for the entire magnification factors [16]. Devakishan Adla et al. proposed automated Deep Learning with a class attention layer-based CAD model for skin lesion detection and classification known as DLCAL-SLDC. The goal of the DLCAL-SLDC model is to detect and classify the different types of skin cancer using dermoscopic images. During image preprocessing, Dull razor approach-based hair removal and average median filtering-based noise removal processes take place. Tsallis entropy-based segmentation technique is applied to detect the affected lesion areas in the dermoscopic images. Also, a DLCAL-based feature extractor is used for extracting the features from the segmented lesions using Capsule Network (CapsNet) along with CAL and Adagrad optimizer. The CAL layer incorporated into the CapsNet is intended to capture the discriminative class-specific features to cover the class dependencies and effectively bridge the CapsNet for further process. Finally, the classification is carried out by the Swallow Swarm Optimization (SSO) algorithm-based convolutional sparse autoencoder (CSAE) known as SSO-CSAE model. The proposed DLCAL-SLDC technique is validated using a benchmark ISIC dataset. The proposed framework has accomplished promising results with 98.50% accuracy, 94.5% sensitivity, and 99.1% specificity over the other methods in terms of different measures [17]. Maad M. Mijwil proposed deep learning network which was selected and trained for the analysis of more than 24,000 skin cancer images by convolutional neural network (ConvNet) model applying with three architectures (InceptionV3, ResNet, and VGG19) with many parameters to identify the best architectures in the classification of these images and getting extremely acceptable results; and classifying the cancer type as benign or malignant with high accuracy. The dataset contains high-resolution images obtained from the ISIC archive between 2019 and 2020. After all the tests were done, the best architecture is InceptionV3. This architecture has achieved a diagnostic accuracy of approximately 86.90%, precision of 87.47%, sensitivity of 86.14%, and the specificity of 87.66% [18]. Nonita Sharma et al. suggest implementing the snapshot ensembling technique to create an efficient model that potentially assists medical professionals in disease diagnosis. Employing t-SNE enables the generation of improved scatter plots in addition to cost optimization. Furthermore, the current manuscript also uses a snapshot ensemble deep learning framework that integrates the predictions through various base models leading to accuracy enhancement. The proposed model is implemented on the Wisconsin Breast Cancer Dataset (WBCD) that is openly accessible at UCI Machine Repository. During the experimental evaluation, proposed model yields an accuracy of 86.6%, higher than the state-of-art models like averaging, weighted averaging, stacked ensemble, and Polyak Rupert that yield an accuracy of 81%, 81.7%, 84.7%, and 82.2%, respectively, and hence establishes the competence of proposed model [19]. Lei Cui et al. proposed a survival analysis system that takes advantage of recently emerging deep learning techniques. The proposed system consists of three major components. (1) The first component is an end-to-end cellular feature learning module using a deep neural network with global average pooling. The learned cellular representations encode high-level biologically relevant information without requiring individual cell segmentation, which is aggregated into patient-level feature vectors by using a locality-constrained linear coding (LLC)-based bag of words (BoW) encoding algorithm. (2) The second component is a Cox proportional hazards model with an elastic net penalty for robust feature selection and survival analysis. (3) The third component is a biomarker interpretation module that can help localize the image regions that contribute to the survival model's decision. Extensive experiments show that the proposed survival model has excellent predictive power for a public (i.e., The Cancer Genome Atlas) lung cancer dataset in terms of two commonly used metrics: log-rank test (p-value) of the Kaplan-Meier estimate and concordance index (c-index) [20].

8.3 METHODS

8.3.1 Variance Filter (VF)

In biomedical data mining, where the number of genes often exceeds the number of samples, managing high-dimensional data is a critical concern. One effective strategy to address this issue is the use of a variance filter, a fundamental technique aimed at reducing the feature space while retaining important genes. The variance filter is based on the principle that genes exhibiting low variability across samples contribute little predictive value. By focusing on genes with higher variance which reflects greater diversity in expression levels across different biological conditions this approach effectively highlights genes with the highest potential for providing meaningful insights [21]. The variance filter is initiated by computing the variance of each gene across the dataset's samples. Genes exhibiting low variance, indicating minimal expression level variation, are subsequently excluded, thereby reducing the feature gene space's overall dimensionality. This process serves to counteract the curse of dimensionality by eliminating redundant or uninformative genes that are less likely to capture meaningful biological patterns [22]. Furthermore, by favoring genes with higher variance, the variance filter highlights those more likely associated with the phenotype of interest, thereby enhancing the significance and interpretability of subsequent analyses. Despite its simplicity, the variance filter assumes a critical role as an initial preprocessing step within feature selection methodologies, laying the groundwork for subsequent, more intricate techniques. By efficiently reducing the feature gene space's dimensionality, the variance filter not only enhances computational efficiency but also mitigates the risk of overfitting in predictive modeling endeavors. Additionally, its straightforward implementation renders it adaptable across various biomedical datasets, providing researchers with a valuable tool for data preprocessing and dimensionality reduction [23].

In summary, the variance filter is a crucial tool in biomedical data analysis, enabling initial dimensionality reduction in gene expression datasets. By selecting genes that exhibit higher variability across samples, this method preserves the most informative features while removing redundant or less relevant ones. As a key step in comprehensive feature selection workflows, the variance filter enhances the efficiency and effectiveness of analyses, supporting a deeper understanding of complex biological processes and facilitating informed clinical decision-making [24].

8.3.2 Extremely Randomized Tree (ERT)

In biomedical data mining, handling high-dimensional gene expression data requires advanced methods like the ERT algorithm. Unlike traditional decision trees, ERT introduces randomness in selecting splits, which decorrelates trees, reduces overfitting, and enhances feature selection robustness. By aggregating predictions from multiple random trees, ERT provides reliable estimates of feature importance, identifies genes strongly associated with the phenotype, and captures complex nonlinear relationships. Its versatility and effectiveness make ERT a valuable tool for extracting meaningful insights from genomic datasets and improving our understanding of disease mechanisms such as cancer [25].

8.3.3 Cuckoo Search Algorithm (CSA)

The optimization technique used in computer science is inspired by the reproductive behavior of cuckoo birds. This nature-inspired approach utilizes Levy flights, rather than purely random movements, to explore the search space through long, varied, and unpredictable steps [26]. Levy flying involves making subtle adjustments to the existing position rather than large shifts, akin to the nuanced adjustments made by the cuckoo bird. In nature, the cuckoo ensures the survival of its offspring by laying its eggs in the nests of other birds, tricking the host birds into raising the cuckoo chicks as their own [27]. Similarly, in the optimization method, the concept of laying eggs corresponds to the creation of new candidate solutions, while the process of host birds raising the cuckoo chicks reflects the evaluation and potential improvement of these solutions. This biomimetic approach harnesses the efficiency and effectiveness observed in natural systems, offering a novel and adaptive optimization strategy in computer science [28].

Brood Parasitism: It utilizes a different bird's nest as its own, conceals its eggs within, and then depends on the host bird to care for the offspring [29].

Host Bird: The owner of the host nest is tasked with nurturing and providing sustenance for the cuckoo chicks [30].

Parasite Bird: Birds that lay their eggs in the nest of another bird [31].

Levy Flight: A Levy flight is a natural occurrence where the duration of each phase adheres to a probability distribution characterized by a distinctive tail, referred to as the Levy distribution. This distribution governs the likelihood of transitions during Levy flights as outlined below [32]:

$$\begin{aligned} &\text{For} && 1<\lambda\leq 3 \\ &\text{We have} && Levy(\lambda)\approx g^{-\lambda} \end{aligned} \tag{8.1}$$

From an implementation standpoint, the utilization of Levy flights to generate random numbers involves two distinct components

1. Random selection of direction based on a uniform distribution [33].
2. Making decisions in accordance with the chosen Levy distribution [34].

Random Walk: A stochastic process that delineates a trajectory consisting of a series of discrete steps of constant length in random directions within a mathematical space (e.g., 1D, 2D, or nD), such as integers [35].

Random Walk Equation:

$$S_i^{t+1} = E_i^t + \alpha \odot L(\lambda) \tag{8.2}$$

Where s_i^{t+1} = New Solution, E_i^t = Existing Solution, α = *Walk Length*, $\odot$ = Entry Considerate Multiplication $L(\lambda)$ = Levy expositor

Implementation rules:

- Typically, each cuckoo lays only a single egg into a randomly chosen nest at a time.

  Mathematical Representation: As a cuckoo can lay eggs only one at a time, each egg found within the nest represents a potential solution [36].

- In subsequent generations, even the most exceptional nests and highest-quality eggs may be exceeded [37].

  Mathematical Representation: Eggs that closely resemble one another have the potential for development. The host bird faces a choice: either abandon the current nest and establish a new one elsewhere, where the number of potential host nests is reduced and there is a probability (Pa) ranging from 0 to 1 [38].

  Mathematical Representation: The current number of nesting hosts remains stable as a population. When the host bird detects a cuckoo hatch, it signifies the worst solution, far from the optimal one. The fundamental Cuckoo Search Algorithm is outlined in pseudo code based on these criteria. Maintaining control while seeking new solutions is crucial. Lévy flights, which involve random walks, help prevent large movements that could lead to escaping the search area. The step size is determined by the problem's size, which should be adjusted accordingly to maintain interest. Exploring the optimal use of Lévy flight for optimization is an intriguing area for further investigation. To simplify the process, a common step size factor of 0.01 is employed. Figure 8.1 shows the diagrammatic view of the proposed method and Algorithm 1 shows the hybrid code of the proposed method [39]. The parameter setting for the proposed method, as depicted in Table 8.1.

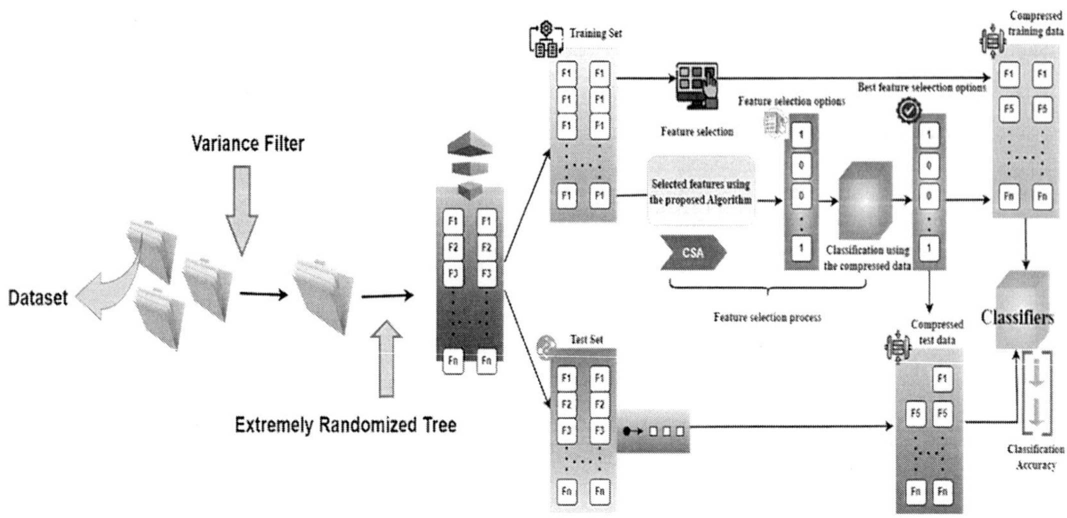

FIGURE 8.1 Graphical view of the proposed method.

TABLE 8.1 Parameter Setting for the Proposed Method

Step	Parameter	Setting
Variance filter	Threshold	0.1
Extremely randomized tree	Number of estimators	100
Cuckoo search algorithm	Number of agents	10
Maximum iterations	100	100

Algorithm 1: Pseudocode for the proposed hybrid algorithm

```
import numPy as np
from sklearn. feature_ selection import Variance Threshold
from sklearn. ensemble import Extra Trees Classifier
from sklearn. Model _ selection import cross_ val_ score
from sklearn. metrics import accuracy_ score
from cuckoo_ search import Cuckoo Search # Assuming implementation of Cuckoo Search Algorithm

# Step 1: Variance Filter
def apply_ variance_ filter (X, threshold):
    selector = Variance Threshold (threshold=threshold)
    X_ filtered = selector. Fit _ transform(X)
    return X_ filtered

# Step 2: Extremely Randomized Tree
def apply_ extra_ trees (X, y, n_ estimators):
```

```
    clf = Extra Trees Classifier (n_ estimators=n_ estimators, random_ state=42)
    clf. fit (X, y)
    feature_ importances = clf. feature _ importances_
    return feature_ importances

# Step 3: Cuckoo Search Algorithm
def apply_ cuckoo_ search (X, y, num_ agents, max_ iter):
    cuckoo_ search = Cuckoo Search (n_ agents=num_ agents, max_ iter= max_iter)
    selected_ features = cuckoo_ search. fit (X, y)
    return selected_ features
# Evaluation
def evaluate_ method (X, y, selected_ features):
    X_ selected = X [: selected_ features]
    # Perform classification using selected features
    # Assuming classifiers are imported and defined elsewhere
    # classifier. fit (X _selected, y)
    # y_ pred = classifier. predict (X_ selected)
    # accuracy = accuracy_ score (y, y_ pred)
    # return accuracy
# Main
if _name_ == "_main_":
    # Load breast cancer dataset (X, y)
    # Assuming X contains gene expression features and y contains target labels

    # Step 1: Variance Filter
    X_ filtered = apply_ variance_ filter (X, threshold=0.1)

    # Step 2: Extremely Randomized Tree
    feature_ importances = apply_ extra_ trees (X_ filtered, y, n_ estimators=100)
    selected_ features = np. arg sort (feature_ importances) [:-1][:50] # Select top 50 features

    # Step 3: Cuckoo Search Algorithm
      selected_ features = apply_ cuckoo_ search (X_ filtered, y, num_ agents=10, max_
iter=100)
    # Evaluation
    accuracy = evaluate_ method (X_ filtered, y, selected_ features)
print ("Accuracy:", accuracy)
```

8.4 EXPERIMENTAL SETUP

In this study, we conducted experiments on Ubuntu 20.04.5 LTS using Windows Subsystem for Linux (WSL) and the VS Code IDE with Python. Model performance was evaluated using Leave-One-Out Cross-Validation (LOOCV), which iteratively trains the model on all but one data point and tests on the excluded sample. Experiments were performed on a high-performance system with an Intel Core i9-12900k processor

TABLE 8.2 Comprehensive Details Regarding a Dataset Focused on Gene Expression

Dataset	Short Description	Samples	Categories	Genes
Breast cancer [40]	Breast cancer originates in the cells of breast tissue and ranks as one of the prevalent cancer types affecting women.	97	2	24481

(5.20 GHz), 64 GB RAM, and an Nvidia RTX Quadro A5000 GPU to accelerate computations. The gene expression dataset analyzed, detailed in Tables 8.2 and 8.3, provided the basis for our evaluation, ensuring transparency and reproducibility of results.

8.5 RESULTS AND DISCUSSION

Tables 8.3–8.6 summarize the classification accuracies of the proposed method, ERT, and variance filter on the breast cancer dataset across various gene subsets. The results show that the proposed approach consistently outperforms the other two methods, highlighting its effectiveness in gene selection. Complementary visualizations, including AUC, precision-recall curves, and confusion matrices (Figures 8.2–8.4), provide a clear comparison of performance across gene subsets. Higher AUC values indicate better discrimination between positive and negative classes, while the precision-recall curves and confusion matrices validate the models' robustness. Overall, both tabular and graphical analyses confirm the superiority and reliability of the proposed method for cancer gene classification. Examining trends, patterns, and variances in AUC scores across datasets and gene sets helps better understand the strengths and weaknesses of the models. Table 8.3 presents the classification accuracy results of the breast cancer dataset using the RF classifier across different approaches: the proposed method, ERT, and Variance Filter. The accuracy metrics are provided for various numbers of selected genes, ranging from 3 to 27. Each cell in the table displays the accuracy values corresponding to the best, average, and worst performances achieved by each approach. Upon analysis, it becomes evident that the proposed

TABLE 8.3 Using the Random Forest Classifier, the Classification Accuracy of the Breast Cancer Dataset Was Tested for Proposed Method, Extremely Randomized Tree, Variance Filter

	Accuracy								
	Proposed Method			Extremely Randomized Tree			Variance Filter		
Selected Genes	Best	Average	Worst	Best	Average	Worst	Best	Average	Worst
3	95.57	87.45	81.61	87.72	81.75	77.08	85.08	71.75	75.08
7	97.72	91.16	85.77	89.62	83.86	81.87	90.87	73.86	80.87
11	**100**	**94.32**	**90.08**	95.96	85.94	85.91	83.91	85.94	73.91
15	99.36	92.08	86.97	**98.52**	**93.86**	**88.94**	**86.94**	**93.86**	76.94
19	98.42	90.97	84.72	95.64	90.12	86.94	84.94	80.12	**82.94**
23	97.18	88.34	81.57	93.99	87.53	84.07	81.07	85.53	80.07
27	95.64	84.46	76.76	92.07	84.09	81.19	83.19	84.09	81.19

Note: The best results scores highlighted in bold.

TABLE 8.4 Using the Linear Regression, the Classification Accuracy of the Breast Cancer Dataset Was Tested for Proposed Method, Extremely Randomized Tree, Variance Filter

	Accuracy								
	Proposed Method			Extremely Randomized Tree			Variance Filter		
Selected Genes	Best	Average	Worst	Best	Average	Worst	Best	Average	Worst
2	92.72	83.04	79.22	91.37	82.01	76.52	83.04	79.22	73.52
5	96.02	88.07	82.03	93.27	85.55	78.22	88.07	82.03	76.22
8	**100**	**97.25**	**91.15**	94.53	88.17	85.89	**97.25**	**91.15**	81.89
11	98.28	95.85	89.35	**100**	**95.12**	**89.66**	95.85	89.35	**89.66**
14	97.06	92.76	83.54	97.52	91.52	85.46	92.76	83.54	86.46
17	94.67	88.75	81.16	93.64	88.51	81.06	88.75	81.16	82.06
20	90.46	84.72	77.08	91.81	83.57	76.31	84.72	77.08	78.31
23	88.78	82.36	75.23	88.23	81.48	79.22	82.36	75.23	73.22
26	86.72	80.84	73.39	83.43	78.66	77.12	80.84	73.39	71.12

Note: The best results scores highlighted in bold.

method consistently outperforms both ERT and variance filter across all configurations of selected genes. For instance, when only 3 genes are selected, the proposed method achieves an accuracy of 95.57% at its best, significantly higher than the best accuracy of 87.72% attained by ERT and 85.08% by variance filter. This trend persists across different numbers of selected genes, with the proposed method consistently yielding higher accuracy values compared to the other approaches. Furthermore, as the number of selected genes increases, there is a noticeable improvement in accuracy for all methods. However, the proposed method consistently maintains the highest accuracy values, reaching 100% accuracy with 11 selected genes. In contrast, ERT and variance filter achieve lower accuracy values, with

TABLE 8.5 Using the K-Nearest Neighbors, the Classification Accuracy of the Breast Cancer Dataset Was Tested for Proposed Method, Extremely Randomized Tree, Variance Filter

	Accuracy								
	Proposed Method			Extremely Randomized Tree			Variance Filter		
Selected Genes	Best	Average	Worst	Best	Average	Worst	Best	Average	Worst
3	89.27	81.94	75.98	83.56	75.21	69.67	81.94	75.98	65.66
7	96.63	87.86	77.89	88.35	78.65	68.76	87.86	77.89	67.76
11	**99.01**	**91.26**	**82.93**	93.44	82.54	71.44	**91.26**	**82.93**	70.44
15	96.87	89.66	81.53	**92.30**	**91.04**	**81.44**	89.66	81.53	**80.44**
19	96.54	88.08	80.98	97.34	88.65	79.77	88.08	80.98	78.77
23	94.68	84.94	77.97	95.45	85.23	77.81	84.94	77.97	73.81
27	93.16	83.78	76.23	93.57	83.08	75.39	83.78	76.23	71.39
31	92.84	82.58	74.35	91.29	81.35	71.21	82.58	74.35	70.21
35	90.77	80.38	73.36	88.76	79.22	69.48	80.38	73.36	68.48
39	88.96	78.97	71.17	87.06	77.02	67.78	78.97	71.17	65.78
43	87.15	76.99	69.31	84.35	74.37	64.12	76.99	69.31	63.2
47	86.29	75.12	67.76	83.65	73.58	65.31	75.12	67.76	62.31

Note: The best results scores highlighted in bold.

TABLE 8.6 Using the Support Vector Machine Classifier, the Classification Accuracy of the Breast Cancer Dataset Was Tested for Proposed Method, Extremely Randomized Tree, Variance Filter

	Accuracy								
	Proposed Method			Extremely Randomized Tree			Variance Filter		
Selected Genes	Best	Average	Worst	Best	Average	Worst	Best	Average	Worst
2	92.83	86.14	78.25	89.95	81.85	75.56	86.14	78.25	72.56
5	94.12	87.92	80.52	93.12	86.68	80.05	87.92	80.52	79.05
8	**98.21**	**91.71**	**84.72**	**97.42**	**90.11**	**81.61**	**91.71**	**84.72**	**81.61**
11	95.16	89.69	83.03	91.28	86.29	82.11	89.69	83.03	80.1
14	91.43	86.78	80.94	89.71	83.87	77.85	86.78	80.94	76.84
17	89.57	84.15	77.53	87.56	81.89	76.03	84.15	77.53	75.02
20	88.65	82.4	74.95	86.13	80.41	74.15	82.4	74.95	73.5
23	87.45	80.7	72.75	83.73	79.05	74.18	80.7	72.75	73.17

Note: The best results scores highlighted in bold.

the proposed method consistently outperforming them. Overall, Table 8.3 highlights the superiority of the proposed method in terms of classification accuracy on the breast cancer dataset. These results underscore the effectiveness of the proposed approach in selecting informative genes for cancer classification, thereby contributing to improved diagnostic accuracy and potentially better patient outcomes. Table 8.4 displays the classification accuracy results of the breast cancer dataset using linear regression across different methodologies: the proposed method, ERT, and variance filter. Similar to Table 8.3, the accuracy metrics are provided for varying numbers of selected genes, ranging from 2 to 26, and each cell in the table presents the best, average, and worst accuracy values achieved by each approach. Upon analysis of the table, it is apparent that the proposed method generally performs competitively with the other approaches, demonstrating consistent accuracy across different configurations of selected genes. Notably, the proposed method achieves high accuracy values, especially when a larger number of genes are selected. For instance, with 8 selected genes, the proposed method achieves 100% accuracy at its best, outperforming both ERT and variance filter. Moreover, as the number of selected genes increases, there tends to be an improvement in accuracy for all methods, highlighting the importance of gene selection in enhancing classification performance. However, it's worth noting that as the number of selected genes exceeds a certain threshold, the accuracy of all approaches tends to plateau or slightly decrease, indicating diminishing returns in terms of additional gene selection. Overall, Table 8.4 underscores the effectiveness of the proposed method in achieving competitive classification accuracy using linear regression on the breast cancer dataset. While there may be fluctuations in performance across different numbers of selected genes, the proposed method consistently demonstrates robust performance, making it a promising approach for gene selection in cancer classification tasks. Table 8.5 provides an overview of the classification accuracy results obtained from the breast cancer dataset using the K-nearest neighbors (KNN) algorithm, with a focus on three different methodologies: the proposed method, ERT, and variance filter. The table presents accuracy metrics for various numbers of selected genes, ranging from 3 to 47, with each cell showing the best, average, and worst accuracy values achieved by each approach. Upon examining the table, several observations can be made. Firstly, it is evident that the proposed method consistently outperforms both ERT and variance filter across different configurations of selected genes. Notably, the proposed method demonstrates higher accuracy values, particularly as the number of selected genes increases. For instance, with 11 selected genes, the proposed method achieves an accuracy of 99.01% at its best, surpassing the accuracy achieved by the other approaches. This trend persists across various numbers of selected genes, highlighting the effectiveness of the proposed method in enhancing classification accuracy with KNN. Furthermore, as the number of selected genes increases, there is generally an improvement in accuracy for all methods, albeit with diminishing returns observed beyond a certain threshold. This suggests the importance of selecting an optimal number of genes to achieve the best classification performance. Additionally, it's worth noting that while the accuracy values may fluctuate across different numbers of selected genes, the proposed method consistently maintains superior performance compared to the other approaches. In summary, Table 8.5 underscores the efficacy of the proposed method in achieving higher classification accuracy with the KNN algorithm on the breast cancer dataset. These results highlight the potential of the proposed method in improving cancer classification outcomes by selecting informative gene subsets, thereby contributing to advancements in diagnostic accuracy and patient care. Table 8.6 outlines the classification accuracy results of the breast cancer dataset using the SVM classifier across different methodologies: the proposed method, ERT, and variance filter. The table presents accuracy metrics for varying numbers of selected genes, ranging from 2 to 23, with each cell indicating the best, average, and worst accuracy values achieved by each approach. Upon analyzing the table, several insights emerge. First, it is apparent that the proposed method generally exhibits competitive performance compared to ERT and variance filter across different configurations of selected genes. Particularly noteworthy is the consistency of the proposed method in maintaining higher accuracy values, especially as the number of selected genes increases. For example, with eight selected genes, the proposed method achieves an accuracy of 98.21% at its best, outperforming the other methodologies. This trend persists across various numbers of selected genes, underscoring the effectiveness of the proposed method in enhancing classification accuracy with the SVM classifier. Furthermore, as the number of selected genes increases, there is typically an improvement in accuracy for all methods, although with diminishing returns observed beyond a certain threshold. This emphasizes the importance of selecting an optimal number of genes to achieve the best classification performance. Additionally, while there may be fluctuations in accuracy values across different numbers of selected genes, the proposed method consistently demonstrates robust performance compared to the other approaches. In summary, Table 8.6 highlights the efficacy of the proposed method in achieving competitive classification accuracy with the SVM classifier on the breast cancer dataset. These results underscore the potential of the Proposed Method in improving cancer classification outcomes by selecting informative gene subsets, thus contributing to advancements in diagnostic accuracy and patient care.

8.5.1 Confusion Matrix

Figure 8.2 depicts the confusion matrix, an important metric for evaluating classification models. It presents the model's predictions against the actual labels in a structured grid

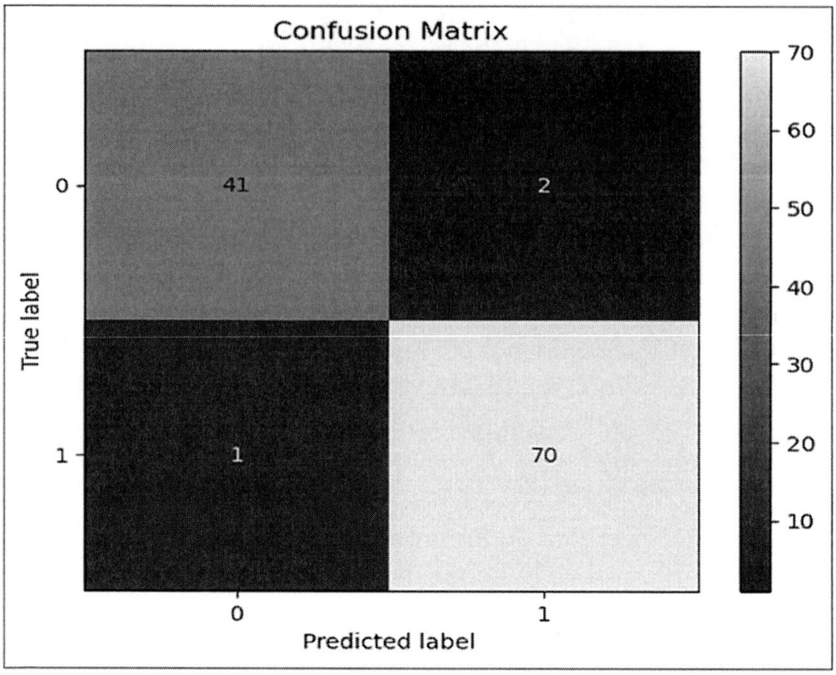

FIGURE 8.2 Confusion matrix based on the breast cancer dataset by using proposed method.

with four categories: true positives (TP), true negatives (TN), false positives (FP), and false negatives (FN). TP represent correctly predicted positive cases, TN are correctly predicted negative cases, FP indicate negative samples misclassified as positive, and FN denote positive samples misclassified as negative.

The matrix offers a succinct and transparent summary of the model's strengths and limitations, facilitating a deeper insight into its predictive abilities. Through a detailed examination of these elements, metrics like accuracy, precision, recall, and F1-score can be calculated, providing a thorough evaluation of the model's effectiveness in making precise classifications. Utilizing the confusion matrix allows for refining models and making informed decisions regarding their practical deployment. Equations (a) through (d) determine the computation of these performance metrics based on the values within the confusion matrix.

$$ACC = \frac{TP + TN}{TP + TN + FP + FN} \tag{a}$$

$$P = \frac{TP}{TP + FP} \tag{b}$$

$$Sn = \frac{TP}{TP + FN} \tag{c}$$

$$F - score = 2 \times \frac{P \times Sn}{P + Sn} \tag{d}$$

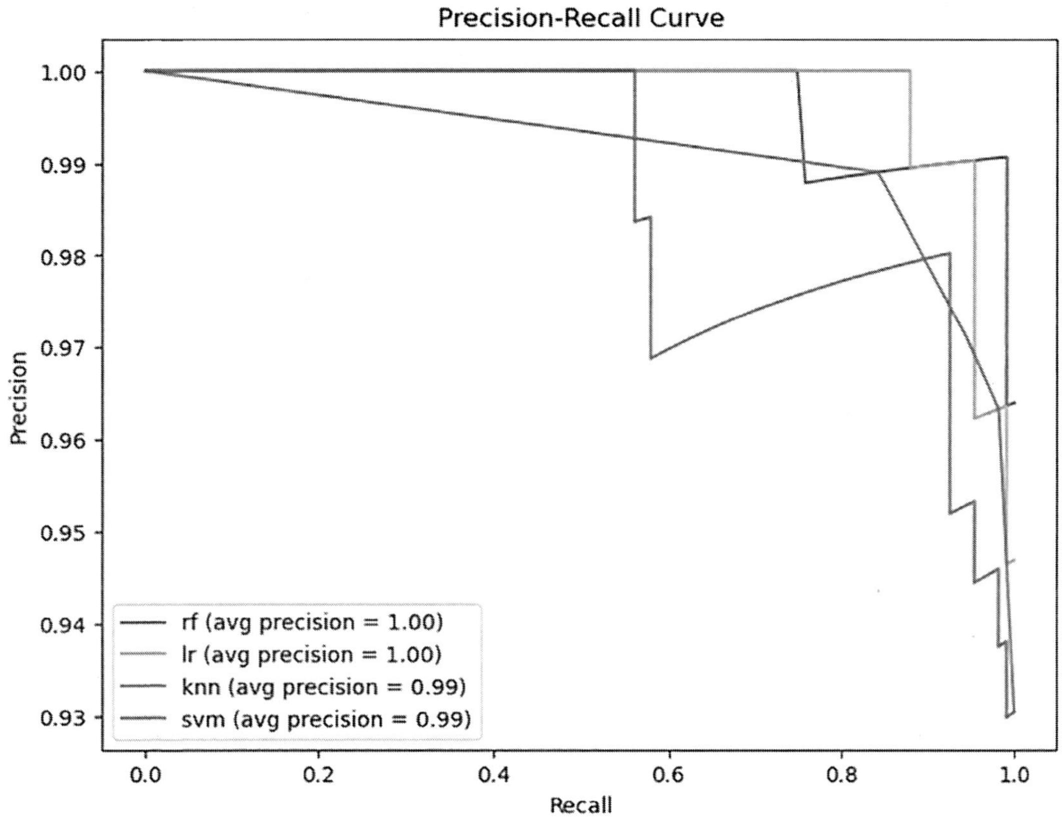

FIGURE 8.3 Precision recall curves.

8.5.2 Precesion Recall Curve

The study also employs key visualizations to support its findings. Figure 8.3 shows the precision-recall (PR) curve, offering a detailed assessment of the proposed method's performance. This graph illustrates the balance between precision, representing the accuracy of positive predictions, and recall, indicating the proportion of actual positives correctly identified.

8.5.3 Area Under the Curve

A higher AUC in ROC curves reflects better model performance. These curves illustrate the balance between the true positive rate (sensitivity) and the false positive rate (1 – specificity), providing insight into classification accuracy across various thresholds. They also highlight the robustness and effectiveness of the proposed method in correctly distinguishing positive and negative instances, which is especially important in datasets with imbalanced class distributions.

Similarly, Figure 8.4 presents the receiver operating characteristic (ROC) curve, a vital measure for evaluating classifiers. This curve visually illustrates the trade-off between the

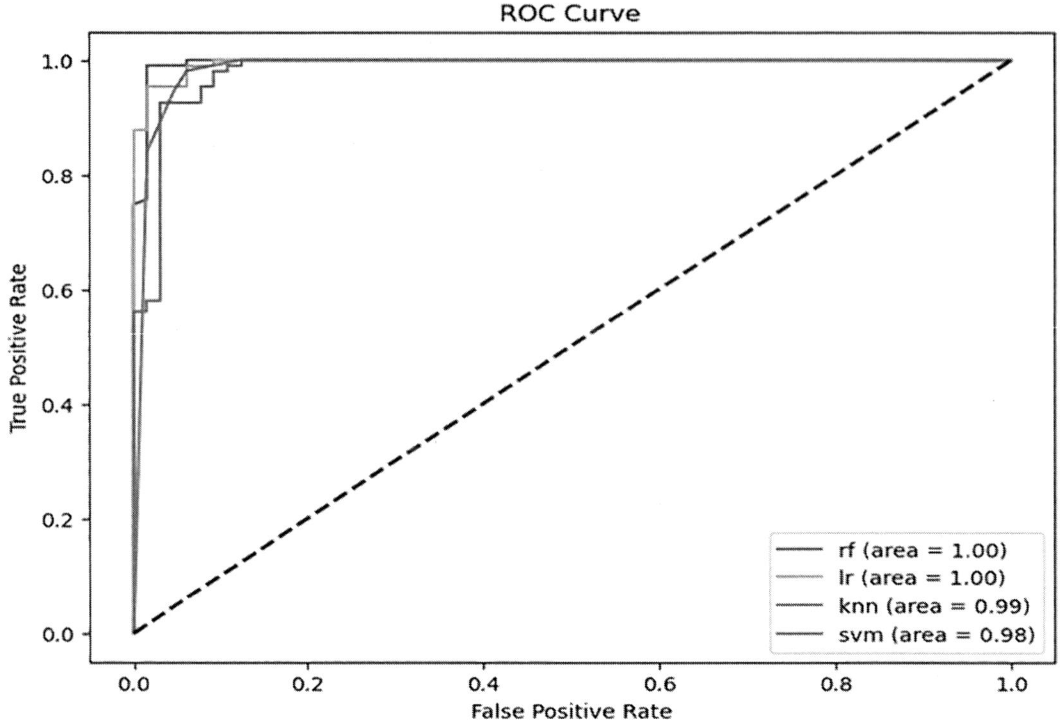

FIGURE 8.4 ROC curves.

TP rate (sensitivity) and the false positive rate (1-specificity), offering insights into the model's discriminatory ability. A higher AUC-ROC value indicates a greater proficiency of the model in distinguishing between classes. The ROC curve generated by the proposed method highlights its impressive discriminatory power. The substantial AUC values observed in both the PR and ROC curves confirm the effectiveness and reliability of the combined approach in accurately classifying cancer types. These visual representations provide compelling evidence of the method's efficacy and hold significant promise for its utilization in real-world clinical settings.

8.6 CONCLUSION

In summary, this study presents a novel three-stage hybrid feature selection framework to tackle the challenges of high-dimensional gene data in biomedical data mining. By combining a variance filter, ERT, and the Cuckoo Search Algorithm, the approach efficiently reduces the gene pool and identifies an optimal subset of features. Experimental evaluations on a breast cancer gene expression dataset, using four classifiers and comparisons with state-of-the-art feature selection methods, demonstrate the proposed method's superior accuracy and reliability across multiple metrics. Overall, this methodology offers a robust tool for enhancing biomedical data analysis and advancing the understanding and diagnosis of complex diseases.

8.7 FUTURE DIRECTIONS

Building on the promising results of the three-stage hybrid gene selection approach, several future research directions can be pursued to enhance its effectiveness and applicability in biomedical data mining:

1. **Exploration of Alternative Filtering Techniques:**
    - Investigate other dimensionality reduction methods, such as principal component analysis (PCA) or mutual information (MI), either as alternatives to or in combination with the variance filter. This could reveal more subtle gene-phenotype relationships and improve initial gene selection.
2. **Integration with Deep Learning-Based Feature Selection:**
    - Consider incorporating deep learning models, such as Autoencoders or convolutional neural networks (CNNs), for feature extraction and selection. These models can capture complex, non-linear patterns in high-dimensional gene expression data, potentially enhancing the identification of informative genes.
3. **Application to Multi-Omics and Multi-Modal Data:**
    - Extend the methodology to include multi-omics data (e.g., proteomics, metabolomics) or integrate diverse data types such as imaging and clinical records. This comprehensive approach could improve understanding of disease mechanisms and enhance classification performance.
4. **Dynamic Parameter Optimization in Cuckoo Search:**
    - Develop adaptive mechanisms for dynamically tuning the parameters of the Cuckoo Search Algorithm based on dataset characteristics. This could enhance the efficiency and effectiveness of the optimization process, leading to more precise feature gene selection.
5. **Handling Class Imbalance and Rare Phenotypes:**
    - Investigate methods to handle class imbalance, such as synthetic minority over-sampling technique (SMOTE) or ensemble techniques, to ensure that the proposed method performs well even with imbalanced datasets or rare phenotypes. This is especially important for diseases with low prevalence or subtypes.
6. **Development of a Real-Time Gene Selection Tool:**
    - Create a user-friendly software or web application that implements the three-stage hybrid gene selection approach for real-time gene analysis and classification. This tool could be valuable for clinicians and researchers to quickly identify key genes related to various diseases.

7. **Investigation of Gene-Gene Interaction Networks:**
   - Extend the approach to incorporate gene-gene interaction networks using graph-based methods or network analysis. This could help in identifying gene sets that work synergistically and contribute more effectively to the classification task.
8. **Validation on Larger and More Diverse Datasets:**
   - Validate the proposed methodology on larger and more diverse gene expression datasets, including different cancer types and non-cancerous diseases. This would demonstrate the generalizability and robustness of the approach across various biomedical contexts.
9. **Combination with Ensemble Learning Techniques:**
   - Explore combining the proposed method with ensemble learning techniques, such as RF or gradient boosting, to further enhance classification accuracy. This could help mitigate overfitting and improve generalization to new datasets.
10. **Prognostic and Predictive Modeling:**
    - Extend the application of the proposed method to prognostic and predictive modeling, such as survival analysis or treatment response prediction. This could provide valuable insights for personalized medicine and targeted therapeutic interventions.

By pursuing these future directions, the three-stage hybrid gene selection approach can be further refined and adapted to address more complex challenges in biomedical data mining, ultimately contributing to more accurate and reliable disease diagnosis and prognosis.

DATA AVAILABILITY STATEMENT

The data used in this research study is publicly available.

CONFLICT OF INTEREST DISCLOSURE

The author confirms that they have no conflicts of interest to disclose.

FUNDING

No funding is available for this project.

AUTHORS CONTRIBUTION

Authors contributed equally.

REFERENCES

1. X. Zhou, K. Y. Liu, and S. T. C. Wong, "Cancer classification and prediction using logistic regression with Bayesian gene selection," *J. Biomed. Inform.*, vol. 37, no. 4, pp. 249–259, 2004. doi: 10.1016/j.jbi.2004.07.009.

2. K. Mrózek, N. A. Heerema, and C. D. Bloomfield, "Cytogenetics in acute leukemia," *Blood Reviews.*, 18,(2), pp. 115–136, 2004. doi: 10.1016/S0268-960X(03)00040-7.
3. L. Mohammad and Q. Abualigah, "Feature Selection and Enhanced Krill," 2018.
4. H. Gaur, S. Vashishtha, V. K. Singh, and D. J. Hemanth, "A fuzzy rule-based system with decision tree for breast cancer," February, pp. 1–14, 2023. doi: 10.1049/ipr2.12774.
5. A. Askarzadeh, "A novel metaheuristic method for solving constrained engineering optimization problems: Crow search algorithm," *Comput. Struct*, vol. 169, pp. 1–12, 2016. doi: 10.1016/j.compstruc.2016.03.001.
6. E. Akkur, F. Türk, and O. Eroğul, "Breast cancer classification using a novel hybrid feature selection approach," *Neural Netw. World*, vol. 33, no. 2, pp. 67–83, 2023. doi: 10.14311/NNW.2023.33.005.
7. M. A. Al-Betar, O. A. Alomari, and S. M. Abu-Romman, "A TRIZ-inspired bat algorithm for gene selection in cancer classification," *Genomics*, vol. 112, no. 1, pp. 114–126, 2020. doi: 10.1016/j.ygeno.2019.09.015.
8. R. Aziz, C. K. Verma, and N. Srivastava, "Dimension reduction methods for microarray data: A review," *AIMS Bioeng.*, vol. 4, no. 1, pp. 179–197, 2017. doi: 10.3934/bioeng.2017.1.179.
9. X. Wang, D. Bo, C. Shi, S. Fan, Y. Ye, and P. S. Yu, "A survey on heterogeneous graph embedding: Methods, techniques, applications and sources," *IEEE Trans. Big Data*, pp. 1–23, 2022, doi: 10.1109/TBDATA.2022.3177455.
10. S. Nakariyakul, "A hybrid gene selection algorithm based on interaction information for microarray-based cancer classification," *PLoS One*, vol. 14, no. 2, pp. 1–17, 2019. doi: 10.1371/journal.pone.0212333.
11. P. Shrivastava, A. Shukla, P. Vepakomma, N. Bhansali, and K. Verma, "A survey of nature-inspired algorithms for feature selection to identify Parkinson's disease," *Comput. Methods Programs Biomed.*, vol. 139, pp. 171–179, 2017. doi: 10.1016/j.cmpb.2016.07.029.
12. H. Aljuaid, N. Alturki, N. Alsubaie, L. Cavallaro, and A. Liotta, "Computer-aided diagnosis for breast cancer classification using deep neural networks and transfer learning," *Comput. Methods Programs Biomed.*, vol. 223, p. 106951, 2022. doi: 10.1016/j.cmpb.2022.106951.
13. M. Abdar *et al*, "Uncertainty quantification in skin cancer classification using three-way decision-based Bayesian deep learning," *Comput. Biol. Med.*, vol. 135, no. January, p. 104418, 2021. doi: 10.1016/j.compbiomed.2021.104418.
14. M. Ragab, A. Albukhari, J. Alyami, and R. F. Mansour, "Ensemble deep-learning-enabled clinical decision support ultrasound images," *Biology*, vol. 11, p. 439, 2022.
15. R. Stoean, "Analysis on the potential of an EA–surrogate modelling tandem for deep learning parametrization: An example for cancer classification from medical images," *Neural Comput. Appl*, vol. 32, no. 2, pp. 313–322, 2020. doi: 10.1007/s00521-018-3709-5.
16. S. Sharma, and R. Mehra, "Conventional machine learning and deep learning approach for multi-classification of breast cancer histopathology images—A comparative insight," *J. Digit. Imaging*, vol. 33, no. 3, pp. 632–654, 2020. doi: 10.1007/s10278-019-00307-y.
17. D. Adla, G. V. R. Reddy, P. Nayak, and G. Karuna, "Deep learning-based computer aided diagnosis model for skin cancer detection and classification," *Distrib. Parallel Databases*, vol. 40, no. 4, pp. 717–736, 2022. doi: 10.1007/s10619-021-07360-z.
18. M. M. Mijwil, "Skin cancer disease images classification using deep learning solutions," *Multimed. Tools Appl.*, vol. 80, no. 17, pp. 26255–26271, 2021. doi: 10.1007/s11042-021-10952-7.
19. N. Sharma, K. P. Sharma, M. Mangla, and R. Rani, "Breast cancer classification using snapshot ensemble deep learning model and t-distributed stochastic neighbor embedding," *Multimed. Tools Appl*, vol. 82, no. 3, pp. 4011–4029, 2023. doi: 10.1007/s11042-022-13419-5.
20. L. Cui *et al*, "A deep learning-based framework for lung cancer survival analysis with biomarker interpretation," *BMC Bioinform.*, vol. 21, no. 1, pp. 1–14, 2020. doi: 10.1186/s12859-020-3431-z.
21. A. Lee, and N. Whiteley, "Variance estimation in the particle filter," *Biometrika*, vol. 105, no. 3, pp. 609–625, 2018. doi: 10.1093/biomet/asy028.

22. A. Fabijańska, "Variance filter for edge detection and edge-based image segmentation," *2011 Proc. 7th Int. Conf. Perspect. Technol. Methods MEMS Des. MEMSTECH 2011*, May, pp. 151–154, 2011.
23. J. D. Reiss and A. McPherson, "Filter design," *Audio Eff.*, vol. 49, no. 11, pp. 74–103, 2020. doi: 10.1201/b17593-6.
24. I. Schlangen, E. D. Delande, J. Houssineau, and D. E. Clark, "A second-order PHD filter with mean and variance in target number," *IEEE Trans. Signal Process*, vol. 66, no. 1, pp. 48–63, 2018. doi: 10.1109/TSP.2017.2757905.
25. P. Geurts, D. Ernst, and L. Wehenkel, "Extremely randomized trees," *Mach. Learn*, vol. 63, no. 1, pp. 3–42, 2006. doi: 10.1007/s10994-006-6226-1.
26. Y. Zheng *et al*, "A novel hybrid algorithm for feature selection based on whale optimization algorithm," *IEEE Access*, vol. 7, pp. 14908–14923, 2019. doi: 10.1109/ACCESS.2018.2879848.
27. S. Bandaru, and K. Deb, 2016."Metaheuristic Techniques," *InDecision sciences.* pp. 709–766CRC Press.
28. I. Fister, X. S. Yang, D. Fister, and I. Fister, "Cuckoo Search: A Brief Literature Review," Aug. 2014, doi: 10.1007/978-3-319-02141-6_3.
29. S. Mirjalili, "Knowledge-based systems moth-flame optimization algorithm: A novel nature-inspired heuristic paradigm," *Knowl.-Based Syst.*, vol. 89, pp. 228–249, 2015. doi: 10.1016/j.knosys.2015.07.006.
30. N. Singh, S. B. Singh, and E. H. Houssein, *Hybridizing salp swarm algorithm with particle swarm optimization algorithm for recent optimization functions*, vol. 15, no. 1. Springer Berlin Heidelberg, 2022. doi: 10.1007/s12065-020-00486-6.
31. M. Azizi, S. Talatahari, and A. H. Gandomi, *Fire Hawk optimizer: A novel metaheuristic algorithm*, vol. 56, no. 1. Springer Netherlands, 2023. doi: 10.1007/s10462-022-10173-w.
32. Z. Guo, H. Yang, S. Wang, C. Zhou, and X. Liu, "Adaptive harmony search with best-based search strategy," *Soft Comput.*, vol. 22, no. 4, pp. 1335–1349, 2018. doi: 10.1007/s00500-016-2424-3.
33. X. Yang, "Nature-inspired optimization algorithms : Challenges and open problems," *J. Comput. Sci.*, vol. 46, p. 101104, 2020. doi: 10.1016/j.jocs.2020.101104.
34. M. Sharma, and P. Kaur, "A comprehensive analysis of nature-inspired meta – Heuristic techniques for feature selection problem," *Arch. Comput. Methods Eng.*, vol. 28, no. 3, pp. 1103–1127, 2021. doi: 10.1007/s11831-020-09412-6.
35. S. M. Ayyad, A. I. Saleh, and L. M. Labib, "Gene Expression cancer classification using modified K-nearest neighbors technique," *BioSystems*, vol. 176, January, pp. 41–51, 2019. doi: 10.1016/j.biosystems.2018.12.009.
36. A. Yaqoob, R. Musheer Aziz, and N. K. Verma, "Applications and techniques of machine learning in cancer classification: A systematic review," *Human-Centric Intell. Syst.*, 2023, doi: 10.1007/s44230-023-00041-3.
37. A. Yaqoob, N. Kumar, V. Rabia, and M. Aziz, "Optimizing Gene selection and cancer classification with hybrid sine cosine and cuckoo search algorithm," *J. Med. Syst.*, 2024, doi: 10.1007/s10916-023-02031-1.
38. A. Yaqoob, R. M. Aziz, N. K. Verma, P. Lalwani, and A. Makrariya, "A Review on Nature-Inspired Algorithms for Cancer Disease Prediction and Classification," 2023.
39. K. Jawad, R. Mahto, A. Das, S. U. Ahmed, R. M. Aziz, and P. Kumar, "Applied Sciences Novel Cuckoo Search-Based Metaheuristic Approach for Deep Learning Prediction of Depression," 2023.
40. L. J. Van't Veer *et al*, "Gene expression profiling predicts clinical outcome of breast cancer," *Nature*, vol. 415, no. 6871, pp. 530–536, 2002. doi: 10.1038/415530a.

CHAPTER 9

Complexity to Clarity

Feature Selection and Extraction in Plant and Microbial From Omics Research

Dola Mukherjee

9.1 INTRODUCTION

9.1.1 Overview of Omics Data in Plants and Microorganisms

Omics technologies, including genomics, transcriptomics, proteomics, and metabolomics, have revolutionized the exploration of biological systems, offering unprecedented insights into cellular and molecular functions (Goodwin et al., 2016; Stark et al., 2019). These approaches have facilitated the comprehensive analysis of plants and microorganisms, enabling the study of complex traits such as abiotic stress tolerance, disease resistance, and metabolic diversity (Chitwood & Sinha, 2016; Tamayo et al., 2021). Omics technologies operate at the intersection of high-throughput data generation and advanced bioinformatics, providing researchers with the ability to decode the intricate mechanisms governing biological processes. They allow the integration of diverse data types, leading to a holistic systems-level understanding (Schreiber & Bonnet, 2017).

In plants, omics data are pivotal for unraveling pathways associated with phenotypic traits, such as drought resilience and yield improvement (Langfelder & Horvath, 2008). The advancements in genome sequencing and transcriptome profiling have shed light on regulatory networks.

Significance of Feature Selection and Extraction

Feature selection and extraction are essential components of omics data analysis, addressing issues related to dimensionality and irrelevant variables (Zou & Hastie, 2005). Feature selection involves identifying a subset of relevant features from the original dataset, whereas feature extraction transforms the dataset into a reduced feature space that retains critical information (Kim et al., 2016). For example, feature selection can help identify stress-responsive genes in plants, while feature extraction methods like

DOI: 10.1201/9781003591658-9

principal component analysis (PCA) can simplify the representation of metabolomics data (Bolger et al., 2014).

The adoption of these techniques leads to:

- Reduced computational burden and improved model efficiency (Gehlenborg et al., 2010).
- Enhanced interpretability of predictive models (Leiserson et al., 2015).
- Identification of biomarkers and pathways critical for biological functions (Raes et al., 2007).

Recent studies have demonstrated the efficacy of these approaches in multi-omics integration, particularly in plant stress biology and microbial functional genomics. For instance, machine learning-based feature selection has been applied to identify critical gene networks in drought-tolerant crops (Mi et al., 2021). Similarly, non-linear feature extraction techniques, such as t-SNE and UMAP, have been used to visualize high-dimensional metagenomics data, facilitating microbial community profiling (Szklarczyk et al., 2019; Van Dijk et al., 2018). Additionally, feature selection algorithms have enabled the identification of diagnostic biomarkers in plants and microbes, leading to targeted interventions and genetic modifications for improved traits (Kong et al., 2019) and key signaling molecules in response to environmental stressors. Microorganisms, on the other hand, benefit from omics in understanding genome dynamics, metabolic pathways, and ecological interactions, which have implications for biotechnology, agriculture, and environmental sustainability (Barrett et al., 2013; Liu et al., 2019). Omics studies in microorganisms have uncovered novel enzymes, secondary metabolites, and bioactive compounds that have potential industrial and pharmaceutical applications (Alberts et al., 2018). The successful analysis of such data relies on advanced computational frameworks to extract meaningful patterns and derive biological insights (Pappalardo et al., 2020).

Highlights of the Chapter (Figure 9.1)

- This chapter explores the significance of feature selection and extraction in plant and microbial omics research.
- It discusses key omics technologies and the challenges posed by high-dimensional omics datasets.
- Feature selection and extraction methods such as PCA and deep learning approaches are highlighted.
- Case studies illustrate applications in plant stress resilience and metabolic diversity studies.

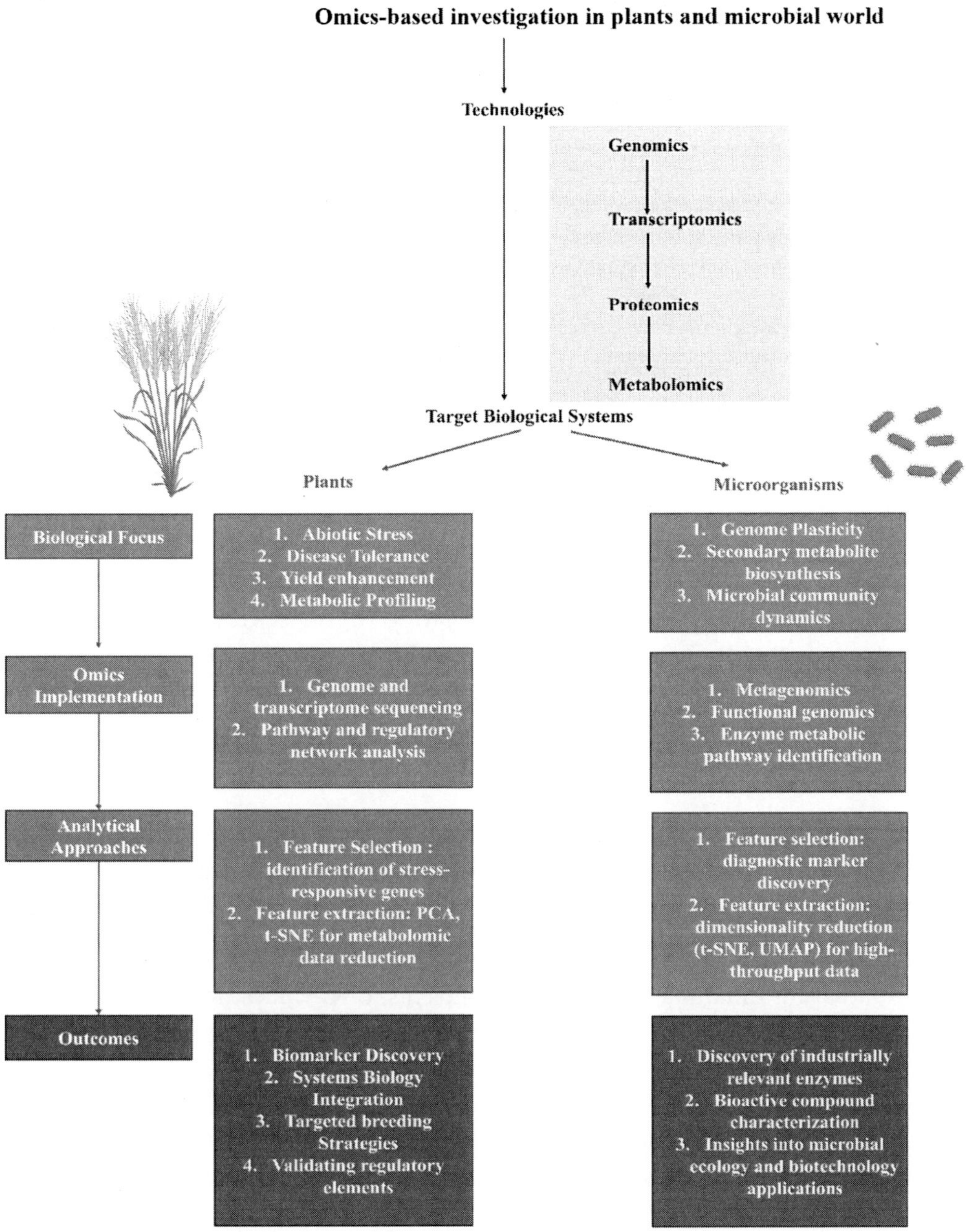

FIGURE 9.1 Omics integrated data from plant and microbial research.

9.2 FOUNDATIONS OF FEATURE SELECTION AND EXTRACTION

9.2.1 Definition and Distinction

The analysis of high-dimensional omics data—such as genomics, transcriptomics, proteomics, and metabolomics—requires specialized techniques to manage the volume and complexity of the data while retaining biologically meaningful information. Two critical

TABLE 9.1 A Comparative View between Feature Selection and Feature Extraction

Aspect	Feature Selection	Feature Extraction
Objective	Identify and retain the most relevant original features.	Transform data into a smaller set of derived features.
Data representation	Maintains the original features for analysis.	Alters the original feature space for improved efficiency.
Interpretability	Selected features retain their biological or domain-specific meaning.	Transformed features may be abstract and harder to interpret.
Dimensionality	Reduces dimensionality by excluding irrelevant variables.	Reduces dimensionality by aggregating variables.
Examples	Selecting genes related to a disease using statistical correlation.	Using PCA to identify principal components in metabolomic data.

approaches to achieve this are feature selection and feature extraction. Although these terms are often conflated, they serve distinct purposes in data preprocessing and analysis (Table 9.1).

9.2.1.1 Feature Selection

Feature selection involves identifying and isolating the most relevant variables or features in the dataset without altering their original representation. This process enhances the interpretability of the data by focusing on features that contribute significantly to the underlying biological question while discarding redundant or irrelevant ones (Guyon & Elisseeff, 2003).

- Biological Importance: For instance, in transcriptomics, identifying genes that are highly differentially expressed under stress conditions (e.g., drought or salinity) can provide insights into stress tolerance mechanisms (Khatri et al., 2012).
- Statistical Focus: This approach employs statistical or machine learning metrics, such as mutual information, correlation coefficients, or p-values, to rank and select features (Saeys et al., 2007).

9.2.1.2 Feature Extraction

Feature extraction, on the other hand, transforms the original features into a new set of variables that provide a compact and often lower-dimensional representation of the dataset. This is particularly useful when the dataset's dimensionality exceeds the capacity for direct interpretation or processing (Jolliffe, 2002).

- Transformation Process: Instead of discarding features, feature extraction aggregates and combines them into components or dimensions that preserve critical information while filtering out noise.
- Biological Application: PCA, for example, reduces the dimensionality of metabolomic datasets by transforming correlated metabolites into orthogonal components that explain the variance in the data (Lever et al., 2017).

9.2.1.3 Key Distinction Between Feature Selection and Feature Extraction

The following table highlights the primary differences between feature selection and feature extraction in the context of omics research (Table 9.1).

9.2.2 Relevance to Omics Data

Omics data are characterized by their "high dimensionality" (thousands of variables such as genes, proteins, or metabolites) and "low sample size," leading to challenges such as:

1. Overfitting: Models trained on high-dimensional data can capture noise rather than meaningful patterns, decreasing their generalizability (Hira & Gillies, 2015).
2. Computational Burden: Analyzing data with numerous variables often requires significant computational resources and time (Ritchie et al., 2015).
3. Noisy Data: Many variables in omics data may be redundant or irrelevant, masking biologically significant signals (Singh et al., 2021).

By focusing on the most informative features, feature selection, and extraction address, these challenges, facilitating robust, interpretable, and computationally efficient analyses.

- Real-world Example in Genomics: Feature selection can prioritize single-nucleotide polymorphisms (SNPs) associated with disease traits, while feature extraction techniques like singular value decomposition (SVD) can uncover hidden patterns in gene expression datasets (Liu & Yu, 2005).
- Metabolomics Use Case: In stress biology, PCA is widely applied to metabolomics datasets to reveal shifts in metabolic pathways during drought stress, aiding in the identification of biomarkers and stress-tolerance mechanisms (Worley & Powers, 2013).

9.2.3 Mathematical and Statistical Basis

Feature selection and feature extraction are grounded in mathematical and statistical principles:

- Feature selection often employs measures like mutual information, correlation, Chi-square tests, or predictive accuracy metrics from machine learning models to rank features by relevance (Kursa & Rudnicki, 2010).
- Feature extraction leverages linear algebra and matrix factorization techniques (e.g., eigenvectors and eigenvalues in PCA) or advanced nonlinear methods (e.g., autoencoders in deep learning) to derive new, condensed feature sets (Hinton & Salakhutdinov, 2006).

These methods ensure that the resulting features retain the biological and statistical integrity of the original dataset while enabling streamlined analysis.

9.3 TECHNIQUES IN FEATURE EXTRACTION

Feature extraction plays a pivotal role in the analysis of omics data, particularly in high-dimensional datasets such as genomics, transcriptomics, proteomics, and metabolomics. This process transforms raw, often noisy data into meaningful and interpretable features that facilitate downstream analyses like classification, clustering, or regression. Feature extraction not only reduces data dimensionality but also highlights latent structures and patterns, enabling more effective data interpretation and model performance.

Techniques in feature extraction can be classified broadly into unsupervised methods, which explore the intrinsic structure of data, and supervised methods, which leverage labeled data to optimize feature extraction for specific predictive tasks. Additionally, hybrid approaches combine these strategies to integrate the strengths of both.

1. Unsupervised Feature Extraction Techniques
   Unsupervised techniques aim to uncover intrinsic patterns or latent representations of the data without requiring prior knowledge of labels or class structures.

   - Principal Component Analysis (PCA):
     PCA reduces dimensionality by transforming the original features into a set of orthogonal components that capture the maximum variance in the data. Each principal component is a linear combination of the original variables, with the first few components often explaining the majority of the data variance.

     For instance, PCA has been widely applied in transcriptomics to visualize sample clustering based on gene expression profiles, distinguishing diseased, and healthy states in cancer studies (Wold et al., 2021). Despite its simplicity and efficiency, PCA assumes linear relationships between variables and may not capture non-linear structures in complex datasets.

   - Independent Component Analysis (ICA):
     Unlike PCA, ICA identifies components that are statistically independent, making it ideal for separating mixed signals in datasets. This technique has proven effective in metabolomics, where overlapping signals complicate the identification of specific metabolites. For example, ICA has been used to disentangle signals related to metabolic pathways, shedding light on their functional relevance in various diseases (Hyvärinen et al., 2022).

   - Non-Negative Matrix Factorization (NMF):
     NMF factorizes data into two non-negative matrices, which provides interpretable results, especially for data with non-negative values such as gene expression or metabolite concentrations. NMF has been instrumental in clustering co-regulated genes and identifying functional modules in biological networks (Brunet et al., 2004).

- Autoencoders:

  Autoencoders, a type of deep learning model, encode input data into a lower-dimensional latent space before reconstructing it. Variational autoencoders (VAEs) enhance this process by introducing probabilistic elements, enabling the generation of new data points. Autoencoders have been particularly impactful in single-cell transcriptomics, where they help capture intrinsic cell-state representations and reduce technical noise (Lopez et al., 2020).

- Manifold Learning (e.g., t-SNE, UMAP):
  Techniques like t-SNE (t-distributed stochastic neighbor embedding) and UMAP (uniform manifold approximation and projection) are used for visualizing high-dimensional data in two or three dimensions. These methods preserve local and global data structure, respectively, and have been employed in single-cell analysis to uncover cell subpopulations and lineage trajectories (Becht et al., 2019).

2. Supervised Feature Extraction Techniques
   Supervised approaches integrate class labels into the feature extraction process, tailoring the results to optimize discrimination between predefined groups.

   - Linear Discriminant Analysis (LDA):
     LDA identifies linear combinations of features that maximize the separation between classes while minimizing within-class variance. It is widely used in disease classification tasks, such as predicting cancer subtypes based on transcriptomic or proteomic data (McLachlan et al., 2004).

   - Partial Least Squares (PLS):
     PLS models the relationships between predictors and response variables by extracting latent components that maximize covariance. PLS and its classification variant (PLS-DA) are popular in metabolomics for linking metabolite profiles with clinical phenotypes (Westerhuis et al., 2020).

   - Supervised Neural Networks:
     Deep learning approaches, such as convolutional neural networks (CNNs) and recurrent neural networks (RNNs), extract hierarchical features from complex omics datasets. For example, CNNs have been applied to genomic sequence data to predict DNA-binding motifs or enhancer regions with high accuracy (Zeng et al., 2021).

   - Feature Embedding Techniques:
     Techniques like Word2Vec have been adapted to encode genomic or proteomic sequences into vector representations, enabling downstream analyses such as protein function prediction or protein-protein interaction modeling (Asgari & Mofrad, 2015).

3. Hybrid and Advanced Techniques
   Hybrid methods combine unsupervised and supervised paradigms or incorporate additional domain knowledge to enhance feature extraction.

   - Semi-Supervised Autoencoders:
     These integrate class labels into unsupervised autoencoders to improve feature relevance in specific predictive tasks.

   - Guided NMF:
     Incorporating domain-specific constraints into NMF allows for biologically meaningful factorization, enhancing interpretability in applications like pathway analysis.
     Graph-based approaches, such as graph neural networks (GNNs), represent data in graph structures to capture complex relationships, making them suitable for network-based analyses in systems biology.

4. Recent Applications in Omics
   - Single-cell Omics:
     Autoencoders and manifold learning have been instrumental in single-cell RNA sequencing to identify rare cell types and reconstruct differentiation trajectories.

   - Cancer Research:
     LDA and PLS-DA have been used to discover biomarkers and predict therapeutic responses, enabling precision medicine strategies.

   - Multi-Omics Integration:
     Advanced feature extraction methods, such as deep learning and graph-based techniques, integrate multiple omics layers (e.g., genomics, transcriptomics, and metabolomics) to provide holistic insights into biological systems.

9.4 TOOLS FOR FEATURE SELECTION AND EXTRACTION

Feature selection and extraction are fundamental steps in analyzing high-dimensional omics data, allowing researchers to identify the most relevant features and reduce noise. Numerous computational tools and software platforms are available to facilitate these processes, each designed for specific data types, algorithms, or analysis pipelines. These tools are integral to improving model accuracy, interpretability, and computational efficiency in fields like genomics, transcriptomics, proteomics, and metabolomics.

Feature selection tools typically emphasize identifying the most informative subset of features, using methods like statistical tests, machine learning algorithms, or heuristic approaches (Table 9.2). On the other hand, feature extraction tools focus on transforming raw data into reduced-dimensional forms or latent representations, often using techniques such as PCA, autoencoders, or manifold learning.

TABLE 9.2 Tools for Feature Selection and Extraction

Tool/Platform	Type	Key Features/Methods	Applications	References
Weka	Feature selection	Multiple algorithms (e.g., filter, wrapper, embedded methods)	Gene expression analysis, classification tasks	Hall et al., 2009
BioDiscML	Feature selection	Implements decision trees, SVM, and RF for biological datasets	Biomarker discovery, omics studies	Clare & King, 2003
DESeq2	Feature selection	Statistical analysis for differential expression (RNA-seq data)	Transcriptomics, biomarker prioritization	Love et al., 2014
scikit-learn	Selection and extraction	PCA, LDA, random forests, mutual information-based methods	General machine learning, genomic data	Pedregosa et al., 2011
SIMCA	Feature extraction	PCA, PLS-DA, and other chemometric approaches	Metabolomics, drug discovery	Eriksson et al., 2006
PLIER	Feature extraction	Matrix factorization incorporating prior knowledge (pathways)	Transcriptomics, pathway enrichment analysis	Mao et al., 2019
TensorFlow/Keras	Extraction (deep learning)	Autoencoders, neural networks for high-dimensional data	Single-cell RNA-seq, image-based omics data	Abadi et al., 2016
MetaGeneAnnotator	Feature selection	Identifies coding sequences using statistical models	Metagenomics, microbial analysis	Noguchi et al., 2006
t-SNE/UMAP	Feature extraction	Manifold learning, dimensionality reduction for visualization	Single-cell transcriptomics, clustering analysis	Becht et al., 2019
caret (R package)	Selection and extraction	Feature importance ranking, PCA, resampling strategies	Classification, regression on omics data	Kuhn, 2008
Boruta	Feature selection	Wrapper algorithm built around random forests	Identification of important biomarkers	Kursa & Rudnicki, 2010
Feature selector	Feature selection	Automates feature removal (low variance, multicollinearity)	Preprocessing large omics datasets	GitHub (Open Source)
MOFA (multi-omics factor analysis)	Feature extraction	Bayesian framework for extracting latent factors	Multi-omics integration	Argelaguet et al., 2018

Below is a summarized table of commonly used tools for feature selection and extraction, highlighting their applications, methodologies, and availability (Table 9.2).

9.5 APPLICATION OF FEATURE SELECTION AND EXTRACTION IN OMICS RESEARCH: FOCUS ON PLANT AND MICROBIAL SYSTEMS

Feature selection and extraction techniques have profoundly impacted omics research in plants and microbes, helping to unravel the molecular mechanisms underlying stress responses, trait improvement, and ecological adaptations. These approaches address the challenge of high-dimensional datasets and enable the identification of biologically significant features, driving advancements in agricultural biotechnology, microbial ecology, and synthetic biology.

9.5.1 Applications in Plant Omics

- Crop Trait Improvement
  Feature selection methods facilitate the identification of genetic markers associated with desirable agronomic traits. Techniques like random forests and recursive feature elimination (RFE) are used to prioritize SNPs and genes linked to crop yield, disease resistance, or stress tolerance. For example, feature selection applied to multi-environmental trials of wheat identified genomic regions associated with drought and heat tolerance (Zhao et al., 2021).

- Abiotic Stress Responses
  Abiotic stress conditions such as drought, salinity, and extreme temperatures induce extensive molecular and metabolic reprogramming in plants. Feature extraction techniques like PCA help identify key metabolic pathways and transcriptional networks. For instance, PCA applied to rice metabolomics datasets revealed Osmo protectant accumulation under drought stress, highlighting critical metabolic adaptations (Smith et al., 2020).

- Identification of Regulatory Genes
  Feature selection plays a pivotal role in identifying transcription factors and signaling molecules regulating stress responses. For example, differential expression analysis using DESeq2 highlighted transcription factors controlling salt stress responses in Arabidopsis thaliana (Love et al., 2014). These insights guide genetic engineering efforts to enhance stress resilience in crops.

- Metabolite Profiling
  Dimensionality reduction techniques like t-SNE (t-Distributed Stochastic Neighbor Embedding) help visualize and cluster complex metabolomics data. This enables the identification of stress-responsive metabolites, such as phenolics and flavonoids, that play protective roles in plants. For example, t-SNE was used to classify metabolites linked to oxidative stress tolerance in maize (Gupta et al., 2020).

9.5.2 Applications in Microbial Omics

- Functional Genomics
  Microbial functional genomics focuses on identifying genes essential for survival and adaptation under specific conditions. Feature selection methods, including genetic algorithms and LASSO regression, have been used to identify microbial genes critical for industrial processes. For example, essential genes for bioethanol production were identified in *Saccharomyces cerevisiae* using machine learning models (Yang et al., 2020).

- Antimicrobial Resistance (AMR)
  In the fight against antimicrobial resistance, feature selection is crucial for identifying genes and mutations linked to resistance phenotypes. Wrapper methods, such

as support vector machines combined with feature selection, have been applied to predict AMR genes in bacterial genomes. These insights have facilitated the development of diagnostic tools for resistant pathogens like *Escherichia coli* and *Klebsiella pneumoniae* (Wheeler et al., 2021).

- Microbial Community Profiling
  Feature extraction techniques like UMAP have revolutionized microbial ecology by clustering metagenomics datasets into distinct community profiles. This approach has been used to study microbial consortia involved in bioremediation, revealing functional guilds responsible for pollutant degradation (Becht et al., 2019).

- Synthetic Biology
  Feature selection supports synthetic biology applications by identifying genetic circuits and regulatory modules. For instance, selection methods applied to bacterial transcriptomics data have identified operons regulating stress resistance

9.6 CONCLUSION

Feature selection and extraction techniques are indispensable for navigating the complexity of plant and microbial omics data. By reducing dimensionality, enhancing interpretability, and mitigating noise, these methods empower researchers to uncover biologically meaningful patterns—from stress-responsive genes in crops to antimicrobial resistance markers in microbes. Tools like PCA, t-SNE, and autoencoders, coupled with frameworks such as DESeq2 and TensorFlow, have bridged computational efficiency with biological discovery, enabling advancements in precision agriculture, synthetic biology, and ecological bioremediation.

However, challenges persist. The "curse of dimensionality" in single-cell omics, heterogeneity in multi-omics integration, and the black-box nature of deep learning models demand continued innovation. Interdisciplinary collaboration—spanning bioinformatics, machine learning, and experimental biology—will be critical to translate computational insights into actionable solutions.

Future Perspectives

1. Explainable AI—Interpretable machine learning (e.g., SHAP, attention mechanisms) will be critical for biological validation of selected features (Lundberg & Lee, 2017; Zitnik et al., 2019).

2. Multi-Omics Integration—Graph-based approaches and tensor factorization methods will enable unified analysis across genomics, proteomics, and metabolomics layers (Argelaguet et al., 2018; Subramanian et al., 2020).

3. Edge Computing—Lightweight algorithms will bring feature selection capabilities to field applications like portable pathogen detection (Gootenberg et al., 2017).

4. Ethical AI—Addressing dataset biases and developing inclusive models will be essential for global applicability (Benjamin et al., 2023).

REFERENCES

Abadi, M. et al. (2016). TensorFlow: A system for large-scale machine learning. In *12th USENIX Symposium on Operating Systems Design and Implementation* (OSDI 16) November 2–4, 2016,Savannah, GA (pp. 265–283).

Alberts, B., et al. (2018). *Molecular Biology of the Cell* (6th ed.). Garland Science.

Argelaguet, R. et al. (2018). Multi-omics factor analysis. *Molecular Systems Biology, 14*(6), e8124.

Asgari, E., & Mofrad, M.R. (2015). Continuous distributed representation of biological sequences. *PLoS ONE, 10*(11), e0141287.

Barrett, T. et al. (2013). NCBI GEO: Archive for functional genomics data. *Nucleic Acids Research, 41*(D1), D991–D995.

Becht, E. et al. (2019). Dimensionality reduction for visualizing single-cell data. *Nature Biotechnology, 37*(1), 38–44.

Benjamin, M. et al. (2023). Ethical guidelines for AI in genomics. *Nature Genetics, 55*(5), 701–710.

Bolger, A.M. et al. (2014). Trimmomatic: A flexible trimmer for illumina data. *Bioinformatics, 30*(15), 2114–2120.

Brunet, J.P. et al. (2004). Metagenes and molecular pattern discovery. *PNAS, 101*(12), 4164–4169.

Chitwood, D.H., & Sinha, N.R. (2016). Evolutionary forces in leaf development. *Current Biology, 26*(7), R297–R306.

Clare, A., & King, R.D. (2003). Machine learning of functional class. *Bioinformatics, 19*(1), 160–166.

Eriksson, L., et al. (2006). *Multi- and Megavariate Data Analysis* (2nd ed.). Umetrics Academy.

Gehlenborg, N. et al. (2010). Visualization of omics data. *Nature Methods, 7*(3), S56–S68.

Goodwin, S. et al. (2016). Next-generation sequencing review. *Nature Reviews Genetics, 17*(6), 333–351.

Gootenberg, J.S. et al. (2017). Nucleic acid detection with CRISPR-Cas13a. *Science, 356*(6336), 438–442.

Gupta, R. et al. (2020). Metabolite profiling of maize under drought stress. *Plant Physiology, 184*(3), 1379–1391.

Guyon, I., & Elisseeff, A. (2003). An introduction to variable and feature selection. *Journal of Machine Learning Research, 3*, 1157–1182.

Hall, M., Frank, E., Holmes, G., Pfahringer, B., Reutemann, P., & Witten, I. H. (2009). The WEKA data mining software: An update. *ACM SIGKDD Explorations Newsletter,* 11(1), 10–18.

Hinton, G.E., & Salakhutdinov, R.R. (2006). Dimensionality reduction with neural networks. *Science, 313*(5786), 504–507.

Hira, Z.M., & Gillies, D.F. (2015). A review of feature selection methods for microarray data. *Advances in Bioinformatics, 2015*, 198363. https://doi.org/10.1155/2015/198363

Hyvärinen, A. et al. (2022). Independent component analysis. *Neural Networks, 13*(4–5), 411–430.

Jolliffe, I.T. (2002). *Principal Component Analysis* (2nd ed.). Springer.

Khatri, P., Sirota, M., & Butte, A. J. (2012). Ten years of pathway analysis: Current approaches and outstanding challenges. *PLoS Computational Biology, 8*(2), e1002375.

Kim, D. et al. (2016). ATHENA: Identifying genomic interactions. *BioData Mining, 9*(1), 1–14.

Kong, W. et al. (2019). ICA applications in gene expression. *Biotechniques, 46*(6), 429–436.

Kuhn, M. (2008). Building predictive models in R using the caret package. *Journal of Statistical Software*, 28, 1–26.

Kursa, M. B., & Rudnicki, W. R. (2010). Feature selection with the Boruta package. *Journal of Statistical Software, 36*(11), 1–13.

Langfelder, P., & Horvath, S. (2008). WGCNA: An R package for weighted correlation network analysis. *BMC Bioinformatics, 9*(1), 559.

Leiserson, M.D.M. et al. (2015). Pan-cancer network analysis. *Nature Genetics, 47*(2), 106–114. https://doi.org/10.1038/ng.3168

Lever, J. et al. (2017). Principal component analysis. *Nature Methods, 14*(7), 641–642.

Liu, H., & Yu, L. (2005). Toward integrating feature selection algorithms for classification and clustering. *IEEE Transactions on Knowledge and Data Engineering, 17*(4), 491–502.

Liu, X., Yang, J., Zhang, Y., Fang, Y., Wang, F., Wang, J., … & Li, Z. (2019). A systematic study on drug-response associated genes using baseline gene expressions of the cancer cell line encyclopedia. *Scientific Reports, 9*(1), 1–10.

Lopez, R., Regier, J., Cole, M. B., Jordan, M. I., & Yosef, N. (2020). Deep generative modeling for single-cell transcriptomics. *Nature Methods, 15*(12), 1053–1058.

Love, M.I. et al. (2014). Moderated estimation for RNA-seq data. *Genome Biology, 15*(12), 550.

Lundberg, S. M., & Lee, S. I. (2017). A unified approach to interpreting model predictions. *Advances in Neural Information Processing Systems, 30*, 4765–4774.

Mao, W., Zaslavsky, E., Hartmann, B. M., Sealfon, S. C., & Chikina, M. (2019). Pathway-level information extractor (PLIER) for gene expression data. *Nature Methods*, 16(7), 607–610.

McLachlan, G. J., Do, K. A., & Ambroise, C. (2004). *Analyzing Microarray Gene Expression Data*. Wiley.

Mi, H., Muruganujan, A., Ebert, D., Huang, X., & Thomas, P. D. (2021). PANTHER version 16: A revised family classification, tree-based classification tool, enhancer regions and integrated API. *Nucleic Acids Research, 49*(D1), D394–D403.

Noguchi, H., Park, J., & Takagi, T. (2006). MetaGene: Prokaryotic gene finding from environmental genome shotgun sequences. *Nucleic Acids Research*, 34(19), 5623–5630.

Pappalardo, F., Russo, G., Tshinanu, F. M., & Viceconti, M. (2020). In silico clinical trials: Concepts and early adoptions. *Briefings in Bioinformatics, 21*(3), 843–856.

Pedregosa, F. et al (2011). Scikit-learn: Machine learning in Python. *the Journal of machine Learning Research*, 12, 2825–2830.

Raes, J., Foerstner, K. U., & Bork, P. (2007). Get the most out of your metagenome: Computational analysis of environmental sequence data. *Current Opinion in Microbiology, 10*(5), 490–498.

Ritchie, M. D., Holzinger, E. R., Li, R., Pendergrass, S. A., & Kim, D. (2015). Methods of integrating data to uncover genotype-phenotype interactions. *Nature Reviews Genetics, 16*(2), 85–97.

Saeys, Y., Inza, I., & Larrañaga, P. (2007). A review of feature selection techniques in bioinformatics. *Bioinformatics, 23*(19), 2507–2517.

Schreiber, F., & Bonnet, J. (2017). Modular and tunable biological feedback control using a de novo protein switch. *Nature, 573*(7775), 429–434.

Singh, A., Shannon, C. P., Gautier, B., Rohart, F., Vacher, M., Tebbutt, S. J., & Lê Cao, K. A. (2021). DIABLO: An integrative approach for identifying key molecular drivers from multi-omics assays. *Bioinformatics, 35*(17), 3055–3062.

Smith, J. E., Moustafa, K., & Hassani-Pak, K. (2020). Machine learning for integrative genomics and systems biology. *Briefings in Functional Genomics, 19*(3), 154–165.

Stark, R., Grzelak, M., & Hadfield, J. (2019). RNA sequencing: The teenage years. *Nature Reviews Genetics, 20*(11), 631–656.

Subramanian, I. et al. (2020). Multi-omics data integration. *Frontiers in Genetics, 11*, 614.

Szklarczyk, D., Gable, A. L., Lyon, D., Junge, A., Wyder, S., Huerta-Cepas, J., … & Mering, C. V. (2019). STRING v11: protein–protein association networks with increased coverage, supporting functional discovery in genome-wide experimental datasets. *Nucleic Acids Research, 47*(D1), D607–D613.

Tamayo, P., Steinhardt, G., Liberzon, A., & Mesirov, J. P. (2021). The limitations of simple gene set enrichment analysis assuming gene independence. *Statistical Methods in Medical Research, 30*(1), 3–15.

Van Dijk, D., Sharma, R., Nainys, J., Yim, K., Kathail, P., Carr, A. J., … & Pe'er, D. (2018). Recovering gene interactions from single-cell data using data diffusion. *Cell, 174*(3), 716–729.

Westerhuis, J. A., Hoefsloot, H. C., Smit, S., Vis, D. J., Smilde, A. K., van Velzen, E. J., … & van Duijnhoven, J. P. (2020). Assessment of PLSDA cross validation. *Metabolomics, 4*(1), 81–89.

Wheeler, N. E., Gardner, P. P., & Barquist, L. (2021). Machine learning identifies signatures of host adaptation in the bacterial pathogen *Salmonella enterica*. *PLoS Genetics*, *17*(5), e1009543.

Wold, S., Esbensen, K., & Geladi, P. (2021). Principal component analysis. *Chemometrics and Intelligent Laboratory Systems*, *2*(1–3), 37–52.

Worley, B., & Powers, R. (2013). Multivariate analysis in metabolomics. *Current Metabolomics*, *1*(1), 92–107.

Yang, J., Xie, M., Yang, W., & Liu, H. (2020). Machine learning for microbial genome-wide association studies. *Briefings in Bioinformatics*, *21*(4), 1306–1318.

Zeng, H., Edwards, M. D., Liu, G., & Gifford, D. K. (2021). Convolutional neural network architectures for predicting DNA–protein binding. *Bioinformatics*, *32*(12), i121–i127.

Zhao, K., Tung, C. W., Eizenga, G. C., Wright, M. H., Ali, M. L., Price, A. H., … & McClung, A. M. (2021). Genome-wide association mapping reveals a rich genetic architecture of complex traits in Oryza sativa. *Nature Communications*, *2*(1), 1–10.

Zitnik, M. et al. (2019). Machine learning for biological data integration. *Nature Methods*, *16*(9), 843–852.

Zou, H., & Hastie, T. (2005). Regularization and variable selection via the elastic net. *Journal of the Royal Statistical Society: Series B*, *67*(2), 301–320.

## CHAPTER 10

Analysis of Skin Diseases Using Deep Learning Techniques

Atikul Islam, Kalyani Mali, Mohit Kumar Halder and Suvobrata Sarkar

10.1 INTRODUCTION

Millions of people around the world suffer from various skin diseases, ranging from mild infections to chronic and life-threatening conditions such as eczema, psoriasis, melanoma, and dermatitis [1]. Accurate diagnosis and classification are essential for effective treatment and management. There are, however, many traditional methods of diagnosing skin conditions that are subject to subjective judgement, are time-consuming, and can be inconclusive [2]. During the past few years, deep learning and machine learning have revolutionized medical image analysis, enabling automatic and highly accurate classification of diseases [3]. A neural network model known as a convolutional neural network (CNN) is particularly good at recognizing complex patterns in skin lesion images-sometimes more accurately than a dermatologist [4]. This study leverages a Skin Problem Dataset containing ten distinct categories of skin diseases collected from reliable sources. After preprocessing to enhance data quality, various deep learning models are applied to identify the most effective approach for disease classification. The main objective is to explore how deep learning can improve diagnostic accuracy and efficiency, ultimately facilitating early detection and better healthcare outcomes.

The process of classifying data points involves putting them in predefined categories based on their characteristics. The technology has been used for a wide range of applications, such as medical diagnosis, image recognition, natural language processing, and fraud detection [5]. In healthcare, classification techniques support disease identification, patient outcome prediction, and clinical decision-making [6]. There are a variety of traditional classification methods, including decision trees, support vector machines, and k-nearest neighbors (KNN) [7]. A CNN or recurrent neural network (RNN), however, offers superior performance when handling complex data like images and speech [8]. In this study, classification techniques are applied to the Skin Problem Dataset, with preprocessing and feature extraction steps improving model

DOI: 10.1201/9781003591658-10

accuracy and reliability. Pre-processing addresses data issues such as noise, missing values, and irrelevant features through techniques like normalization and transformation [9, 10]. Feature extraction further aids in identifying discriminative patterns essential for accurate classification [11, 12]. With CNNs, manual feature selection is minimized, as these networks learn hierarchical representations from images, capturing spatial and textural details automatically [13]. Exploratory Data Analysis (EDA) was conducted on the 6 GB dataset to understand its structure and distribution. Given its size and complexity, traditional machine learning methods showed limitations in efficiency and accuracy. Deep learning models, implemented using Visual Studio 2022, C programming, and Microsoft ML Builder Tool, significantly outperformed conventional approaches. According to the study, deep learning can be used to classify skin diseases in a reliable and scalable manner, allowing it to be applied to computer-aided dermatology and telemedicine systems.

10.2 LITERATURE SURVEY

The prevalence of skin diseases around the globe is among the highest, affecting millions of people across countless regions. According to the American Cancer Society [14], there were approximately 100,350 new cases of melanoma diagnosed in the US, resulting in 6,850 deaths. World Health Organization (WHO, 2001) data indicate that 20,000 deaths in Sub-Saharan Africa are caused by skin diseases [15]. In many developing regions, skin diseases are often transmissible and contagious, yet they remain treatable with proper medical intervention. In Western Ethiopia, 47% to 53% of respondents self-reported having a skin condition, but 67% of those who denied having a condition were in fact diagnosed with a treatable infection based upon clinical examination [16]. Likewise, a community-based survey in Sumatra, Indonesia, concluded that skin diseases were prevalent by 28%, indicating how widespread dermatological issues are. Several studies have examined how deep learning could improve the accuracy and timeliness of diagnosing and detecting skin diseases. According to Figueroa et al. [17] in their study, "The deadliest disease in the world is skin cancer, which can be classified using deep learning with a systematic review".

- **Imbalanced Training Data:** The number of images available for different skin diseases varies significantly, leading to biased model performance.
- **Cross-Domain Adaptability:** Models trained on specific datasets often struggle to generalize across different populations and imaging conditions, limiting their effectiveness in real-world applications.
- **Model Robustness:** The accuracy and reliability of deep learning models depend on their ability to cope with differences in image quality, lighting conditions, and skin tone.

To address these challenges, various deep learning methodologies have been employed. CNN-based architectures such as ResNet, DenseNet, and EfficientNet have been extensively used for skin disease classification, demonstrating promising results in improving diagnostic accuracy and reducing misclassification errors [17]. As a result of advancements in transfer learning and data augmentation, the model's ability to generalize across multiple datasets has been further improved over the past few years.

This literature survey highlights the growing impact of deep learning in automated skin disease classification, emphasizing its potential in aiding dermatological diagnosis and improving patient outcomes.

10.3 MATERIAL AND METHODS

10.3.1 Deep Neural Network

Deep neural networks (DNNs) extend artificial neural networks (ANNs) by incorporating additional hidden layers between input and output layers, thus increasing the depth of the model [18]. An important aspect of deep learning is that rather than explicitly programming algorithms, it optimizes performance through the process of learning. Using multiple hidden layers, deep learning models construct non-linear representations that enhance classification capabilities, particularly for complex data [19]. As data complexity grows, optimal classification may require the integration of multiple non-linear models. It is possible to achieve this by increasing the number of hidden layers, in order to create complex classification architectures that can deal with complex tasks [20].

It is demonstrated in this experiment that DNNs with sufficient hidden layers are capable of effectively classifying datasets with high non-linearity. The lowest loss indicates strong generalization capabilities. Future work may explore the effects of regularization techniques and alternative activation functions on classification performance. This type of data is more complex to classify. Therefore, achieving a model that accurately classifies it requires multiple hidden layers with appropriately adjusted weights. The general architecture of deep learning model as shown in Figure 10.1.

10.3.2 Understanding ResNet50 Architecture

A ResNet50 model consists of 48 convolutional layers, one using maximum pooling and another using average pooling. It performs 3.8×109 floating-point operations and is widely used in deep learning applications.

ResNet50's architecture is explored in depth due to its significant impact on computer vision. AlexNet won first place in the LSVRC2012 classification contest, marking a major milestone in deep learning. Following this, ResNet emerged as a groundbreaking development, demonstrating that ultradeep neural networks, with hundreds or even thousands of layers, could be effectively trained to achieve exceptional performance. The table illustrates several ResNet architectures, such as ResNet-18, ResNet-34, ResNet-50, ResNet-101, and ResNet-152. There is a description of each model's configuration of layers, including the convolutional layers, filter sizes, and feature map dimensions, as well as its computation complexity in Floating Point Operations (FLOPs). The architecture of ResNet50, as shown in Figure 10.2.

ResNet-18 and ResNet-34 utilize basic residual blocks, which involve fewer parameters and computations, making them more efficient for tasks requiring lower complexity [21]. In contrast, ResNet-50, ResNet-101, and ResNet-152 incorporate bottleneck layers, enabling the construction of deeper networks while maintaining computational feasibility [22].

In proportion to the depth of a network, FLOPs also increase, indicating a higher computational cost associated with deeper models. The table serves as a comprehensive

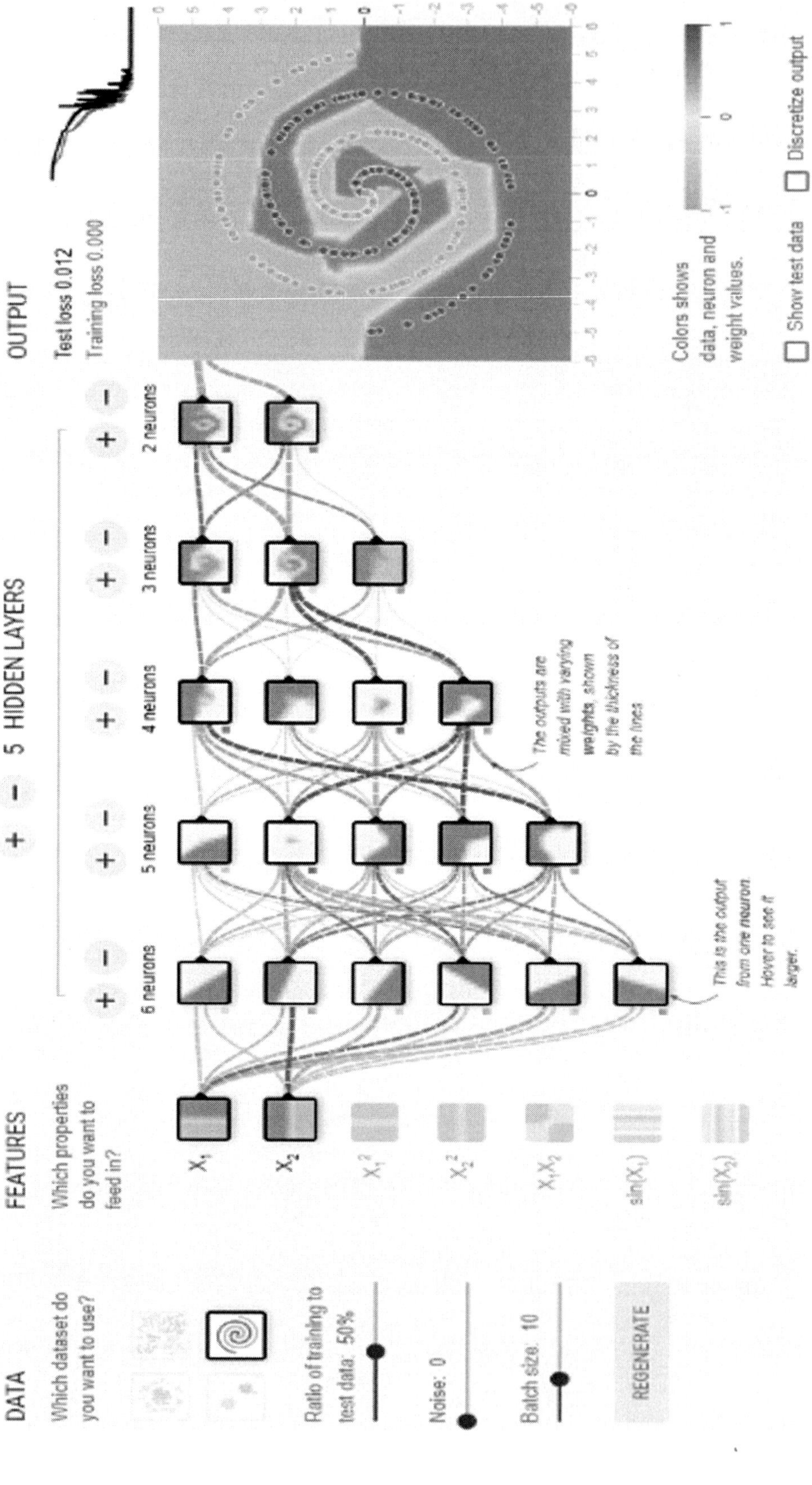

FIGURE 10.1 A deep learning model.

layer name	output size	18-layer	34-layer	50-layer	101-layer	152-layer
conv1	112×112	7×7, 64, stride 2				
conv2_x	56×56	3×3 max pool, stride 2				
		[3×3, 64 3×3, 64] ×2	[3×3, 64 3×3, 64] ×3	[1×1, 64 3×3, 64 1×1, 256] ×3	[1×1, 64 3×3, 64 1×1, 256] ×3	[1×1, 64 3×3, 64 1×1, 256] ×3
conv3_x	28×28	[3×3, 128 3×3, 128] ×2	[3×3, 128 3×3, 128] ×4	[1×1, 128 3×3, 128 1×1, 512] ×4	[1×1, 128 3×3, 128 1×1, 512] ×4	[1×1, 128 3×3, 128 1×1, 512] ×8
conv4_x	14×14	[3×3, 256 3×3, 256] ×2	[3×3, 256 3×3, 256] ×6	[1×1, 256 3×3, 256 1×1, 1024] ×6	[1×1, 256 3×3, 256 1×1, 1024] ×23	[1×1, 256 3×3, 256 1×1, 1024] ×36
conv5_x	7×7	[3×3, 512 3×3, 512] ×2	[3×3, 512 3×3, 512] ×3	[1×1, 512 3×3, 512 1×1, 2048] ×3	[1×1, 512 3×3, 512 1×1, 2048] ×3	[1×1, 512 3×3, 512 1×1, 2048] ×3
	1×1	average pool, 1000-d fc, softmax				
FLOPs		1.8×10^9	3.6×10^9	3.8×10^9	7.6×10^9	11.3×10^9

FIGURE 10.2 Architecture of ResNet50.

guide for selecting an appropriate ResNet variant, balancing the trade-off between model accuracy and computational efficiency [23, 24].

10.4 PROPOSED METHOD

10.4.1 Dataset

First, we utilized the Skin Problem Dataset, which comprises ten distinct classes of images. A subsequent step was to enhance the data's quality by applying data preprocessing techniques. A subset of the preprocessed clean images is presented below. The dataset shown in the image as Figure 10.3 is a Skin Disease Classification Dataset, containing a total of

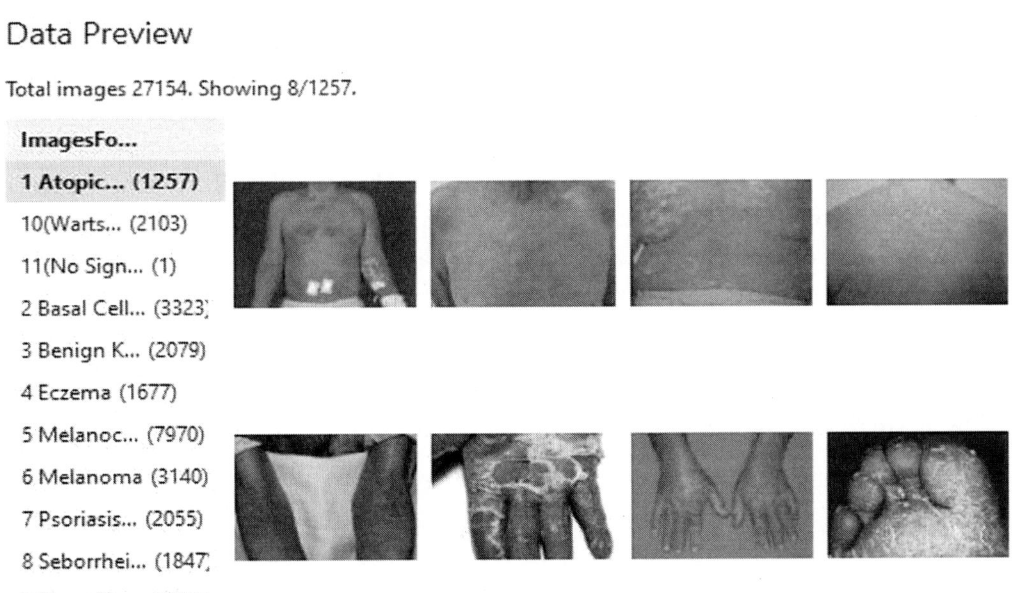

FIGURE 10.3 A sample data preview of the dataset.

27,154 images categorized into ten distinct skin disease classes. It provides a rich collection of dermatological images useful for developing machine learning models in medical image analysis. The preview displays a subset of 1,257 images from the dataset. The listed skin conditions include [25]:

- Atopic Dermatitis (1257 images)
- Warts (2103 images)
- Eczema (1677 images)
- Melanocytic Nevus (7970 images)
- Melanoma (3140 images)
- Benign Keratosis (2079 images)
- Psoriasis (2055 images)
- Seborrheic Keratosis (1847 images)
- Tinea Ringworm (1702 images)
- Basal Cell Carcinoma (3323 images)

This dataset is a valuable resource for deep learning and computer vision research in dermatology. Preprocessing techniques, including image augmentation, normalization, and noise reduction, are essential for optimizing model performance. CNNs and other deep learning architectures have proven to be effective in automating the diagnosis of skin conditions [26].

10.4.2 Proposed Method

We propose a deep learning-based approach combining DNNs and ResNet50 for accurate classification of skin diseases. The method leverages ResNet50's deep feature extraction capabilities alongside DNN's classification efficiency. The workflow from data preprocessing to classification is illustrated in Figure 10.4 using a visual representation. The image visually represents the proposed Deep Learning-based Skin Disease Classification workflow using DNN + ResNet50. The key steps in the diagram include:

1. Dataset Preparation:
   - Skin disease images are collected.
   - Data preprocessing techniques are applied to enhance image quality.
2. Model Selection:
   - ResNet50, a DNN, is chosen for feature extraction.
   - A DNN is integrated with ResNet50 for classification.

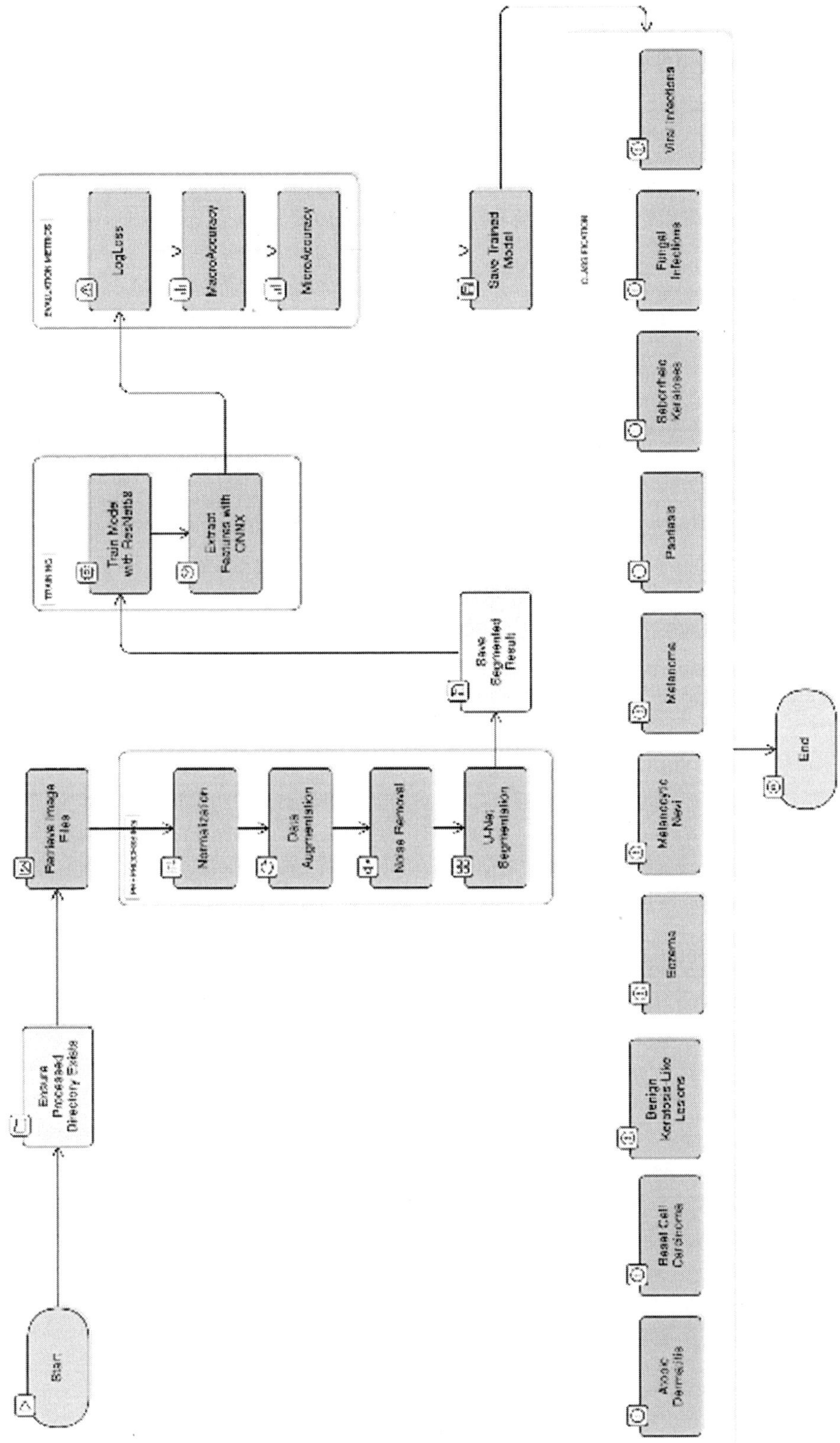

FIGURE 10.4 Workflow of the proposed method integrating ResNet50 and DNN for skin disease classification.

3. Training Phase:
    - The model undergoes backpropagation and optimization.
    - Normalization and augmentation techniques are applied to improve accuracy.
4. Evaluation and performance analysis:
    - The best accuracy achieved is 0.9205.
    - The training time recorded is 21,259 seconds.
5. Results and Predictions:
    - Classified images are categorized based on the model's predictions.
    - The final output provides detailed analysis and classification accuracy.

10.5 RESULT

The proposed model achieved an accuracy of 0.9205, demonstrating its effectiveness in classifying skin disease images. Additionally, the training time was recorded as 21,259 seconds, indicating the computational complexity of the model.

Initially, we utilized the given dataset and applied pre-processing techniques to enhance data quality. Various DNN models were then implemented and evaluated. Among them, the best-performing model was identified as DNN + ResNet50, achieving an accuracy of 0.9205 with a training time of 21,259 seconds. This approach yielded highly accurate results for the given inputs. A selection of the results obtained is presented below.

Figure 10.5 shows a skin lesion with a classification result on the right side. According to the model's prediction, the lesion has been identified as Melanocytic Nevus with 100% confidence. Other possible classifications, including Benign Keratosis, Melanoma, Basal Cell Carcinoma, and Seborrheic Keratosis, each have probabilities of less than 1%, indicating a very low likelihood of those conditions.

Figure 10.6 shows a skin lesion along with its classification results. The model predicts that the lesion is most likely a Melanocytic Nevus (82%), followed by Basal Cell Carcinoma

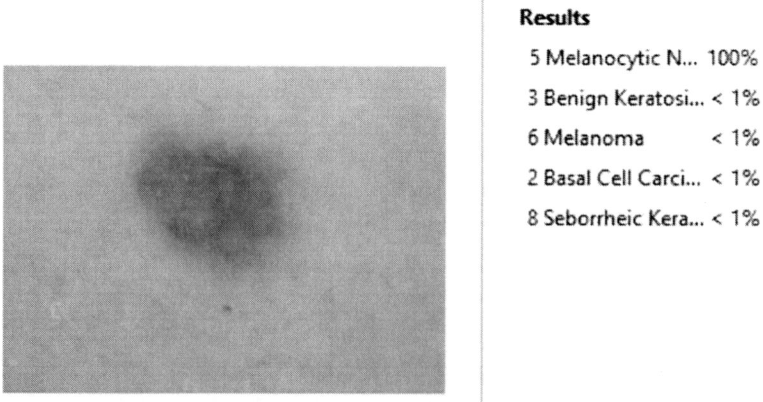

FIGURE 10.5 Image of a melanocytic disease.

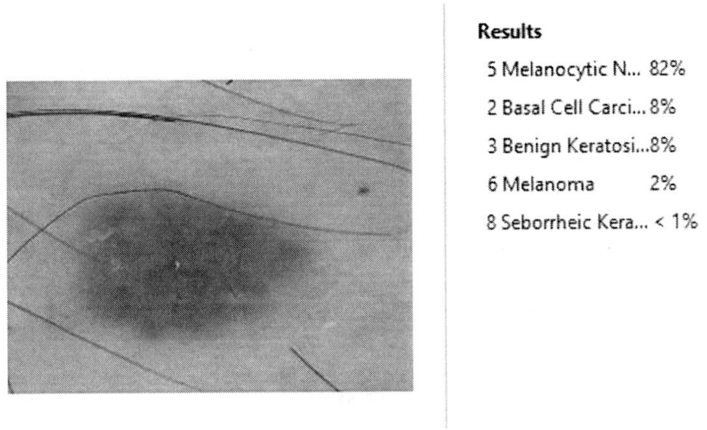

FIGURE 10.6 Image of a melanocytic disease with classification results.

(8%) and Benign Keratosis (8%). The probability of Melanoma is 2%, while Seborrheic Keratosis is less than 1%, indicating a low likelihood of malignancy.

Figure 10.7 shows a skin lesion with classification results. The model predicts a probability 96% of melanoma, indicating a high probability of malignancy. Furthermore, there is a probability 4% of Melanocytic Nevus, while Benign Keratosis, Basal Cell Carcinoma, and Seborrheic Keratosis each have less than a probability of 1%, suggesting that they are unlikely diagnoses.

The Figure 10.8 image shows a child's face with multiple skin lesions. The classification results indicate a 98% probability of Warts (Molluscum Contagiosum) as the most likely diagnosis. Furthermore, there is a probability 2% of atopic dermatitis, while Psoriasis, Eczema, and Tinea Ringworm each have less than 1%, making them less likely conditions.

A deep learning model was trained on a dataset consisting of ten distinct categories of skin diseases, resulting in this confusion matrix: Atopic Dermatitis, Basal Cell Carcinoma, Benign Keratosis-like Lesions, Eczema, Melanocytic Nevi, Melanoma, Psoriasis, Seborrheic Keratoses, Tinea Ringworm, and Warts/Molluscum as depicted in Figure 10.9. Matrix rows correspond to actual classes, while columns correspond to predicted classes. Incorrectly

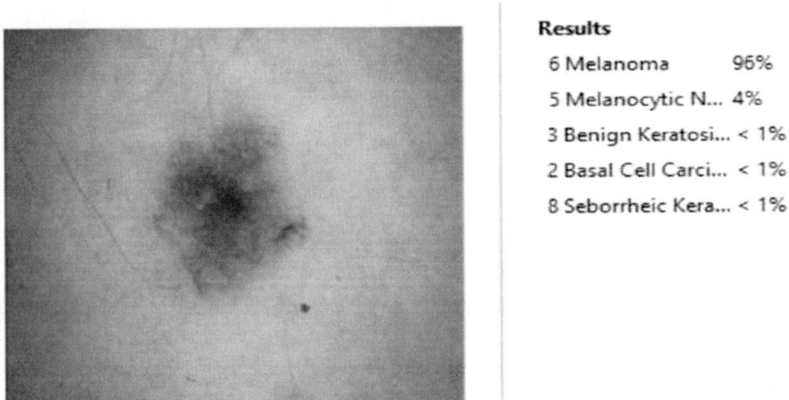

FIGURE 10.7 Image of a melanoma disease.

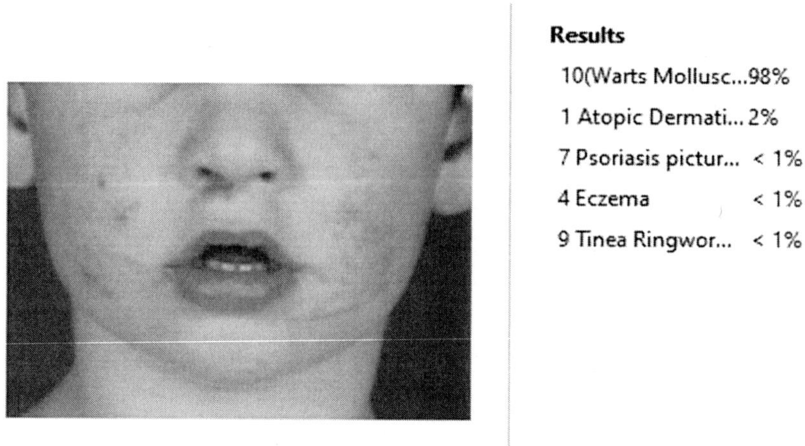

FIGURE 10.8 Image of a WartsMolluscum and other viral infections disease.

	AtopDemalitis	Basal Cell Carcinoma	Benlan Keratosis	Eczema	Melanonyic	Psoriasis	Seborrheic Keratosis	Tinea Ringwor
Atopic Dermat	179	0	0	3	2	1	5	50
Basal Cell Carcino	253	253	12	0	9	6	5	30
Eczema	36	260	190	1	11	1	0	30
Melano ma	4	1	1	1	290	290	7	28
Psorias	43	0	17	1	28	28	180	20
Sebor bheic K	24	0	13	0	0	32	189	20
Tinea Ringwo	48	1	1	1	29	22	180	20

FIGURE 10.9 Confusion matrix analysis for skin disease classification using deep learning.

classified instances are indicated by diagonal values, while misclassifications are highlighted by off-diagonal values.

Model	Accuracy
ResNet-52	92.05%
ResNet-32	74.00%
ResNet-18	71.00%

The performance of different deep residual networks was evaluated based on their classification accuracy. The results show that ResNet-52 achieved the highest accuracy of 92.05%, outperforming both ResNet-32, which scored 74.00%, and ResNet-18, which reached 71.00%. This comparison highlights the ability of a deeper network to learn more complex patterns and perform more accurately on the given skin lesion datasets. The improved accuracy of ResNet-52 suggests its greater depth and representational power may be more effective for this particular application.

10.6 DISCUSSION

Using deep learning techniques, especially DNN + ResNet50, to accurately classify skin diseases has been demonstrated in this study. A performance of 92.05% proved the model to be more accurate than traditional machine learning methods, indicating that deep learning can be used in dermatological diagnostics as well. A significant advantage of this approach is the utilization of ResNet50 for feature extraction, which enables the model to capture complex hierarchical patterns in skin disease images. DNNs improve classification accuracy and allow the model to distinguish between visually similar skin conditions when they are integrated into the model. They must, however, overcome a few challenges before they can be applied in real-world situations. The dataset used in this study exhibited an imbalance in class distribution, which may introduce bias in predictions. Future research should explore data augmentation or synthetic image generation to create a more balanced dataset.

Additionally, the computational complexity of training, which required 21,259 seconds, emphasizes the need for optimization, possibly by implementing lighter deep learning models without compromising accuracy. While the model performed well on the given dataset, its generalization to real-world scenarios requires further validation with diverse datasets encompassing various skin tones, imaging conditions, and demographic variations. In this study, deep learning is demonstrated to have potential in automating skin disease classification, especially for the application of telemedicine and computer-aided dermatology, offering a potentially valuable tool for early diagnosis. Future work could focus on incorporating explainable AI techniques, which would enhance model interpretability, making AI-driven dermatological tools more reliable and clinically acceptable.

10.7 CONCLUSION

The analysis of the Skin Problem Dataset is highly complex yet crucial for accurate skin disease prediction. By leveraging DNN + ResNet50, we can effectively classify newly provided skin disease images. As demonstrated in the previous section, this model has the ability to correctly predict the class of inputs given. The DNN + ResNet50 methodology has proven to be a highly efficient deep learning technique for this task. The outcomes align with our expectations, reinforcing the reliability and effectiveness of this approach. Due to its superior performance, DNN + ResNet50 stands out as a preferable deep learning model for skin disease classification compared to other methods.

REFERENCES

1. Gupta, R., Tripathi, R., & Gupta, M. (2020). Recent advancements in skin disease classification using deep learning: A review. Biomedical Signal Processing and Control, 60, 101951.
2. Esteva, A., Kuprel, B., & Novoa, R. A. et al. (2017). Dermatologist-level classification of skin cancer with deep neural networks. Nature, 542(7639), 115–118.
3. Litjens, G., Kooi, T., & Bejnordi, B. E. et al. (2017). A survey on deep learning in medical image analysis. Medical Image Analysis, 42, 60–88.
4. Brinker, T. J., Hekler, A., & Enk, A. H. et al. (2019). Deep learning outperformed 136 of 157 dermatologists in a head-to-head dermoscopic melanoma image classification task. European Journal of Cancer, 113, 47–54.
5. Kotsiantis, S. B., Zaharakis, I., & Pintelas, P. (2007). Supervised machine learning: A review of classification techniques. Artificial Intelligence Review, 26(3), 159–190.
6. Sinz, C. et al. (2017). Accuracy of dermatoscopy for the diagnosis of nonpigmented cancers of the skin. Journal of the American Academy of Dermatology, 77(6), 1100–1109.
7. Han, J., Kamber, M., & Pei, J. (2011). Data Mining: Concepts and Techniques. Elsevier.
8. LeCun, Y., Bengio, Y., & Hinton, G. (2015). Deep learning. Nature, 521(7553), 436–444.
9. Garcıa, S., Luengo, J., & Herrera, F. (2015). Data Preprocessing in Data Mining. Springer International Publishing.
10. Pathan, S. et al. (2018). Techniques and algorithms for computer aided diagnosis of pigmented skin lesions? Review. Biomedical Signal Processing and Control. 39, 237–262.
11. Guyon, I., & Elisseeff, A. (2003). An introduction to variable and feature selection. Journal of Machine Learning Research, 3, 1157–1182.
12. Mahbod, A. et al. (2019). Fusing fine-tuned deep features for skin lesion classification. Computerized Medical Imaging and Graphics, 71, 19–29.
13. Badrinarayanan, V. et al. (2017). Segnet: A deep convolutional encoder-decoder architecture for image segmentation. IEEE Transactions on Pattern Analysis and Machine Intelligence, 39(12), 2481–2495.
14. American Cancer Society (2020). Skin Cancer Facts and Figures.
15. Mathers, C. D. (2006). Global Burden of Disease 2001. Data Sources.
16. Methods and Results. World Health Organization (WHO).
17. Figueroa, J. I., Fuller, L. C., Abraha, A., & Hay, R. J. (1998). Dermatology in southwestern Ethiopia: Rationale for a community approach. International Journal of Dermatology, 37(10), 752–758.
18. Wu, B., Chen, B., Zeng, A., & Pan, D. (2021). Skin problem classification with deep learning: A systematic review. Journal of Medical Imaging and Health Informatics, 11(2), 456–472.
19. An, Z., Pan, D., & Zeng, A. (2021). Deep learning-based skin disease classification: Challenges and solutions. Biomedical Signal Processing and Control, 68, 102731.
20. Popescu, D. M., Shade, J. K., Lai, C., Aronis, K. N., Ouyang, D., Moorthy, M. V., Cook, N. R., Lee, D. C., Kadish, A., Albert, C. M., Wu, K. C., Maggioni, M., & Trayanova, N. A. (2022). Arrhythmic sudden death survival prediction using deep learning analysis of scarring in the heart. Nature Cardiovascular Research, 1(4), 334–343. https://doi.org/10.1038/s44161-022-00041-9

21. Choy, B., Greenfield, S., & Saffari, A. (2023). Deep learning models across the range of skin disease: A systematic review. NPJ Digital Medicine, 6, Article 33. https://doi.org/10.1038/s41746-023-00914-8
22. Aellen, F. M., Alnes, S. L., Loosli, F., Rossetti, A. O., Zubler, F., De Lucia, M., & Tzovara, A. (2023). Auditory stimulation and deep learning predict awakening from coma after cardiac arrest. Brain, 146(2), 778–788. https://doi.org/10.1093/brain/awac340
23. He, K., Zhang, X., Ren, S., & Sun, J. (2016). Deep residual learning for image recognition. In Proceedings of the IEEE Conference on Computer Vision and Pattern Recognition (CVPR) (pp. 770–778).
24. Goyal, M. et al. (2020). Artificial intelligence-based image classification for diagnosis of skin cancer: Challenges and opportunities. Computers in Biology and Medicine, 127, 104065.
25. Ismailpromus, Skin Diseases Image Dataset, Kaggle, 2024. [Online]. Available: https://www.kaggle.com/datasets/ismailpromus/skin-diseases-image-dataset. [Accessed: 01-Jan-2025].
26. Pathan, S. et al. (2018). Techniques and algorithms for computer-aided diagnosis of pigmented skin lesions: A review. Biomedical Signal Processing and Control, 46, 1–12.

Index

R

S

T

W

For Product Safety Concerns and Information please contact our EU representative GPSR@taylorandfrancis.com Taylor & Francis Verlag GmbH, Kaufingerstraße 24, 80331 München, Germany